MATHEMATICS IN INDUSTRY 7

Editors

Dr. Armin Iske

University of Leicester
Department of Mathematics
University Road
Leicester, LE1 7RH, United Kingdom
Email: iske@mcs.le.ac.uk

Dr. Trygve Randen

Schlumberger Stavanger Research
Risabergveien 3
4068 Stavanger, Norway
Email: trygve.randen@slb.com

Library of Congress Control Number: 2004114140

Mathematics Subject Classification (2000):
65M25, 65M12, 65M50, 76M25, 76M28, 76T99, 65D15, 65D05, 65D07, 65D17

ISBN 3-540-22536-6 Springer Berlin Heidelberg New York

Springer is a part of Springer Science+Business Media

springeronline.com

Printed in Germany

Typeset by the authors using a Springer TEX macro-package
Cover design: *design & production* GmbH, Heidelberg
Printed on acid-free paper 46/3142LK - 5 4 3 2 1 0

Preface

Fossil fuels, including oil and gas, store a part of the energy which the earth has received from the sun during the last several hundred million years. These days, oil and gas account for around 64 % of the total world energy consumption. Despite the efforts in developing new renewable energy sources, oil and gas will continue to play a major role in meeting the world's ever increasing energy demand for the next few decades. Moreover, oil and gas are expected to remain the most cost effective and the most convenient sources of energy that we have at our disposal.

The required exploration and production of hydrocarbons, however, incorporate great technological challenges for the oil and gas industry. Indeed, about 70 % of today's oil and gas production rate comes from hydrocarbon fields that are more than 30 years old. But several of these fields are exhibiting a significant production decline. In order to meet the world's future demand for oil and gas, further technological advances are essentially needed, where new developments should aim at efficiency and accuracy in sub-surface mapping, monitoring of reservoir depletion, and numerical simulation of production scenarios. This requires research across multiple disciplines, including mathematics, geophysics, geology, petroleum engineering, signal processing, and computer science.

This work explains important aspects and fundamental concepts in hydrocarbon exploration and production. Moreover, new developments and recent advances in the relevant research areas are discussed, where special emphasis is placed on mathematical methods and modelling. The book reflects the multi-disciplinary character of the hydrocarbon production workflow, ranging from seismic data acquisition through imaging, seismic analysis & interpretation, and geological model building to numerical reservoir simulation. Various challenges concerning the production workflow are discussed in detail.

The thirteen chapters of this joint work, authored by international experts from academic and industrial institutions, include survey papers of expository character as well as original research articles. The material of this book is arranged in three parts.

Part I. *Seismic Interpretation*;
Part II. *Geological Model Building*;
Part III. *Reservoir Modelling and Simulation.*

The first part of the book, comprising Chapters 1-5, treats the analysis and interpretation of the input seismic data. The first chapter provides a very brief and basic introduction to geology and seismic. The next two chapters present a set of seismic attributes (measures computed from the seismic data) which are useful for the subsequent classification of geological responses. Chapters 4 and 5 propose novel pattern recognition strategies for the automated interpretation of seismic data.

The second part of the book, Chapters 6-9, addresses the construction of models of the sub-surface. Chapter 6 provides a very comprehensive treatment of geological modelling and reservoir simulation. The following Chapters 7 and 8 propose new concepts for 3D geo-body model building. The approach taken in Chapter 7 is based on hierarchical segmentation, whereas the method of Chapter 8 works with level set and marching methods. Velocity estimation is an important task in sub-surface model construction, and techniques from tomography are effective tools in this respect. The complexity of tomographic inversion from seismic data is, however, significant. Chapter 9 uses modern techniques from seismic tomography in order to design efficient tomographic inversion methods.

The last part of the book, Chapters 10-13, is concerning the simulation of oil and gas production from geological models. Chapter 10 presents an integrated workflow and a case study, where advanced concepts from previous chapters of the book are applied. The resulting workflow, ranging from seismic data acquisition to reservoir simulation, is orders of magnitude faster than the conventional workflow. In order to effectively model physical phenomena in reservoir flow, multiscale methods are essentially required for reservoir simulation. To this end, a novel finite volume method over adaptive triangular meshes is explored in Chapter 11, where multivariate scattered data interpolation is an important ingredient. Selected details and principles of optimal multivariate interpolation are discussed in Chapter 12. Finally, Chapter 13 provides a practical approach to history matching the model with time-lapse measurements. This includes an illustrative real-world case study concerning monitoring the injection of CO_2 into the sub-surface.

Large parts of the material presented in this book were developed between November 2000 and April 2004 through the European research and training network NetAGES, "Network for Automated Geometry Extraction from Seismic". The NetAGES project hosted a distributed multi-disciplinary team of young researchers, PhD and postdoctoral students, and senior researchers at one industrial partner (Schlumberger Stavanger Research, Norway) and three academic partners (Munich University of Technology, University of Vienna, and Stavanger University College). Recent research results of the NetAGES young researchers and their supervisors are presented in Chapters 7-12.

In addition, collaborators from other institutions contributed to this book, either through interconnected projects (University of Surrey, University of Oxford and Schlumberger Abingdon Technology Centre, UK) or by extensions of the NetAGES training network (SINTEF Petroleum Research, Trondheim, Norway). Chapters 1-6 and Chapter 13 were written on the basis of such joint work.

All thirteen contributions to this book are invited chapters. In order to ensure their scientific and instructive quality, the chapters went through two stages of reviews. The first stage was organized by the editors in October 2003, where each chapter was carefully reviewed by independent experts external to NetAGES, as well as by experts from the NetAGES consortium and by young researchers from NetAGES. After submission of its prefinal version to Springer-Verlag in March 2003, the entire manuscript was then reviewed by a panel of ten anonymous reviewers. This second review stage was organized by Springer-Verlag.

Finally, we wish to thank those who supported the making of this book. It is in particular our great pleasure to acknowledge Dr. Martin Peters from Springer-Verlag, Heidelberg, for his everlasting support and great enthusiasm towards the book project. Moreover, the friendly and effective collaboration with Ute McCrory (Springer-Verlag) is kindly appreciated. Special thanks go to the authors for their fine contributions, and to the reviewers for their constructive comments and suggestions. Last but not least, partial financial support was granted by the European Commission through the NetAGES network (contract no. IST-1999-29034).

Stavanger, October 2004

Armin Iske
Trygve Randen

Table of Contents

Part III Reservoir Modelling and Simulation

List of Contributors

Per Bergmo
SINTEF Petroleum Research
N-7465 Trondheim, Norway
Per.Bergmo@iku.sintef.no

Hilde G. Borgos
Schlumberger Stavanger Research
P.O. Box 8013
N-4068 Stavanger, Norway
Hilde.Borgos@slb.com

Alexander A. Boukhgueim
University of Vienna
Department of Mathematics
A-1090 Wien, Austria
boukhga5@univie.ac.at

Alexis Carrillat
Schlumberger Stavanger Research
P.O. Box 8013
N-4068 Stavanger, Norway
ACarrillat@slb.com

Geir Vaaland Dahl
Schlumberger Stavanger Research
P.O. Box 8013
N-4068 Stavanger, Norway
gvdahl@slb.com

Maria Faraklioti
University of Surrey
CVSSP, SEPS
Guildford GU2 7XH, UK

Chris L. Farmer
University of Oxford
Centre for Industrial and
Applied Mathematics
Oxford OX1 3LB, UK
farmer5@slb.com

Hans Georg Feichtinger
University of Vienna
Department of Mathematics
A-1090 Wien, Austria
hans.feichtinger@univie.ac.at

Kristine Årland Halvorsen
Schlumberger Stavanger Research
P.O. Box 8013
N-4068 Stavanger, Norway
KHalvorsen@slb.com

Armin Iske
University of Leicester
Department of Mathematics
Leicester LE1 7RH, UK
iske@mcs.le.ac.uk

Martin Käser
University of Trento
Civil & Environmental Engineering
I-38050 Trento, Italy
martin.kaeser@ing.unitn.it

Erik Lindeberg
SINTEF Petroleum Research
N-7465 Trondheim, Norway
Erik.Lindeberg@iku.sintef.no

Magne Lygren
Statoil ASA
Forusbeen 50
N-4035 Stavanger, Norway
mly@statoil.com

Erik Monsen
Schlumberger Stavanger Research
P.O. Box 8013
N-4068 Stavanger, Norway
Erik.Monsen@slb.com

Jan E. Odegard
Rice University
CITI
Houston, TX, U.S.A.
odegard@rice.edu

Stein Inge Pedersen
Schlumberger Stavanger Research
P.O. Box 8013
N-4068 Stavanger, Norway
Stein.Inge.Pedersen@slb.com

Maria Petrou
University of Surrey
CVSSP, SEPS
Guildford GU2 7XH, UK
m.petrou@surrey.ac.uk

Trygve Randen
Schlumberger Stavanger Research
P.O. Box 8013
N-4068 Stavanger, Norway
Trygve.Randen@slb.com

Stine Kjersti Richardsen
Norwegian Defence
Research Establishment
FFI Horten, P.O. Box 115
N-3191 Horten, Norway
Stine-Kjersti.Richardsen@ffi.no

Jürgen Schlaf
ConocoPhillips Norway
P.O. Box 220
N-4098 Tananger, Norway
juergen.schlaf@conocophillips.com

Thorleif Skov
Schlumberger Stavanger Research
P.O. Box 8013
N-4068 Stavanger, Norway
Thorleif.Skov@slb.com

Lars Sønneland
Schlumberger Stavanger Research
P.O. Box 8013
N-4068 Stavanger, Norway
Lars.Sonneland@slb.com

Brice Vallès
RF - Rogaland Research
Thormøhlensgt. 55
N-5008 Bergen, Norway
Brice.Valles@rf.no

Tobias Werther
University of Vienna
Department of Mathematics
A-1090 Wien, Austria
tobias.werther@univie.ac.at

Part I

Seismic Interpretation

Introduction to Seismic Texture

Jürgen Schlaf, Trygve Randen, and Lars Sønneland

Schlumberger Stavanger Research, P.O. Box 8013, N-4068 Stavanger, Norway

Summary. This chapter introduces the concept of seismic texture analysis. Several seismic textures and their geological messages are described with respect to depositional history and reservoir quality. Finally, a strategy of how to automatically map these textures is recommended and limitations of this method are mentioned.

1 Introduction

Most of the hydrocarbons (gas and oil) occur in sedimentary rocks that were generated in different depositional environments (e.g. river channels, delta systems, submarine fans, carbonate mounds and reefs). Seismic waves penetrating into and reflected within sedimentary rock bodies yield a seismic image of their external shape and of their internal texture. Therefore, the analysis of the external shape of seismic bodies and its internal textures, which is called *seismic facies analysis* [2], helps to specify the depositional environment of the investigated sedimentary rocks. An analysis of the seismic facies is a must in seismic interpretation to determine the depositional environment and to locate potential reservoirs, especially in complex oilfields.

Generally, different sedimentary rocks yield different seismic facies. The seismic facies of a buried carbonate mound or reef, for example, differs significantly from a submarine fan or a delta system. Hence, each depositional system has its particular seismic facies.

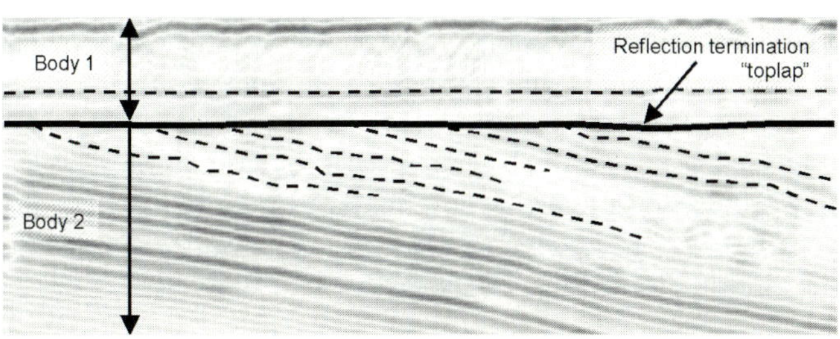

Fig. 1. [Reproduced in colour in Plate 1 on page 421.] Seismic section showing one example of a reflection termination surface or sequence boundary. Reflection termination surfaces are the boundaries of seismic bodies.

For a seismic interpreter, seismic facies analysis is a monotonous and time consuming task because it still has to be done manually by scanning through hundreds to thousands of seismic cross sections. Hence, a process is highly required which makes this interpretation step automatic.

The following of this chapter explains (1) the principals of seismic facies analysis, (2) what kind of principal external shapes and internal textures are desirable to extract, and, (3) what kind of strategy should be developed to achieve an automatic mapping of the specified features. The attention of this chapter is directed to carbonate mounds of the Barents Sea and their seismic facies.

2 Seismic Facies Analysis

In every exploration program it is a must to find out in what kind of depositional environment the investigated rocks were generated. To locate reservoirs it has to be known if they were generated in a deltaic, fluvial or reef environment. Reservoir sands in a reef environment show a different seismic facies than sands of a fluvial environment. If the depositional environment of a sedimentary rock is known then further exploration and production strategies can be refined and optimised. A tool to describe depositional environments out

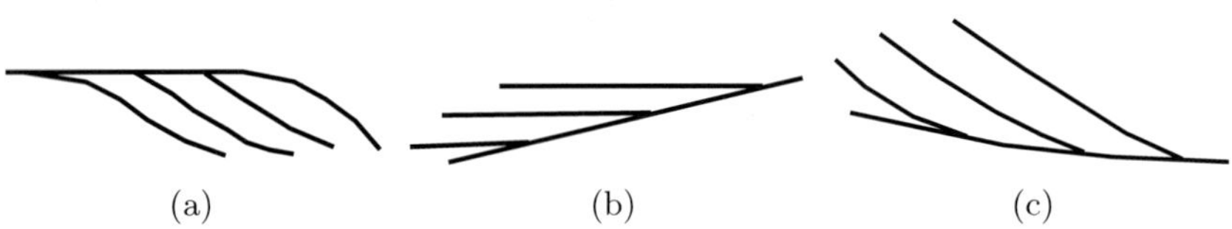

Fig. 2. The three basic types of reflection terminations for upper and lower boundaries of seismic sequences. Upper boundary type: (a) toplap. Lower boundary types: (b) onlap and (c) downlap.

of seismic data is called *seismic facies analysis*. It applies certain techniques that can help to specify depositional environments. This technique is based on (1) the external shapes of seismic bodies, (2) boundary relationships between those bodies, and, (3) the internal texture of those bodies. These three criteria should be taken into account when describing and mapping seismic facies. Therefore, seismic facies analysis subdivides a seismic data set into seismic facies units which are mappable three-dimensional units of reflections whose characteristics differ from that of the adjacent facies unit.

2.1 External Shapes and Boundary Relationships

Eight basic types of external shapes can be differentiated. The boundaries between different shapes are reflection terminations (Figure 1, Figure 2), so that they meet the requirements of seismic sequences. The mapping of reflection terminations is the key to seismic facies analysis. Problems in mapping reflection terminations can arise when terminations laterally pass into concordant relationships as indicated in Figure 3. Normally, termination surfaces are characterised by high reflection strengths. Therefore, following reflection terminations through regions where there is no angularity should be possible by virtue of their amplitude standout. The eight basic shapes that are bounded by reflection terminations are: sheet, sheet drape, wedge, bank, lens, mound, fan, and fill (Figure 4 to Figure 5).

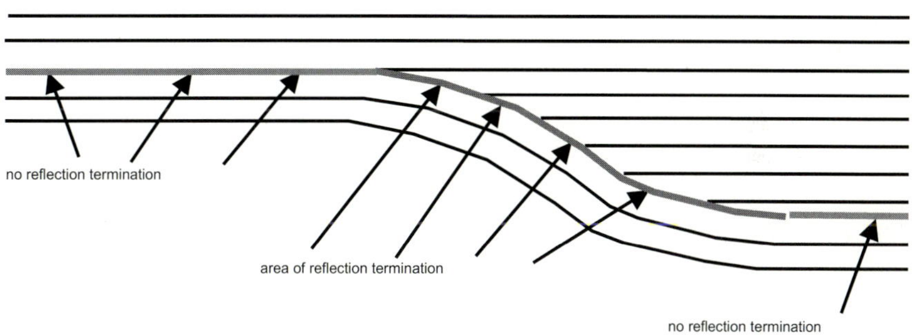

Fig. 3. Scheme explaining the mapping of reflection terminations. The solid line in the middle indicates an area where reflections terminate with an angle onto a surface (sequence boundary, see annotation). This surface, however, continues into areas where no terminations exist (left and right solid line, see annotations). It is important to automatically map also the continuation of the termination surface into areas where no reflection terminations occur.

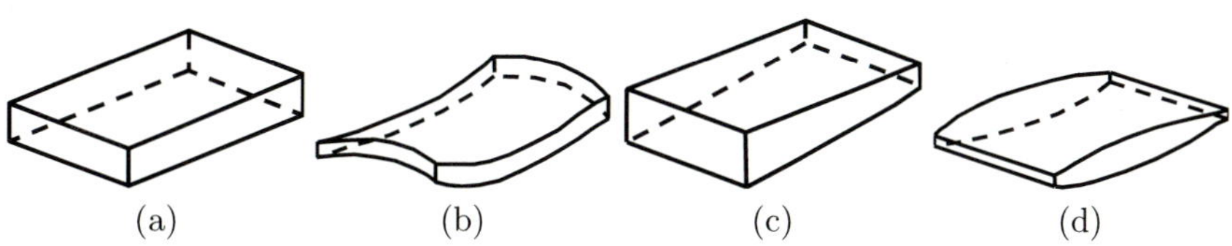

Fig. 4. External shapes of seismic bodies: (a) sheet, (b) sheet drape, (c) wedge, and (d) lens. The boundaries of all of them should be, at least in some parts, reflection terminations.

All the shapes mentioned above can consist of different textures. A sheet geometry, for example, can contain several textures simultaneously, e.g. a chaotic texture and a parallel texture side by side. On the other hand, there is no texture that is restricted only to a certain shape. Prograding textures, for example, can occur in almost all shapes.

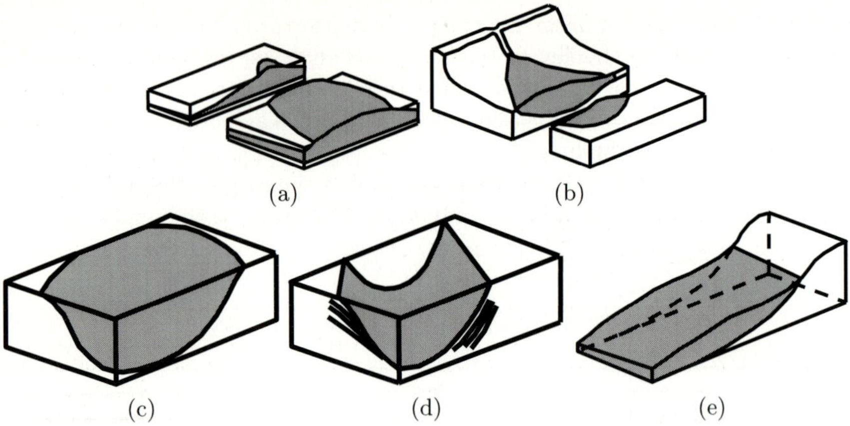

Fig. 5. External shapes of seismic bodies: (a) mound, (b) fan, (c) basin fill, (d) channel fill, and (e) slope front fill.

2.2 Texture: Reflection Configuration and Reflection Continuity

The external shapes mentioned above can be investigated with respect to the configuration of the reflections and their continuity, which is called *texture*. Mentioned below are the most important textures that an automatic system should be able to detect.

Parallel and Subparallel Textures (Figure 6). Parallel seismic events can be even or wavy and occur as sheets, sheet drapes and fill units. Intervals consisting of parallel events can be further subdivided with respect to reflector spacing and continuity. Parallel textures indicate uniform rates of deposition on a uniformly subsiding shelf or stable basin plain setting.

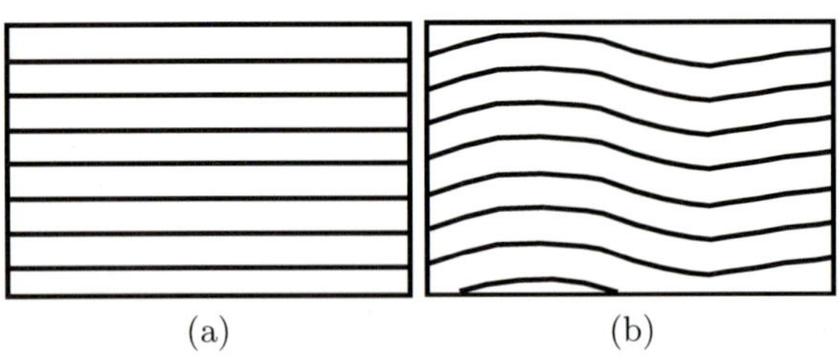

Fig. 6. Parallel textures: (a) even and (b) wavy (modified after Mitchum et al. [2]).

Divergent Textures and Convergent Textures (Figure 7). This texture is characterised by a wedge-shaped unit in which most of the lateral thickening or thinning is accomplished by thickening of individual reflections within the unit, rather than by onlap, toplap, or erosion at the base or top. Lateral terminations of reflections are probably due to progressive thinning of strata to below the resolution of the seismic tool. Divergent and convergent textures suggest lateral variations in the rate of deposition, or progressive tilting of the depositional surface (e.g. growth fault).

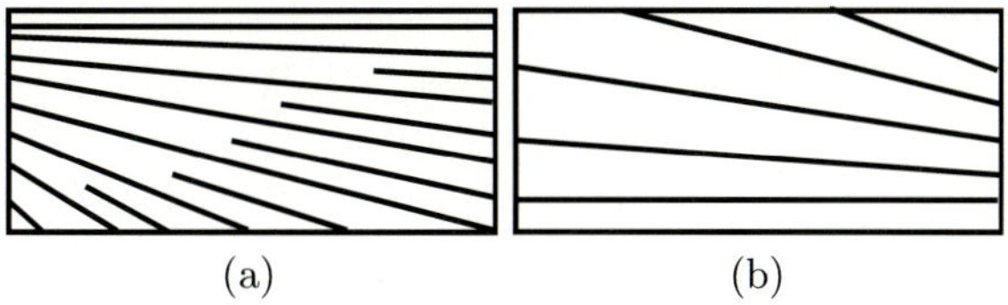

Fig. 7. Divergent and convergent textures: (a) divergent and (b) convergent (modified after Mitchum et al. [2]).

Prograding Textures (Figure 8). This texture indicates lateral outbuilding or progradation of a sedimentary system due to lateral development of gently sloping depositional surfaces, called clinoforms. Prograding textures are among the most common depositional features. They occur in deltas, carbonate platforms, beaches, shelfes and submarine fans. Therefore, the prograding texture should be further subdivided into several sub-textures, each reflecting different depositional modes.

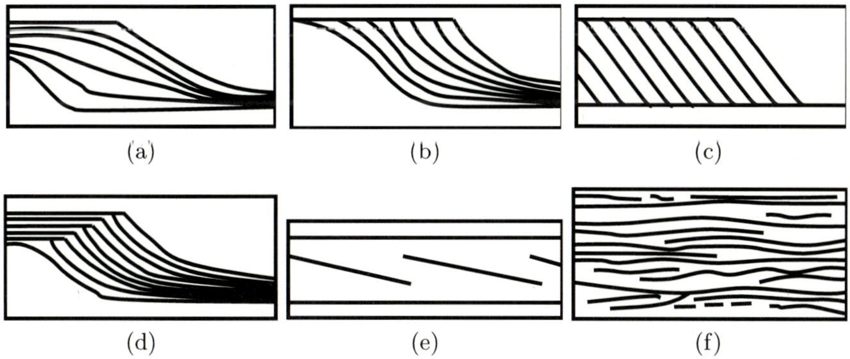

Fig. 8. Different prograding textures: (a) sigmoid, (b) oblique tangential, (c) oblique parallel, (d) complex sigmoid-oblique, (e) shingled, and (f) hummocky (modified after Mitchum et al. [2]).

Prograding sub-Texture Sigmoid. This texture shows S-shaped reflections with thin, gently dipping upper and lower segments. The upper segments show almost horizontal dips. The middle segments are thicker and form lenses.

Depositional angles are low, usually. The lower segments approach the lower surface at very low angles. The most distinctive feature of the sigmoid is the parallelism and concordance of the upper segments coincident with the progradation of the middle segments. A low sedimentary regime can be interpreted, the possibility of reservoir sands being small.

Prograding sub-Texture Tangential-Oblique. The dips of the clinoforms decrease in lower portions forming concave-upward strata which pass into gently dipping bottomset strata. This texture indicates low to moderate energy and points to a decrease in grain size downslope.

Prograding sub-Texture Parallel-Oblique. The relatively steep-dipping parallel foreset strata terminate downdip at a high angle by downlap against the lower surface. This texture indicates coarse grain sizes on the slope because of the steeper angles. The steep angles can also be generated by carbonates and reefs. Furthermore, steep slope angles point also to sedimentary by pass.

Prograding sub-Texture Complex Sigmoid-Oblique. This pattern consists of a combination of variably alternating sigmoids and oblique progradational textures within one single unit. This subdivision allows to differentiate higher energy segments from lower energy segments.

Prograding sub-Texture Shingled. This represents a thin prograding texture, commonly with parallel upper and lower boundaries, and with gently dipping parallel oblique internal reflections that terminate by apparent toplap and downlap. The overall pattern resembles that of the parallel oblique configuration, except that the thickness of the unit is just at the point of seismic resolution of the oblique beds. Shingled seismic textures are one of the most common textures in seismic indicating progradation into shallow water.

Prograding sub-Texture Hummocky Clinoforms. This texture consists of irregular discontinuous subparallel reflections forming a practically random hummocky pattern marked by nonsystematic reflection terminations and splits. Hummocky clinoforms are indicative for deltaic lobes prograding into shallow water.

Chaotic Texture (Figure 9). A chaotic texture is characterised by discontinuous, discordant reflections suggesting a disordered arrangement of reflection surfaces. Chaotic textures can represent a wide range of features like slumped areas, reefs and carbonate mounds, cut-and-fill channel complexes, and highly faulted and folded areas.

Reflection-Free Areas. Reflection-free textures on seismic sections indicate large igneous masses (e.g. buried volcanoes), salt features or shales.

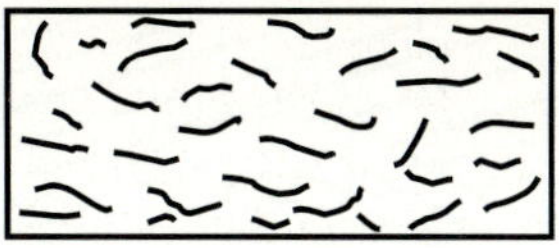

Fig. 9. Chaotic Texture.

2.3 Channels (Figure 10)

Channels are very important shapes because they are indicative of fluvial channels or submarine channels. Generally, channels point to high depositional energy and the possibility of deposited sand, either within a channel or close by.

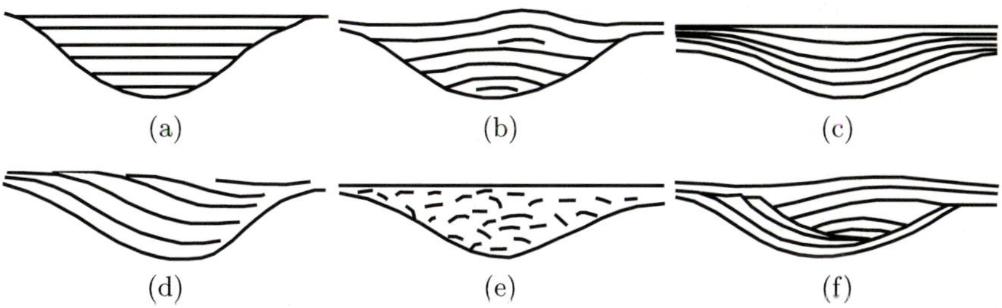

Fig. 10. Different channel fill textures: (a) onlap fill, (b) mounded onlap fill, (c) divergent fill, (d) prograded fill, (e) chaotic fill, and (f) complex fill (modified after Mitchum et al. [2]).

In most of the cases channels show a concave upward base and a more or less flat top. This is due to the filling of negative-relief features due to erosion of a flowing medium in the underlying strata.

Channels can show different types of textures due to various filling mechanisms and size, aspect ratio, sinuosity, etc. We can differentiate onlap fill, prograded fill (progradational clinoforms can have all the shapes described in Subsection 4.2), mounded onlap fill (in this case the channel top is convex upward), chaotic fill, divergent fill, and complex fill.

2.4 Mounds (Figure 11)

Due to the fact that one of the key objectives of the Barents Sea data set is to map carbonate mounds, these features will be described here in more detail. The characteristic of a mound is its elevation above a rather flat surface. Therefore, an automatic system should be able to capture these topographic elevations.

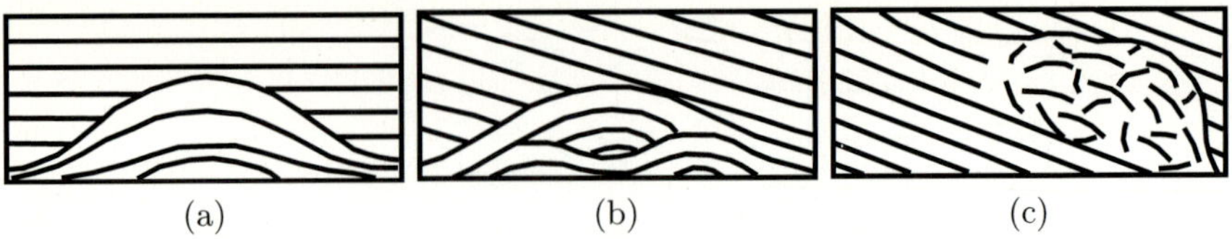

Fig. 11. Different textures of mounds: (a) concave upward simple, (b) concave upward complex, and (c) chaotic.

Internally mounds can consist of different textures each reflecting different environments, ranging from volcanoes to reefs Mitchum et al. [2] differentiate six main textures of mound. However, we will focus here mainly on the external shape and on a few basic textures.

Concave Upward Simple. The texture of the mound consists of a stack of concave upward events. Surrounding events show a pronounced onlap onto the mound surface. This texture can be typical for volcanoes.

Concave Upward Complex. The texture consists of a complex arrangement of parallel and wavy events. The seismic events surrounding the mound show onlap. This texture is characteristic for submarine fans.

Chaotic. The seismic events show a chaotic arrangement. This can be indicative for reefs or slumps.

3 Strategy for Automatic Mapping

To automatically subdivide a seismic cube into shapes and textures, a certain strategy should be applied. It is recommended to follow the proposal of Mitchum and Vail [1]. They propose two steps: (1) subdivide a seismic data set into bodies (sequences) that have a certain external shape and that are separated by surfaces of discontinuity, called *sequence boundaries*, and, (2) to analyse the texture of those bodies.

It is of high importance that the shapes of geological bodies are mapped. This can only be done if we take reflection terminations into account. Different shapes (e.g. wedges, sheets, mounds, channels) represent packages that show pronounced terminations of seismic events at their base and at their top and fulfil, therefore, the requirements of seismic sequences. The mapping of textures should be achieved after having subdivided the seismic data into shapes. An idealised scheme of seismic facies mapping is shown below (Figure 12). An attempt to apply seismic facies mapping on the Barents Sea data is shown in Figure 13 and Figure 14.

If a seismic data set is mapped based on textures only, then we take the risk to subdivide the data neglecting geological meaning. We should, if possible, always focus on a subdivision first into shapes (or sequences) and

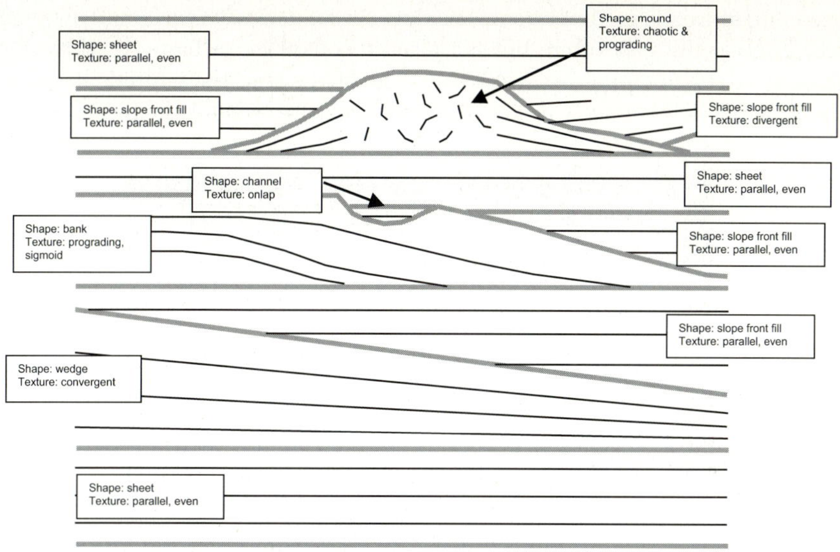

Fig. 12. Simplified scheme of seismic facies analysis. The thick lines indicate the boundaries of different shapes which represent seismic sequences. The different shapes show different textures.

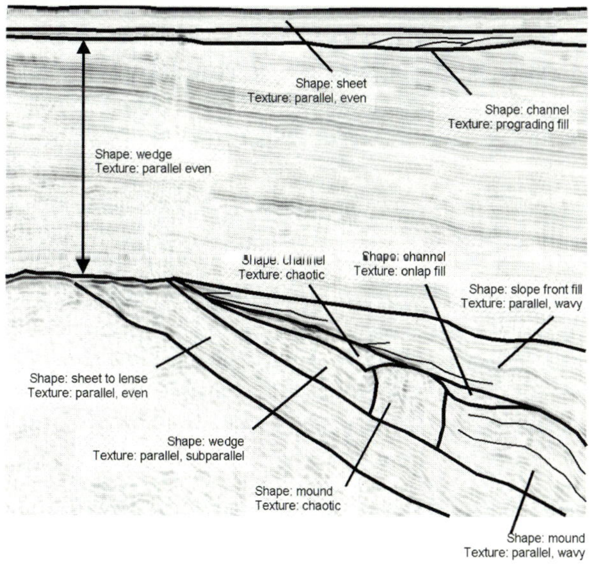

Fig. 13. [Reproduced in colour in Plate 2 on page 421.] Segmentation of a two-dimensional inline of the Barents Sea data set into different shapes and textures. This figure shows that a manual interpretation is challenged when it comes to defining the lateral boundaries of the mound shape in two dimensions, not mentioning three dimensions.

then into textures. In the Barents Sea data set we deal with carbonate mounds which internally very often, but not always, consist of a chaotic texture. The priority of the mapping approach would be (1) to map the mounds based on their shape, and then (2) to detect chaotic textures within them.

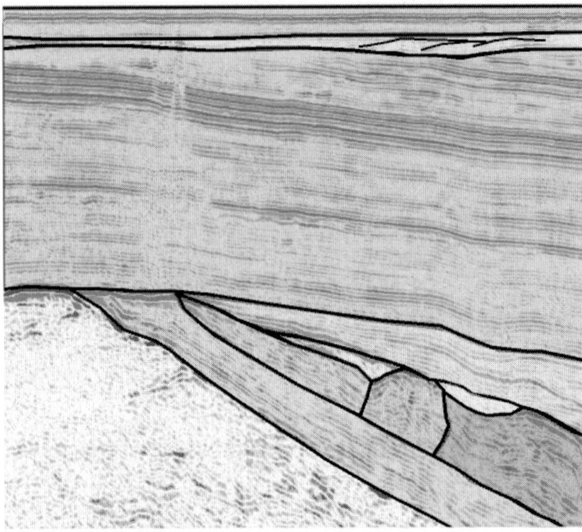

Fig. 14. [Reproduced in colour in Plate 3 on page 422.] Overlay of different colours onto the same inline as for Figure 13. This represents a vision how a seismic data set should be segmented into shapes and textures. Different colours correspond to different shapes and textures. Compare Figure 13 for legend.

3.1 Limitations

The mapping of seismic textures is a very important step in interpreting seismic data because the seismic texture contains information about the reservoir quality. Until recently seismic texture mapping has been done manually which is a time consuming procedure. Automated seismic texture analysis has been proven to be a powerful tool to quickly and precisely map distinct seismic textures. However, some limitations exist that one should be aware of.

Compressional Tectonic Regimes. In compressional tectonic regimes a sedimentary package can be distorted by faults and fractures. Intense faulting and fracturing can generate a chaotic or reflection-free texture that does not represent the primary depositional environment in which the rock was formed. Compressional tectonic regimes can produce stacked thrust systems, called duplexes, which can be confused with prograding clinoforms. Additional features are imbricate fans that resemble downlap geometries. Channel fill patterns can be distorted significantly. When exposed to compressional forces a channel filled with an onlap pattern can easily be turned into a mounded onlap fill or even into a chaotic fill. However, with respect to reservoir quality there is a big difference between an onlap fill and a chaotic fill. The angle of prograding clinoforms can be changed in a compressional regime. Gently dipping clinoforms, normally indicative for lower energy regimes and poor reservoir quality can be transformed into steeply dipping prograding system. Steeper slope angles, however, indicate either carbonates or a larger grain sizes.

Steeply Dipping Reflectors. The amount of seismic energy that is reflected at a boundary depends on many factors [3]. However, important factors are the contrast in acoustic impedance and the angle of incidence. Sedimentary packages consisting of steeply dipping reflectors can generate a reflection free seismic texture even if the acoustic impedance contrasts are high. This is mainly due to the fact that steeply dipping reflectors reflect the seismic energy over such a broad area that it cannot be gathered properly. Due to this dispersion no clear reflection can be produced. An additional disadvantage is that steeply dipping reflectors generate too much refracted energy that cannot be compensated for. Therefore, a sedimentary package that consists of e.g. parallel reflections would give a reflection free texture when this package shows steep dips. In this case an automatic seismic texture program would lead to an erroneous interpretation of the depositional environment.

Recommendations. As discussed above, automatic seismic texture analysis is a very powerful tool in extensional tectonic domains and in tectonically undisturbed areas. There it quickly sorts the data set into a number of seismic facies. Compressional tectonic environments, however, are prone to alter the seismic texture and can lead, when applied uncritically, to wrong conclusions with respect to the reservoir quality. When automatic seismic texture analysis is applied to compressional regimes, the results should be interpreted carefully. To avoid wrong conclusions, the seismic data set under consideration should be interpreted by a geologist before automatic texture algorithms are applied to map out potential areas of error.

4 Summary

A useful automatic seismic facies mapping tool has to combine information about shapes and textures within these shapes. Only a combination of shapes and textures enables a meaningful seismic facies analysis fulfilling the requirements of seismic stratigraphy. Mapping of seismic bodies has to be done on the basis of reflection terminations. As soon as a seismic data set is subdivided into different shapes, texture analysis within each of the shapes can be achieved.

Appendix: Catalog of Seismic Facies Types

4.1 Termination Patterns

Upper Boundary.

Texture	Name	Geological Message
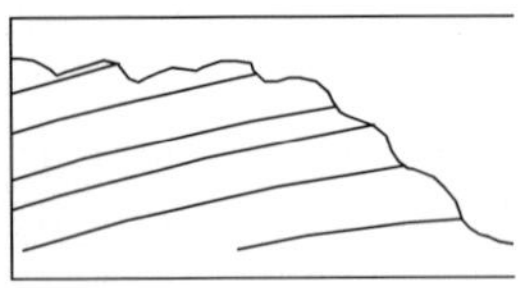	Erosional Truncation	Unconformity; sequence boundary caused by (subaerial or subaqueous).
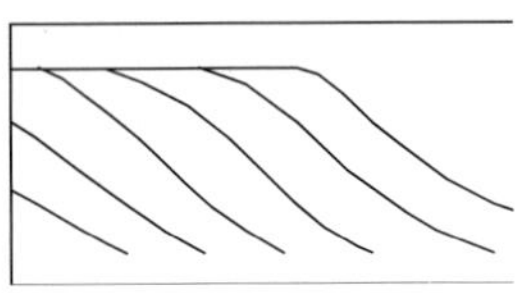	Toplap	Upper boundary caused mainly by nondeposition rather than by erosion.

Lower Boundary.

Texture	Name	Geological Message
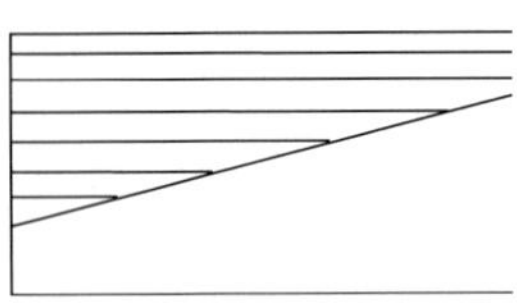	Onlap	(1) In a shelfal environment, rising of relative sea level; (2) In a deep-sea environment, modest rate of deposition, on modest gradient; (3) In an eroded channel, low-energy fill.
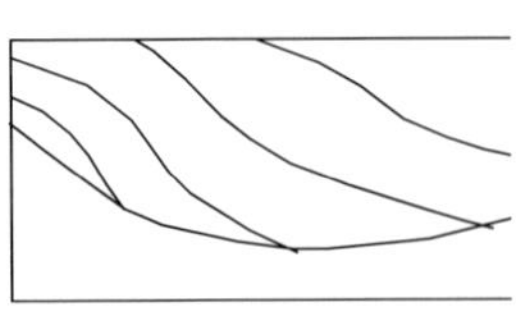	Downlap	Sediment starvation (at least where lower boundary is horizontal).

4.2 Principal Seismic Textures

Texture	Name	Geological Message
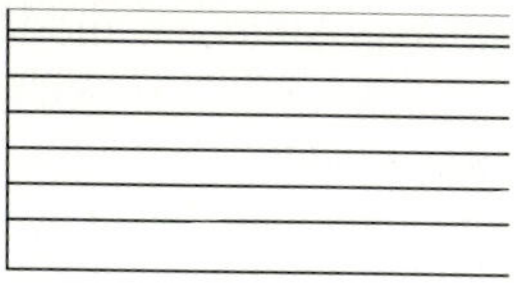	Parallel	Uniform rate of deposition on an uniformly subsiding shelf or stable basin plain.
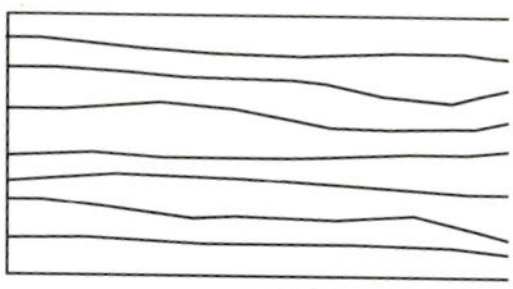	Subparallel	Commonly in zones of fill; also above situation disturbed by marine currents.
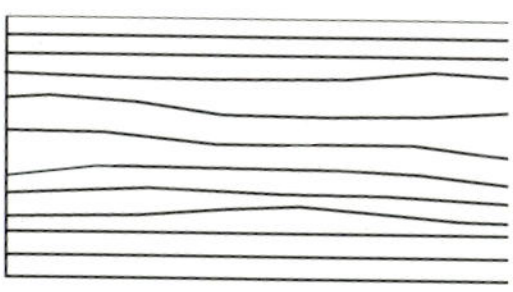	Subparallel between Parallel	General stable tectonic and depositional environment; probably fluvial plain, with dirty medium-grained sediments.
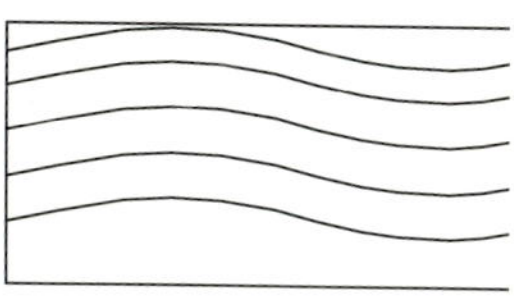	Wavy Parallel	(1) compressional folding of parallel strata over detachment surface, or diapirism (2) sheet drape: very fine-grained deposition, out of suspension.
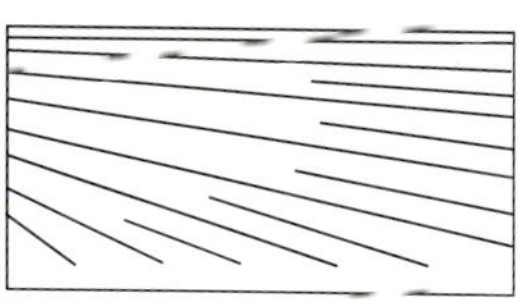	Divergent	During deposition, progressive tilting of a depositional surface above a hinge line.
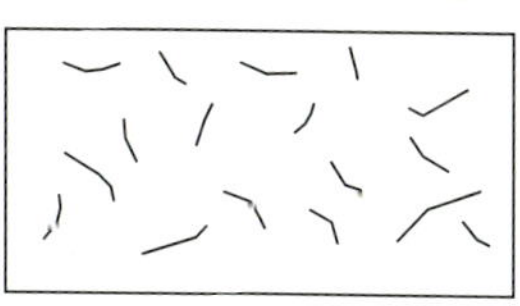	Chaotic	Very variable high-energy deposition (mounding, cut-and-fill channeling), or major post-depositional deformation (faulting, movement of overpressured shale).
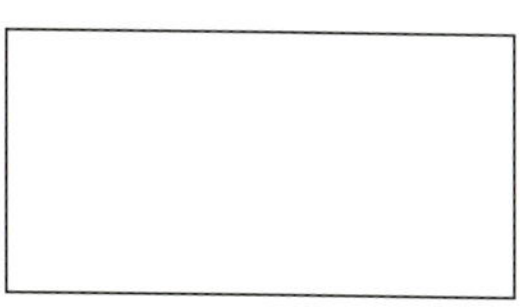	Reflection-free	Igneous rocks, salt, interior of single-stage reefs.
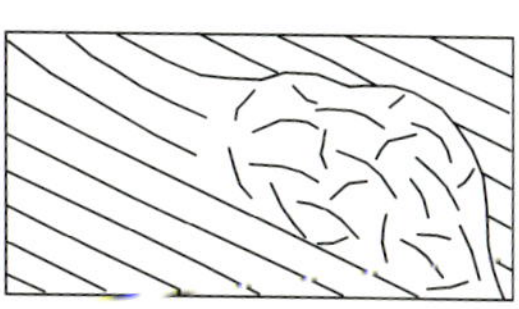	Local Chaotic	Slump (normally deap-sea) triggered by earthquake or gravitational instability; extremely rapid undifferentiated deposition.

4.3 Prograding Textures

Texture	Name	Geological Message
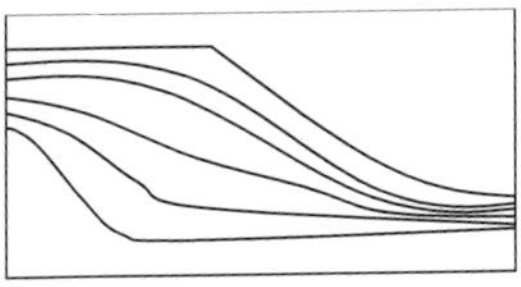	Sigmoid	Outbuilding with upbuilding; modest sediment supply coupled with rapid rise of relative sea level; low-energy depositional regime, such as prograding slope; generally fine-grained sediments.
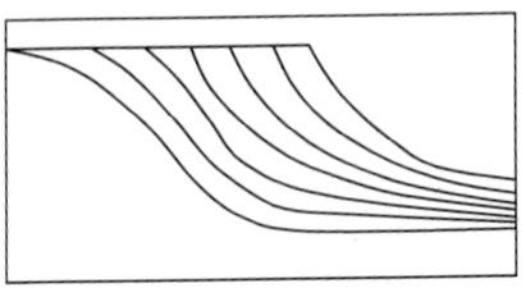	Oblique Tangential	Outbuilding only; modest to large sediment supply; stillstand of relative sea level; higher-energy depositional regime, such as delta; some coarse-grained sediments on delta plain and particular in channels and bars.
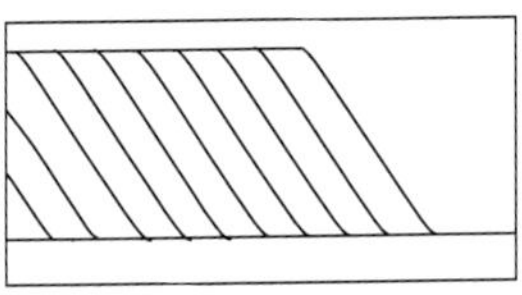	Oblique Parallel	Variant of oblique-tangential; perhaps sediment better sorted.
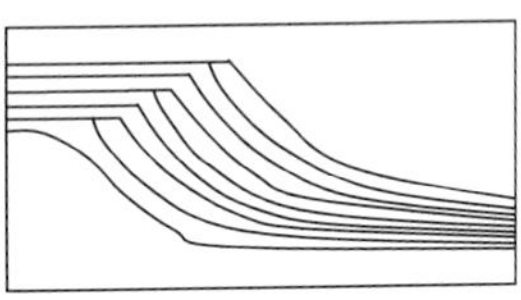	Complex	Locally alternating upbuilding sigmoid-oblique and depositional bypass; typical of section across high-energy deltaic lobes within a low-energy prograding slope.
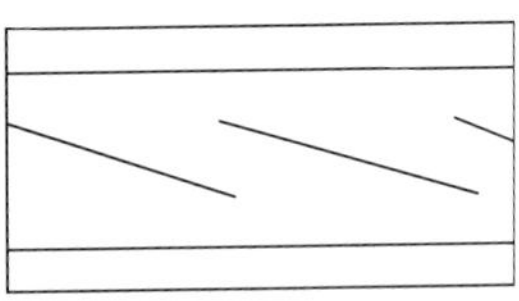	Shingled	Outbuilding into shallow water; generally low-energy regime.
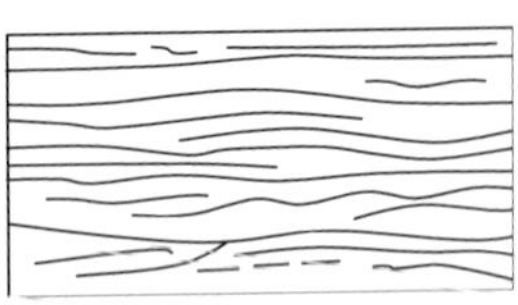	Hummocky	Small interfingering sediment lobes building into shallow water (typically between deltas); modest-energy regime.

4.4 Channel Fill Textures

Texture	Name	Geological Message
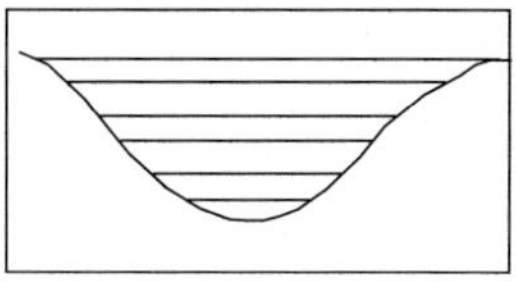	Onlap Fill	Relatively low-energy filling of erosional channel.
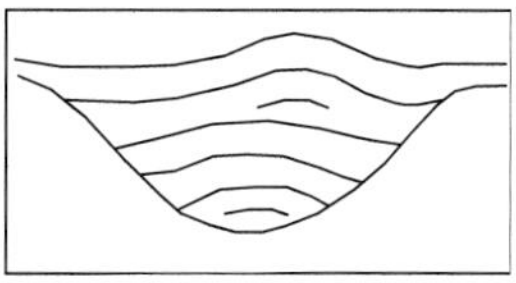	Mounded Onlap Fill	Higher-energy fill in at least two stages.
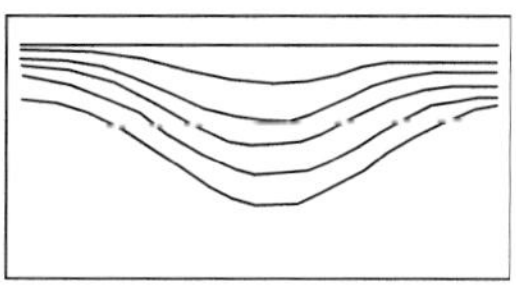	Divergent Fill	Compactible (shale-prone) low-energy sediments; also typical of last stages of graben fill.
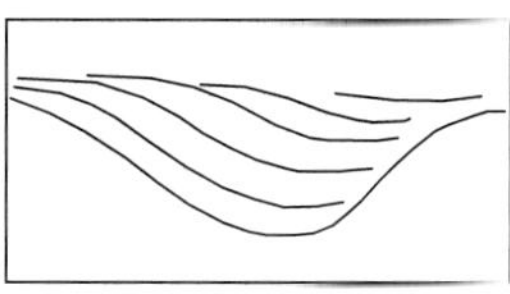	Prograded Fill	Sediment transport over edge, or along channel at a bend.
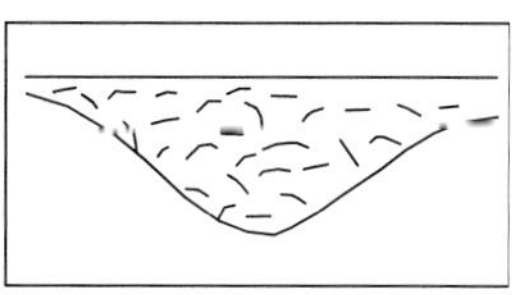	Chaotic Fill	Very high-energy fill.
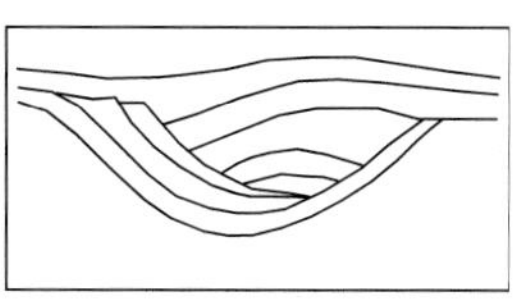	Complex Fill	Changing sediment provenance and/or water flow.

4.5 Mounded Textures in a Carbonate Setting

Texture	Name	Geological Message
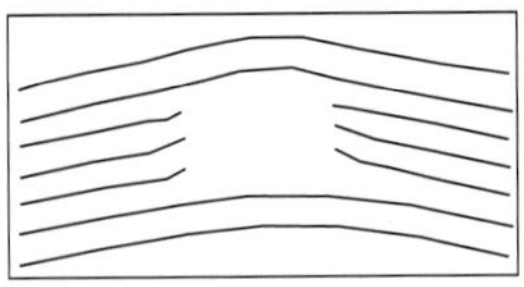	Reflection-free Mound	Patch reef or pinnacle reef; drape shows flanking sediments more compactible (probably shale).
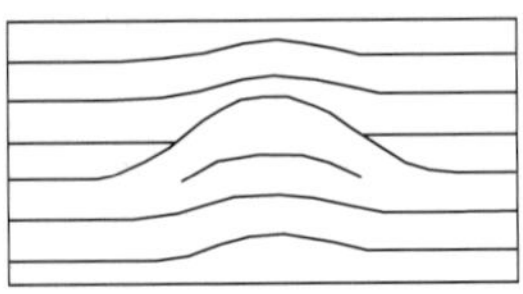	Pinnacle with Velocity Pull-Up	Patch reef or pinnacle reef, multi-stage growth, possibly porous.
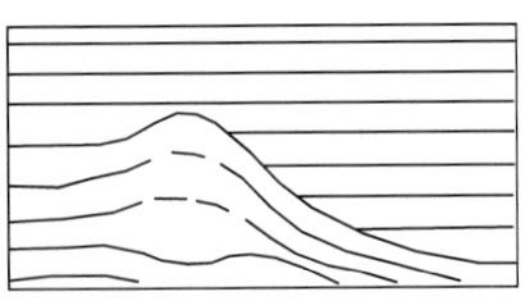	Bank-Edge with Velocity Sag	Shelf-edge reef, with very good porosity; overlying sediments probably carbonate-prone.
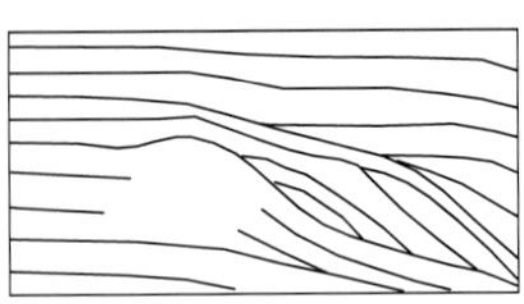	Bank-Edge Prograding Slope	Shelf-edge reef overlain and flanked by clastics; change in sediment supply.

4.6 Mounded Textures in Other Settings

Texture	**Name**	**Geological Message**
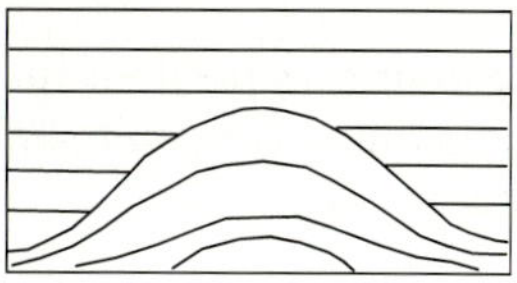	Fan Complex	Lateral section of fan close to sediment entry.
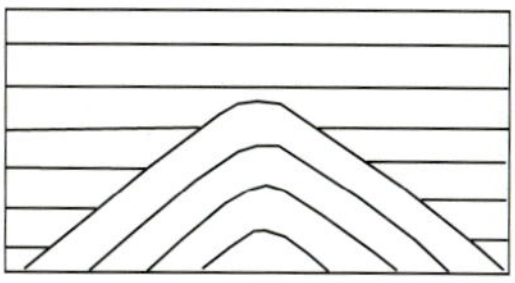	Volcanic Mound	Earlier convergent margin; center of rifting activity in rift basin.
	Compound Fan Complex	Superposed lobes of different fans coalesce laterally and as greater depositional activity moves from one lobe to another.
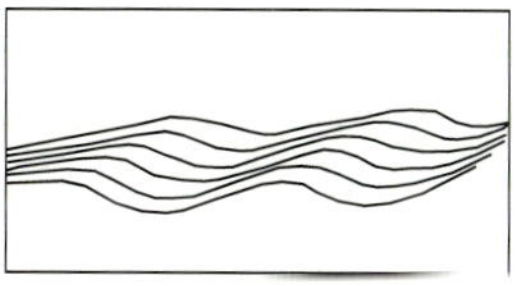	Migrating Wave	Major ocean currents, deep-water.

4.7 Basin Slope and Floor Seismic Facies Types

Texture	Name	Geological Message
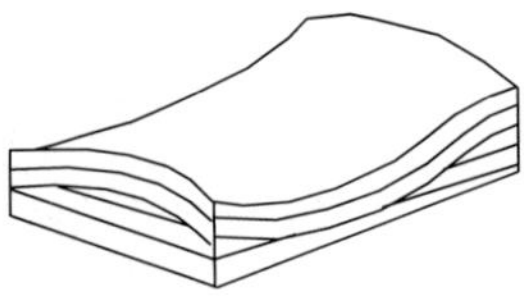	Sheet-Drape (Low Energy)	Uniform, deep-marine deposition independent of bottom relief, uniform lithology, no sand.
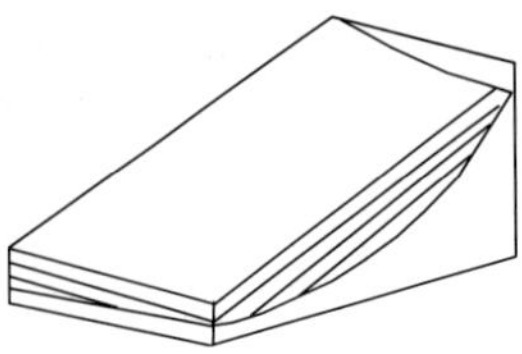	Slope-Front Fill	Deep-water fans, clays and silts (low energy).
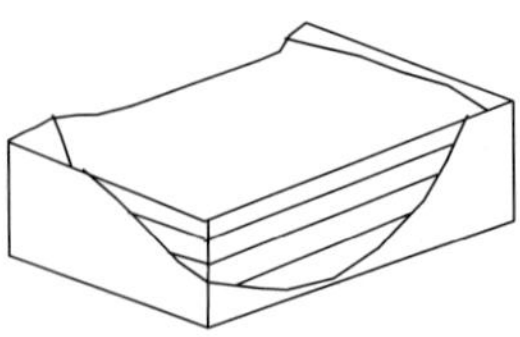	Onlap-Fill (Low Energy)	Deposited by gravity-controlled flows (low velocity turbidity currents) along the bottom.
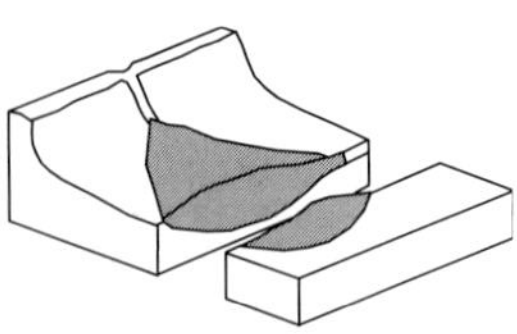	Fan-Complex (High Energy)	Deposited as fans, contourite mounds and slumps; although high energy deposits; may contain good reservoir sandstones.
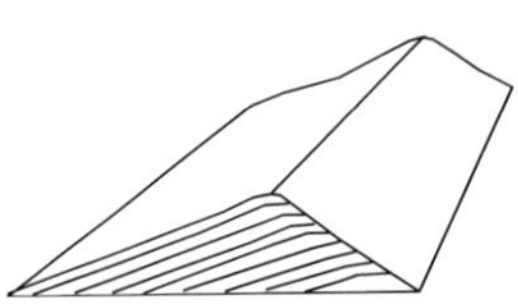	Contourite (Variable Energy)	Usually fine-grained and therefore of no interest for exploration; elongate, asymmetric features; unidirectional flows.

Texture	Name	Geological Message
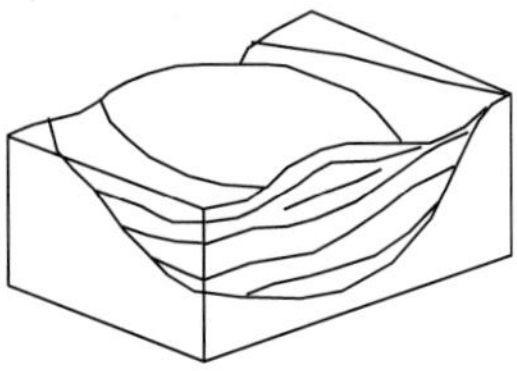	Mounded Onlap-Fill (High Energy)	Intermediate facies between chaotic-fill and onlap-fill; gravity-controlled flow (turbidity currents) along the sea bottom; discontinuous reflections that thicken towards topographic lows indicate high energy deposits.
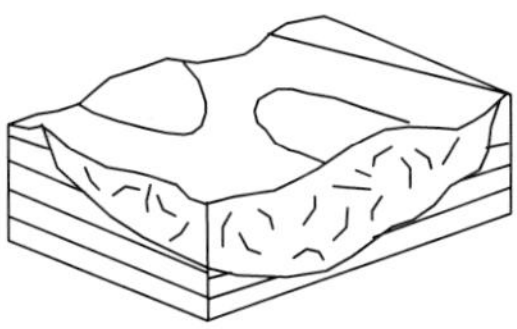	Chaotic Fill (Variable Energy)	Mounded external form; located in topographic lows; slump, creep and high energy turbidites; composition depends on the source; commonly little sand due to lack of winnowing; diffractions are common.

Acknowledgement

The work was partly supported by the European Union through the TriTex project, contract no. IST-1999-20500. We also wish to thank Kai Hogstad, Geir Elvebakk, and the Barents Sea Seismic Area C partners Norsk Hydro, Statoil, AGIP, and Fortum for collaboration and permission to publish data.

References

1. R.M. Mitchum Jr. and P.R. Vail (1977) Seismic stratigraphic interpretation procedure. AAPG Memoir; Seismic Stratigraphy - Applications to Hydrocarbon Exploration **26**, 135–143.
2. R.M. Mitchum Jr., P.R. Vail, and J.B. Sangree (1977) Stratigraphic interpretation of seismic reflection patterns in depositional sequences. AAPG Memoir; Seismic Stratigraphy - Applications to Hydrocarbon Exploration **26**, 117–133.
3. R.E. Sheriff (1975) Factors affecting seismic amplitudes. Geophysical Prospecting **23**, 125–138.

Atlas of 3D Seismic Attributes

Trygve Randen and Lars Sønneland

Schlumberger Stavanger Research, P.O. Box 8013, N-4068 Stavanger, Norway

Summary. Seismic attributes have been a common tool in seismic data analysis since the mid to late seventies. Seismic attributes may highlight geological or geophysical effects, thus leading to a quicker analysis of the data. In the early works, seismic attributes were to a large degree limited to capture 1D effects occurring along the vertical seismic trace. Since the mid nineties, extensions to multi-trace attributes have become more popular. In this chapter we present a set of three-dimensional attributes for seismic data analysis. The attributes are designed to highlight the seismic expression of faults and stratigraphic features, and are designed to be genuine 3D with no implicit directionality bias on the result. The chapter is written to be an introduction to the technology. Feasibility tests of some of the approaches can be found in the chapter [4] of Carrillat and Vallès.

1 Introduction

A *reservoir trap* typically belongs to one of two categories: structural or stratigraphic. Structural traps (anticlines, faults, and domes) have been far easier to detect, and are the dominating traps in most detected reservoirs. It is however anticipated that a large part of the *undiscovered* reservoirs involve stratigraphic traps [15]. Furthermore, most stratigraphic accumulations of hydrocarbons that have been found were found by chance when searching for something else [15]. Proper stratigraphic interpretation of a seismic data set is a huge task requiring months of effort. Still, the seismic contractors around the world introduce more and more acquisition capacity. Hence, if we want to extend seismic exploration to enable detection of these potentially vast reserves, we need tools capable of handling the new trap models.

Stratigraphic interpretation involves the identification and recognition of upper and lower boundaries and internal configuration. *Upper and lower boundaries* typically consist of reflectors and terminations. Reflector tracking solutions are available in most commercial interpretation packages and an approach to automatic termination detection is proposed in Section 4.

Internal configuration is another word for *texture* (some sketches can be seen in Figure 1). In this chapter we propose several new texture attributes with the potential of significantly simplifying the interpretation task.

A few moves towards texture attribute extraction in seismic data are presented by Sheriff et al. [15]. The probably most common seismic texture

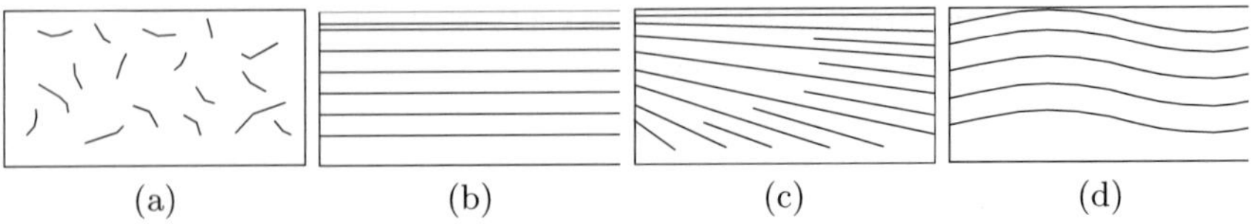

Fig. 1. A few example sketches of internal seismic stratigraphic facies configurations. From left, (a) chaotic, (b) parallel, (c) divergent, and (d) wavy parallel. See the chapter [14] of Schlaf, Randen, and Sønneland for more examples.

attributes are amplitude, instantaneous frequency and phase, and integrated versions of these. These are typically one-dimensional. In the recent years there has been a growing focus on three-dimensional attributes, such as dip and azimuth, see Section 2, and continuity, see Section 5.

This chapter is intended for experts from both the seismic *solution* sciences (e.g. mathematics, signal processing, pattern recognition), as well as from the *application* sciences (e.g. geology). The chapter is more focused towards being a quick reference than an in-depth treatment of the approaches.

More specifically, in Section 2 we discuss the geometrical tensor (the tip and azimuth), in Section 3 we introduce a few texture attributes targeted towards stratigraphic interpretation, in Section 4 we present the termination attribute, designed to detect termination interfaces, in Section 5 attributes for fault detection are presented, and finally in Section 6, techniques for preconditioning the data are discussed.

2 Dip and Azimuth

Key Principle: Estimates the local dominating strata angle; represented by the dip and azimuth [3, 12].

Main Uses: (a) Input to other processes or (b) cubes representing stratigraphic orientation, without requiring a full interpretation of the data.

Key Features:

- Genuine 3D.
- No trace-bias (as opposed to many alternative techniques, such as coherency/semblance techniques).

Any robust seismic attribute must be able to handle dipping layers, a characteristic feature of seismic signals in general, in a consistent way. The traditional approach of extracting attributes along vertical traces, irrespective of any dipping nature of the data, clearly imposes a risk of introducing artifacts. This kind of artifacts may not be apparent on horizon or time-slice views of the attributes, but do still impose risks of poor mapping, wrong positioning of detected events, and false event detections. In order to avoid such artifacts, the attribute extraction must either be invariant to dip and azimuth, or we need to compensate for dip and azimuth.

The approach we present to dip and azimuth estimation consists of three steps:

1. Gradient vector estimation, $\nabla x(t_1, t_2, t_3)$;
2. Local gradient covariance matrix estimation, $C(t_1, t_2, t_3)$;
3. Principal component analysis. The principal eigenvector represents the normal to the local reflection dip and azimuth.

In the sequel, we describe the elements of this setup.

2.1 Gradient Vector Estimation

If a three-dimensional signal, $x(t_1, t_2, t_3)$, has a fixed dip and azimuth, then the loci of the signal's iso-value surfaces, i.e., the points where the signal has constant value, will be parallel planes. Furthermore, the gradient of the signal, ∇x, will be perpendicular to these parallel planes. In real world data, the loci of the iso-value surfaces are not likely to be exact planes and will be varying over the volume. Hence, we must estimate the *local dominating* dip and azimuth. The first element in this is the gradient estimate.

For discrete data, the gradient is a discrete estimate. Several approximations of the derivative are possible, and we have chosen to primarily use the derivative of Gaussian, due to joint optimum resolution in time and frequency of Gaussian filters and nice scalability properties [6]. Adjusting the scale allows tuning of resolution vs. noise sensitivity.

The unit pulse response of a multi dimensional derivative of Gaussian filter is separable, and has the equation

$$h_\kappa(k) = \alpha_\kappa k \exp\left(-\frac{k^2}{2\sigma_\kappa^2}\right) \tag{1}$$

in the differentiated dimension and

$$h_\lambda(l) = \alpha_\lambda \exp\left(-\frac{l^2}{2\sigma_\lambda^2}\right) \tag{2}$$

in the non-differentiated dimensions, where

$$\nabla x(t_1, t_2, t_3) = \begin{bmatrix} \frac{\partial x(t_1,t_2,t_3)}{\partial t_1} \\ \frac{\partial x(t_1,t_2,t_3)}{\partial t_2} \\ \frac{\partial x(t_1,t_2,t_3)}{\partial t_3} \end{bmatrix}, \tag{3}$$

with one partial derivative component for each dimension. The αs are constant scaling factors (not complicated, but not really interesting either).

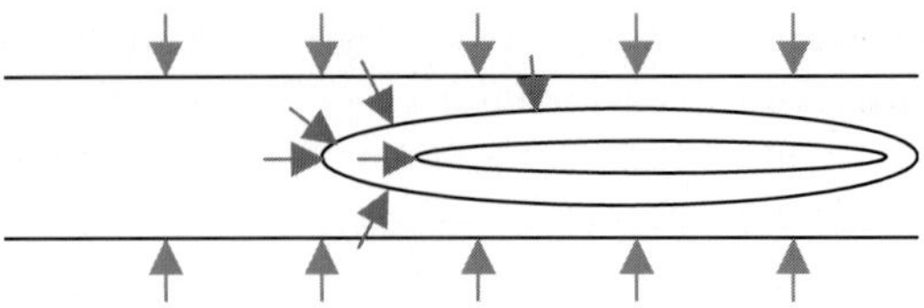

Fig. 2. Gradient estimates for a horizontal reflector with locally varying amplitude. The black lines correspond to reflection amplitude countours and the arrows to the gradient vectors.

2.2 Local Gradient Covariance Matrix Estimation and Principal Component Analysis

The three-dimensional gradient vector represents the local dip and azimuth of the data, but might be contaminated by noise and other artifacts. For example, a horizontal reflector with slight amplitude variations will not be represented by a vertically pointing gradient vector, but with gradient vectors pointing in several directions, as illustrated in Figure 2. Hence it is necessary to smooth the gradient estimate.

Due to wraparound effects (the effect occurring when the angle changes slightly, but its representation changes abruptly, e.g. wrapping from -180 to +180 degrees) of the dip and azimuth estimates, this smoothing is non-trivial. An approach dealing with this is to estimate the covariance matrix of the gradient vectors and determine the dip as the direction of the principal eigenvector of this matrix.

The *covariance matrix* is given by

$$\mathbf{C} = \begin{bmatrix} C_{11}C_{12}C_{13} \\ C_{21}C_{22}C_{23} \\ C_{31}C_{32}C_{33} \end{bmatrix}, \tag{4}$$

where

$$C_{ij} = E\left\{\left(\frac{\partial x}{\partial t_i} - \eta_i\right)\left(\frac{\partial x}{\partial t_j} - \eta_j\right)\right\}$$

and

$$\eta_k = E\left\{\frac{\partial x}{\partial t_k}\right\}.$$

However, the expectation operator, $E\{\cdot\}$, is global in nature. A localized estimate is obtained by replacing the global covariance estimate by a windowed local estimate,

$$C_{ij}(t_1,t_2,t_3) = \sum_{\tau_1,\tau_2,\tau_3} w(\tau_1,\tau_2,\tau_3) \cdot \left[\frac{\partial x(t_1+\tau_1,t_2+\tau_2,t_3+\tau_3)}{\partial t_i} - \eta_i\right] \cdot \left[\frac{\partial x(t_1+\tau_1,t_2+\tau_2,t_3+\tau_3)}{\partial t_j} - \eta_j\right],$$

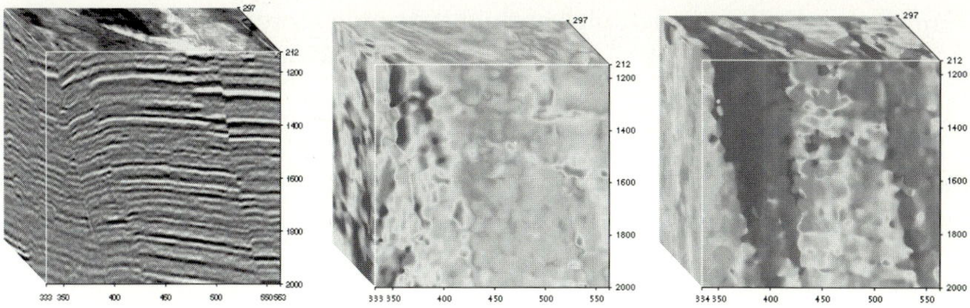

(a) (b) (c)

Fig. 3. [Reproduced in colour in Plate 4 on page 423.] Examples of the dip (b) and azimuth (c) attributes computed from the seismic cube (a). For the dip cube, increasing dip is here encoded from green through red to darker red. For the azimuth cube, yellow/green indicates that the layers are dipping to the right, whereas pink indicate layers dipping to the left.

yielding a localized covariance matrix, $\mathbf{C}(t_1, t_2, t_3)$, at each sample position. The window function, $w(s) = w(\tau_1, \tau_2, \tau_3)$, will typically be a low-pass filter. An implication of the windowing function is a smoothing of the dip and azimuth estimate. We have had good experience using a Gaussian low-pass filter,

$$w(\tau_1, \tau_2, \tau_3) = \frac{1}{\sigma_1\sigma_2\sigma_3\left(\sqrt{2\pi}\right)} \exp\left(-\frac{1}{2}\left(\frac{\tau_1}{\sigma_1}\frac{\tau_2}{\sigma_2}\frac{\tau_3}{\sigma_3}\right)^2\right),$$

for the window function, allowing us to easily adjust the noise sensitivity versus the resolution, and at the same time obtain joint optimum time-frequency resolution [6].

The dip and azimuth is found as the spherical angles of principal eigenvector, $\mathbf{v_1}(t_1, t_2, t_3)$, of the localized covariance matrix, $\mathbf{C}(t_1, t_2, t_3)$; one pair of angle estimates for each voxel.

The dip and azimuth attributes are powerful both for capturing properties of the seismic data and as a basis for compensating for the dip and azimuth. Examples of the dip and azimuth attributes are shown in Figure 3.

3 Stratigraphic Texture Attributes

Stratigraphic texture attributes are attributes capturing various internal configurations properties,which may be related to the stratigraphy.

3.1 Chaos

Key Principle: Maps the *chaoticness* of the local seismic signal, within a 3D window [12].

Main Uses: Fault, stratigraphy, or fluid type indicator, mapping for example

- Gas chimneys;
- Reef internal texture;
- Sink holes;
- Channel infill;
- Faults/discontinuities.

Key Features:

- Genuine 3D.
- No trace-bias (as opposed to many alternative techniques, such as coherency/semblance techniques [1], variance cube [2], etc.).
- Amplitude-invariant; will produce the same response for the same seismic signature, whether in a low- or high-amplitude region.
- Orientation-invariant; will produce the same response for the same seismic signature, whether in tilted or non-tilted strata.

There will be three eigenvectors, $\mathbf{v_i}(t_1, t_2, t_3)$, of each $\mathbf{C}(t_1, t_2, t_3)$-matrix, each of them associated with one eigenvalue, $\lambda_i(t_1, t_2, t_3)$. The larger $\lambda_i(t_1, t_2, t_3)$, the better $\mathbf{v_i}(t_1, t_2, t_3)$ describes the dip and azimuth. The larger the difference between the dominating $\lambda_i(t_1, t_2, t_3)$ and the two other $\lambda_i(t_1, t_2, t_3)$'s, the more reliable the dip and azimuth estimate is. Assuming without loss of generality that $\lambda_1 \geq \lambda_2 \geq \lambda_3$, one possible measure that captures this is the *chaos measure*

$$J = \frac{2\lambda_2}{\lambda_1 + \lambda_3} - 1. \tag{5}$$

By using this measure, regions with low consistency in the estimate typically correspond to regions with chaotic signal patterns. Hence, this measure is a suitable attribute for chaotic texture. This attribute is shown in Figures 4 (d) and 5(b). The chaos texture attribute is inherently dip- and azimuth-invariant, in addition to being amplitude-invariant. This invariance is crucial and allows us to select whether to explicitly accommodate for these properties in the analysis. Dip, azimuth and amplitude may require special treatment.

3.2 Flatness

Key Principle: Maps the *flatness* of the local seismic signal. *Flatness* is the degree to which local reflectors are flat/planar (not necessarily horizontal).

Main Uses: Stratigraphy indicator, mapping for example:

- Reef internal texture;
- Sink holes;
- Channel infill.

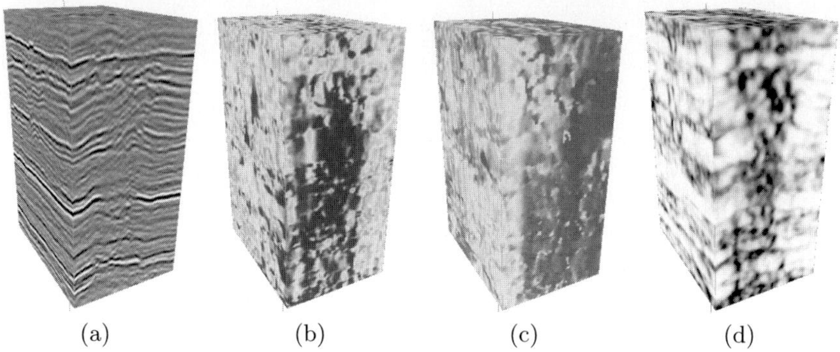

Fig. 4. [Reproduced in colour in Plate 5 on page 423.] An example of a seismic cube with a chaotic pattern representing a gas chimney is shown in (a), its dip and azimuth attributes in (b) and (c) and a chaos texture attribute highlighting the gas migration path in (d).

Key Features:

- Genuine 3D.
- No trace-bias (as opposed to many alternative techniques, such as coherency/semblance techniques, variance cube, etc.).
- Amplitude-invariant; will produce the same response for the same seismic signature, whether in a low- or high-amplitude region.
- Orientation-invariant; will produce the same response for the same seismic signature, whether in tilted or non-tilted strata.

Given that we have a proper orientation estimate for the 3D data cube (Section 2), measuring flatness is a matter of computing statistics of the orientation field. The flatness can thus be measured as the variance in the orientation field, locally. An example can be seen in Figure 5 (c).

3.3 Divergence

Key Principle: Maps the divergence of the local reflections by analysing reflection angles in short traces. A consistent trend in the angles will give a high divergence measure.

Main Uses: Stratigraphy indicator.

Key Features:

- Amplitude-invariant; will produce the same response for the same seismic signature, whether in a low- or high-amplitude region.
- Orientation-invariant; will produce the same response for the same seismic signature, whether in tilted or non-tilted strata.

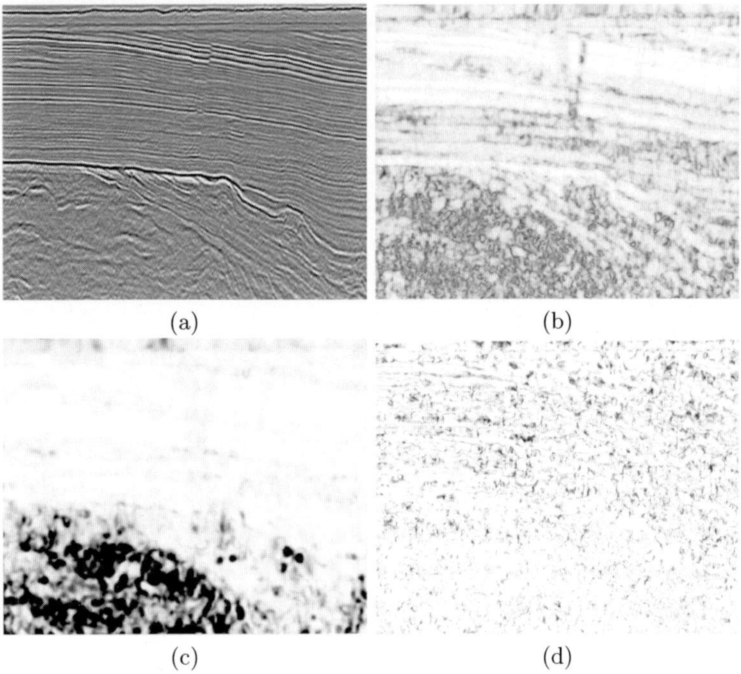

Fig. 5. Examples of the (b) chaos, (c) flatness, and (d) divergence texture attributes.

3.4 Dip Histogram Attributes

Key Principle: Captures an orientation-invariant measure by analysing the dip angles within a small window.

Main Uses: Generic stratigraphy indicator.

Key Features:

- Genuine 3D.
- No trace-bias (as opposed to many alternative techniques, such as coherency/semblance techniques, variance cube, etc.).
- Amplitude-invariant; will produce the same response for the same seismic signature, whether in a low- or high-amplitude region.
- Orientation-invariant; will produce the same response for the same seismic signature, whether in tilted or non-tilted strata.

Kovalev et al. [9] presented an approach based on 3D gradient histograms for 3D texture characterization. Their approach is very attractive in the descriptive power, but sensitive to the tradeoff between bin size and histogram accuracy. However, if we go back to Vail et al. [17], the stratigraphic textures are classically described by their 2D cross sectional views. Assuming validity

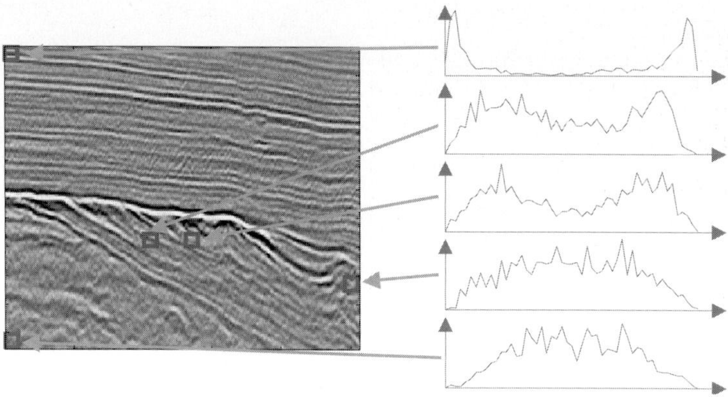

Fig. 6. Illustrations of the dip spectra of different subvolumes (the spectra are computed from 3D subvolumes).

of this approach also on 3D data, we expect that most of the information is captured in the dip angle alone, which opens up for faster and easier description of the orientation field by the histogram of only the dip angle.

Ignoring the amplitude of the gradient vectors when constructing the histogram easily incorporates amplitude-invariance. Furthermore, by using the dip angle from a 3D gradient vector, the method is genuinely three-dimensional. However, a few challenges still remain:

- The angle histogram is not orientation-invariant.
- The classical trade-off between bin sizes vs. numerical accuracy always applies to histogram techniques.

To overcome these, we suggest applying a modification to the scheme. A classical way of obtaining rotational invariance for shapes in image processing is the use of the Fourier descriptor [7]. The idea here is basically that by using a periodic Fourier transform (e.g. discrete Fourier transform, DFT) and discarding the phase (angle) of the Fourier coefficients, the orientation of a shape is not influencing the measure. The same can easily be applied to the histogram curves we have. By assuming the histogram to represent a periodic function and applying a periodic Fourier transform, the magnitude of the Fourier coefficients is representing the histogram at the same time as the phase (i.e., orientation) is discarded.

The next challenge is the bin sizes. However, the Fourier approach also comes in handy here. Consider if we construct the Fourier descriptor by applying the following steps:

- Compute angular histogram, implying fixed bin intervals.
- Compute the DFT of the histogram [5],

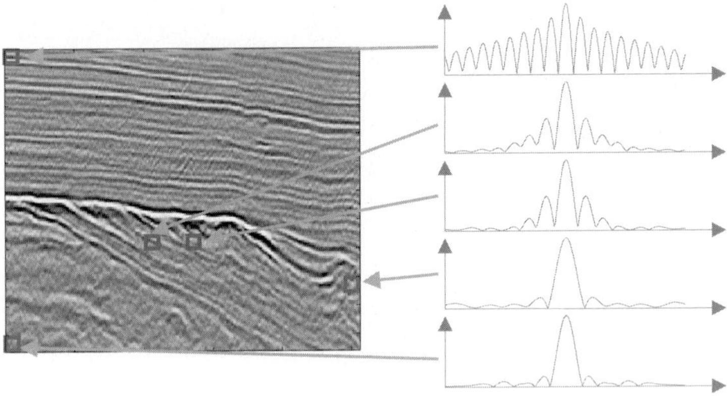

Fig. 7. Illustrations of the dip Fourier descriptors of different subvolumes. We can easily see how different textures are illustrated by different descriptor curves.

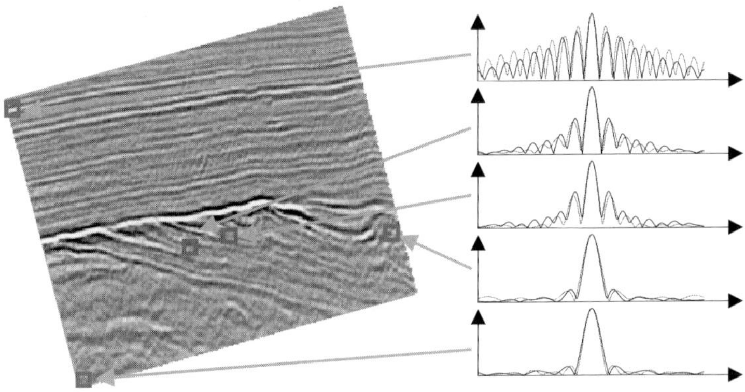

Fig. 8. Illustrations of the dip Fourier descriptors before and after rotation of the subvolumes. The dotted curves represent the descriptors before rotation and the solid curves after. We see that only minor changes can be observed, probably due to numeric effects in the rotation.

$$F(m) = \sum_{k=0}^{N-1} f(k) e^{-jmk2\pi/N} .$$

- Represent the histogram by the magnitudes of the Fourier coefficients.

The DFT computation on the histogram is identical to using the individual angles as input with k representing the angular bin. However, if we use the non-binned angles, we can still use the same framework, simply by replacing the DFT with a continuous and periodical Fourier transform, typically referred to as (continuous time) Fourier series [5]. This hence also avoids the bin size problem. We denote the magnitude of the resulting Fourier coefficients as the dip Fourier descriptors (dFd). The particular attribute cubes

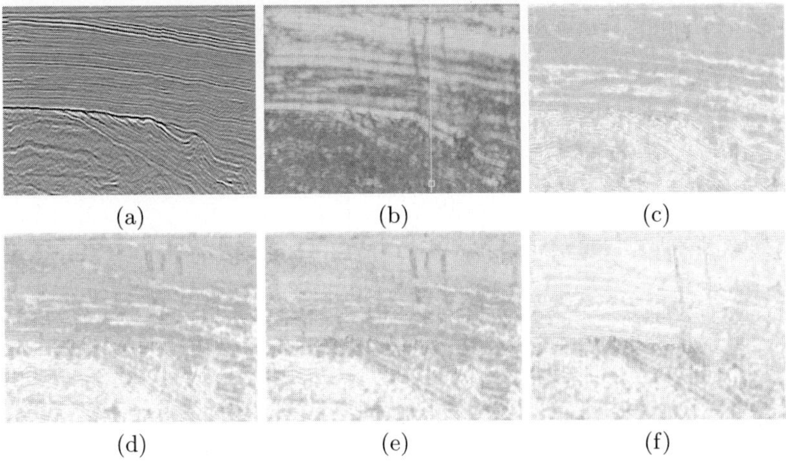

Fig. 9. Example dFd attributes. From the combined attribute response, we see discrimination between at least the 3-4 dominating facies in this section. The *frequencies* represented here are (b) $\omega = 2\pi/8$, (c) $\omega = 6\pi/8$, (d) $\omega = 7\pi/8$, (e) $\omega = 8\pi/8$, and (f) $\omega = 10\pi/8$.

produced are samples of the continuous Fourier series spectrum. Illustrations of the dFd attribute are given in Figure 9.

3.5 Gabor Filter Bank

Key Principle: Maps the frequency characteristic of pseudo-traces orthogonal to the stratification.

Main Uses: Generic stratigraphy indicator.

Key Features:

- Amplitude-invariant; will produce the same response for the same seismic signature, whether in a low- or high-amplitude region.
- Orientation-invariant; will produce the same response for the same seismic signature, whether in tilted or non-tilted strata.

Seismic trace frequency attributes, such as instantaneous frequency [15], are widely used in seismic data analysis. The instantaneous frequency attribute is available in most standard seismic data analysis packages.

An alternative to the instantaneous frequency, with the potential for capturing more than the single dominating frequency, is to use a multi-band frequency decomposition, in other words a *filter bank*. A very popular filter bank in the texture analysis literature is the Gabor filter bank [11]. In this case we use a one-dimensional Gabor filter bank applied along vertical or layer-orthogonal traces. Layer-orthogonal traces means short signal traces extracted orthogonal to the dominating dip and azimuth of the data (see

Section 2). The 1D Gabor filter is a band-pass filter constructed by a cosine modulated Gaussian. Its unit pulse response is

$$h(k) = e^{-\frac{1}{2}\frac{k^2}{\sigma^2}} \cos(2\pi f_0 k), \tag{6}$$

where f_0 is the center frequency of the pass-band and σ is the unit pulse response width. The parameter σ thus determines the pass-band width in the frequency domain (large σ implies small pass-band width and vice versa). The filter has infinite support, but is in practical experiments approximated by a finite discrete time filter through a cut-off at 2.0σ to 2.5σ.

Based on findings in [8], the center frequencies $2^{-3.5}\sqrt{2}$, $2^{-3.0}\sqrt{2}$, $2^{-2.5}\sqrt{2}$, $2^{-2.0}\sqrt{2}$, and $2^{-1.5}\sqrt{2}$ are used, thus constituting a 5-band filter bank. The filters are, along with the approach published in [8], followed by a local energy [11] extraction operation consisting of an absolute value and a Gaussian smoothing operation (see Subsection 6.1).

Finally, the amplitude is normalized such that the square sum of the attributes at a specific voxel position is constant. This last step is important since it will ensure a unified response irrespective of the local seismic contrast level. Thus a specific stratigraphic pattern will be equally well highlighted in low and high amplitude regions.

In the resulting attribute cubes, a large amplitude indicates a high presence of the specific frequency content. In Figure 10 we can see some examples of the Gabor attribute cubes.

3.6 Volume Reflection Spectral (VRS) Decomposition

Key Principle: Maps the VRS spectral frequency characteristic of analytic traces [16].

Main Uses: Generic stratigraphy indicator.

A similar measure to the Gabor filter bank, Subsection 3.5, may be extracted using the volume reflection spectral (VRS) decomposition [16]. In this approach, the seismic traces are decomposed by a Chebychev polynomial, and the polynomial coefficients are used as spectral components, representing an attribute volume. An example for this is shown in Figure 11.

4 Terminations

Key Principle: Traces the orientation field (see Section 2) to detect terminating strata [13].

Main Uses: Termination detection.

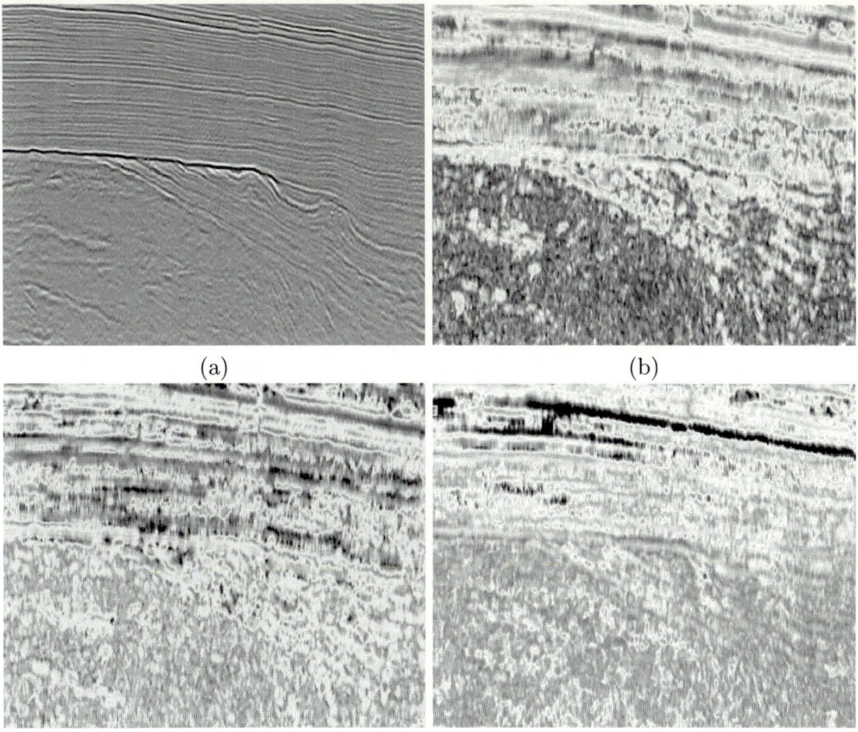

(a) (b)

(c) (d)

Fig. 10. A seismic cube and three of its Gabor attribute cubes. We can see how the different facies generate different attribute signatures.

Key Features:

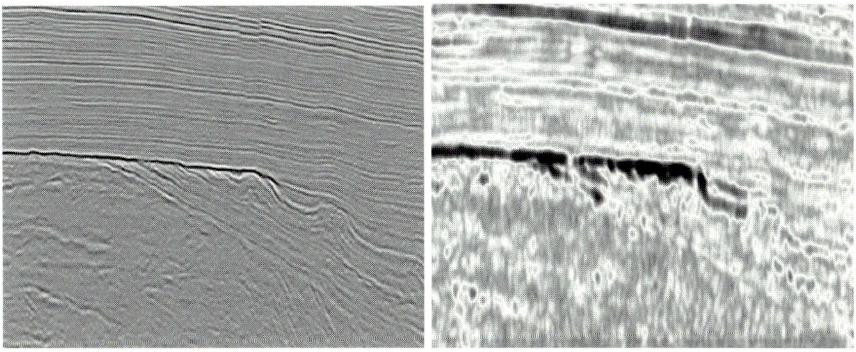

(a) (b)

Fig. 11. Example of a VRS attribute volume.

- Genuine 3D.
- Amplitude-invariant; will produce the same response for the same seismic signature, whether in a low- or high-amplitude region.
- Orientation-invariant; will produce the same response for the same seismic signature, whether in tilted or non-tilted strata.

The primary target of the automated termination detection technique is the identification of stratigraphic boundaries, such as erosional truncation, toplap, onlap, and downlap. In the technique, the local orientation of the seismic cross section is first estimated, see Section 2. Starting at any sample position (seed position) a flow line is generated. The flow line is the curve through that position having the orientation of the local angle estimates in its path.

A sketch of flow lines and terminations is given in Figure 12. The flow lines trace the local orientation of the seismic data, and in the position marked by an $\times$, these lines intersect. This intersection may be likely to correspond to one sedimentary package terminating into another. Several termination surfaces may be identified in the lower picture in Figure 12.

Note that the approach is fully automated. It is not restricted to detection of terminations only at reflectors. The terminations pinpoint the boundaries by virtually continuous lines or zones, provided sufficient quality and clarity of the seismic data. Care is taken to avoid artefacts with noisy data, low resolution, interference between neighbouring reflectors, etc.

The termination surfaces that are highlighted by the method correspond well to the surfaces that may be determined by a tedious manual interpretation done on the seismic cross section.

5 Discontinuities and Faults

Enhancing faults means enhancing discontinuities in the seismic data. This is, however, not straightforward, as the intersections between the different reflection layers constitute great amplitude changes. Hence, we need to enhance changes along the reflection layers and not orthogonal to them. In order to obtain this, the subsequently described attributes use a local dip estimate of the reflection layers as input. It is noted that also the chaos cube, Subsection 3.1, is a good fault attribute.

We would also like to point the attention towards the chapter [10] of Pedersen, Skov, Randen, and Sønneland, presenting a very attractive further processing step beyond the fault attributes for extracting the faults as surfaces in 3D.

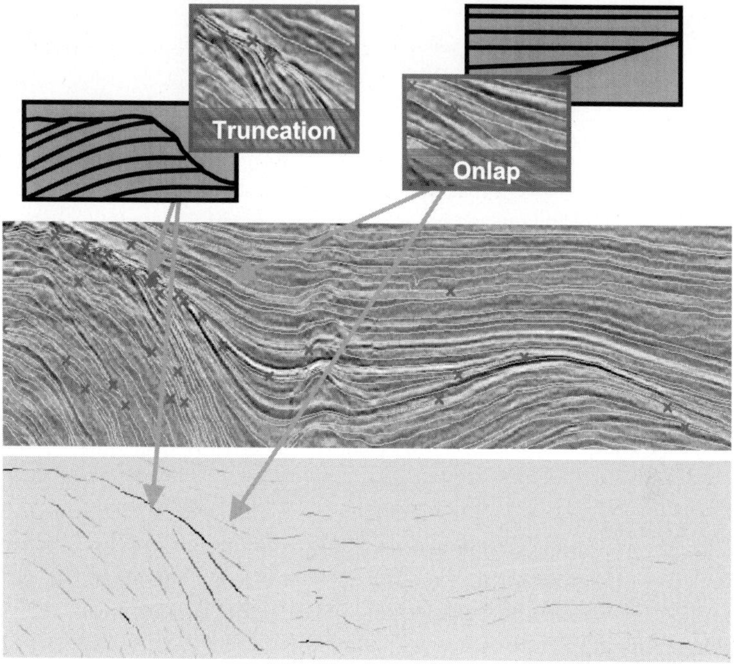

Fig. 12. The principles of termination detection by flow lines.

5.1 Fault Edge

Key Principle: Estimates the *edginess* of the seismic parallel to the stratification, using a derivative edge attribute.

Main Uses: Fault highlighting.

Key Features:

- Genuine 3D.
- Amplitude-variant; will produce different responses depending on low- vs. high-amplitudes. This is a good feature in the cases where amplitude correlates well to signal quality, but may be undesired if there are severe amplitude anomalies in the data.
- Orientation-invariant; will produce the same response for the same seismic signature, whether in tilted or non-tilted strata.

A fault appears as changes in amplitude in the reflectors. We should thus be able to enhance faults by measuring changes in the signal amplitude which is exactly what the edge enhancement attribute does. As previously discussed, the intersections between different layers comprise sharp edges and will produce large outputs by using conventional edge detection techniques. The edge enhancement attribute reduce this problem by using the local dip estimates

Fig. 13. Illustration of the concept of dip guided edge enhancement. (a) Derivatives indicating change along the reflector produces a projection with large magnitude in the dip plane, whereas (b) derivatives indicating change orthogonal to the reflector projects with a small magnitude.

of the reflection layers. The local dip estimate represents a plane, and by projecting the gradient vector with the derivatives,

$$\nabla x(n_1, n_2, n_3) = \begin{bmatrix} \frac{\partial x(n_1,n_2,n_3)}{\partial n_1} \\ \frac{\partial x(n_1,n_2,n_3)}{\partial n_2} \\ \frac{\partial x(n_1,n_2,n_3)}{\partial n_3} \end{bmatrix}, \tag{7}$$

onto this plane, changes that are nearly perpendicular to the reflector will produce projections with small magnitudes, whereas changes in the direction of the reflector will produce projections with larger magnitudes.

Taking the magnitude of the projected vector as the attribute value makes this attribute dependent on the amplitude in the seismic data. Faults in areas of low amplitude will thus have a weak signature which may be hard to detect for a human interpreter. The visual appearance can be corrected by applying some amplitude correction, but with the appropriate subsequent steps this may prove unnecessary in an automated fault extraction setting. This attribute does not introduce artifacts by smoothing and picks up very small amplitude changes (dimming effects of sub-seismic resolution faults). Hence, even very subtle faults, which are very hard to visually detect from the seismic data, are captured. Figure 14 shows an example of an edge enhanced cube.

5.2 Projected Principal Gradient

Key Principle: Estimates the *edginess* of the seismic parallel to the stratification, using the principal component of the derivative.

Main Uses: Fault highlighting.

Key Features:

- Genuine 3D.
- Amplitude-invariant; will produce the same response for the same seismic signature, whether in a low- or high-amplitude region. See Subsection 5.1 for an amplitude variant attribute.
- Orientation-invariant; will produce the same response for the same seismic signature, whether in tilted or non-tilted strata.

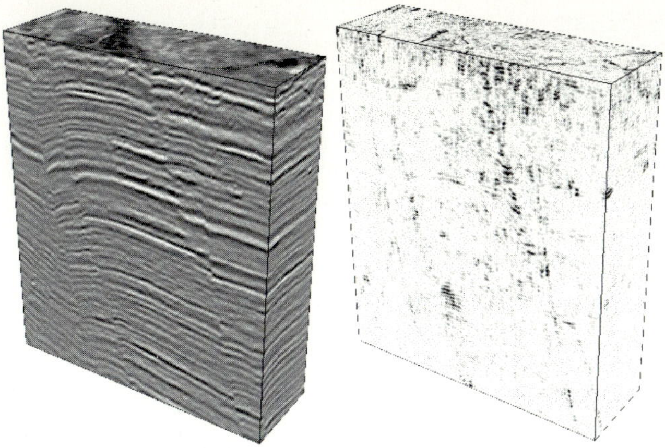

(a) (b)

Fig. 14. Edge enhanced cube (right) generated from the seismic cube (left).

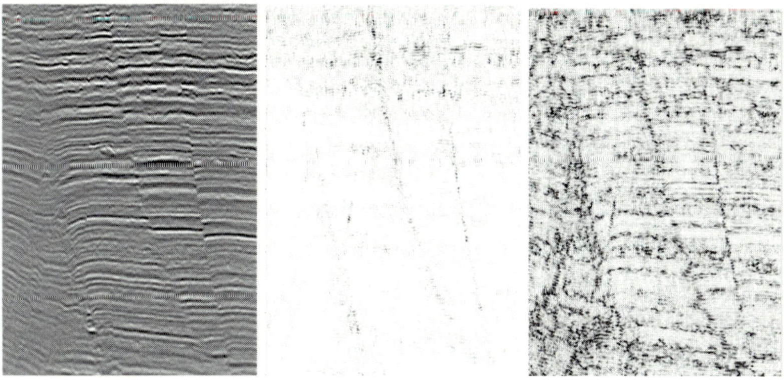

(a) (b) (c)

Fig. 15. The fault edge cube (b) and the projected principal gradient cube (c), showing the improved mapping of faults in low-amplitude regions.

Edges are important in many 3D data types, as they contain significant information on properties of the data. In seismic data, edges describe properties such as faults, horizons, and horizon continuity.

Referring to the previous section, one approach to make that attribute amplitude-invariant, is to project the principal component of the gradients within a small window, instead of projecting the gradient itself. The window can in general be arbitrary, but good results have been obtained by using a Gaussian window. An illustration is given in Figure 15.

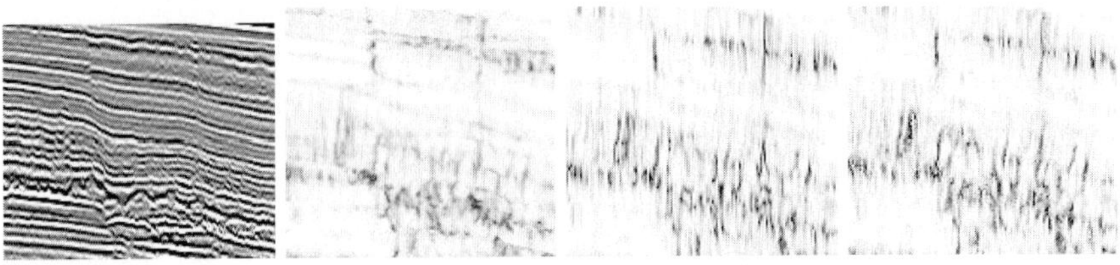

(a) (b) (c) (d)

Fig. 16. (d) Orientation-tuned chaos continuity attribute computed, with the input seismic shown in (a). The result in (b) and (c) shows non-adaptive chaos attributes, with an isotropic computation window (a) and a narrow vertical window (b). By careful examination, it can be observed that the result in (d) has a better definition of the fine features than the results in (b) and (c) on this data set.

5.3 Orientation-Tuned Chaos

Key Principle: Computes the chaos attribute (see Subsection 3.1) in pseudo-traces orthogonal to the local stratification (see Section 2).

Main Uses: Fault highlighting.

Key Features:

- Amplitude-invariant; will produce the same response for the same seismic signature, whether in a low- or high-amplitude region.
- Orientation-invariant; will produce the same response for the same seismic signature, whether in tilted or non-tilted strata.

This attribute, shown in Figure 16, is extracted guided by the dip and azimuth and is thus operating equally well in dipping as in non-dipping strata. It is a variant of the already described chaos attribute, but more tuned to detection of sharp lineaments orthogonal to the stratification. This property makes it attractive for highlighting faults that are more or less orthogonal to the stratification. The attribute will yield continuous and sharp responses irrespective of dipping layers and poor amplitude.

5.4 Variance Attribute

Key Principle: Estimates the variance in small horizontal planes and smoothes the result vertically [2].

Main Uses: Fault highlighting.

Key Features:

- Amplitude-invariant; will produce the same response for the same seismic signature, whether in a low- or high-amplitude region.
- Not orientation-invariant.

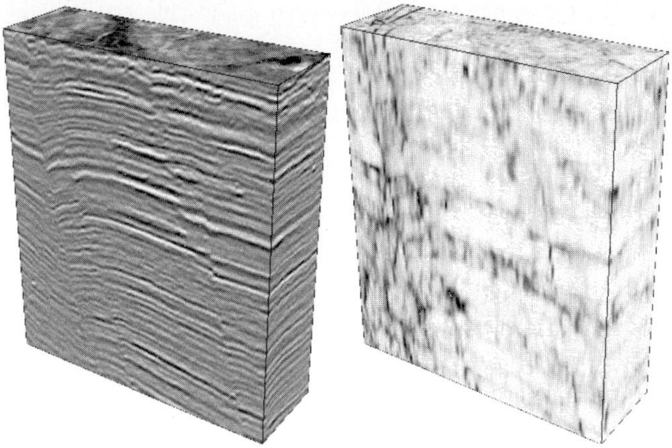

(a) (b)

Fig. 17. Variance cube (b) generated from the seismic cube (a).

The variance cube has the local variance as a measure of signal unconformity. For each voxel, the local variance is computed from horizontal sub-slices. If this slice is within an unbroken reflection layer, the amplitude variance will be small whereas amplitude changes due to a fault will result in a larger variance. Next, the variance estimate is smoothed by a vertical window and amplitude normalized. For more details on the variance attribute, refer to Bemmel and Pepper [2].

5.5 Thinning

Key Principle: Cuts any seismic attribute orthogonal to the key features and retains only the peak values, thus obtaining a thinned (skelotonized) volume [12].

Main Uses: Fault attribute *enhancement.*

Key Features:

- Genuine 3D.
- Orientation-invariant; will produce the same response for the same seismic signature, whether in tilted or non-tilted strata.

The response resulting from the fault enhancement method are not always sharp. Conditioning of the result may be performed by thinning. The thinning operation will for each voxel check whether the current voxel is the peak value of a neighborhood of n voxels forming a line perpendicular or nearly perpendicular to the fault. If so, it is retained, otherwise it will be discarded. As a result, the surfaces will be thinned. In order to identify the neighborhood in which to look for a peak value, an estimate of the normal to the fault in the voxel of interest is needed. One way to make this estimate is to use the orientation of the gradient vector after projection onto the orientation plane, as described in connection with the edge enhancement attribute. This vector will generally be orthogonal to the fault. The estimates should be smoothed, Section 2, in order to increase reliability.

Another approach to fault attribute enhancement is presented in the chapter [10] of Pedersen, Skov, Randen, and Sønneland.

6 Pre- and Post-Conditioning

The noise level of seismic data varies tremendously and noise is a significant problem in many data sets. One way to suppress noise is smoothing. Furthermore, using the principles of scale-space theory, detection of geological features at different resolutions may also take advantage of smoothing; large smoothing leaves primarily the major features, etc. In this section, we present two variants of seismic data smoothing, differing in the preservation of *features of interest.*

6.1 Gaussian Low-Pass Smoothing Filter

Key Principle: Smooths a data set by local averaging with a Gaussian weighted averaging.

Main Uses:

- General smoothing.
- Increased signal to noise ratio.
- Estimating local signal magnitude.

Key Features:

- Genuine 3D.

The Gaussian filter is one of the most popular filters for smoothing data, in particular for its joint optimum time-frequency localization properties [6]. The Gaussian filter is given by

$$h_G(k) = \frac{1}{\sqrt{2\pi}\sigma} \exp\left(-\frac{1}{2}\frac{k^2}{\sigma^2}\right), \tag{8}$$

where the parameter σ defines the width of the filter, or in other words the degree of the smoothing. A larger σ gives more smoothing. Depending on the application, useful values for σ range from 0.5 to 4.0 or higher. The Gaussian filter is separable, meaning that the filter above is applied (convolved) in each dimension separately. Typically, different σ's are allowed for each dimension.

One particular application of smoothing filters is for generating *local energy* cubes. Smoothing the absolute value of a seismic cube, for example,

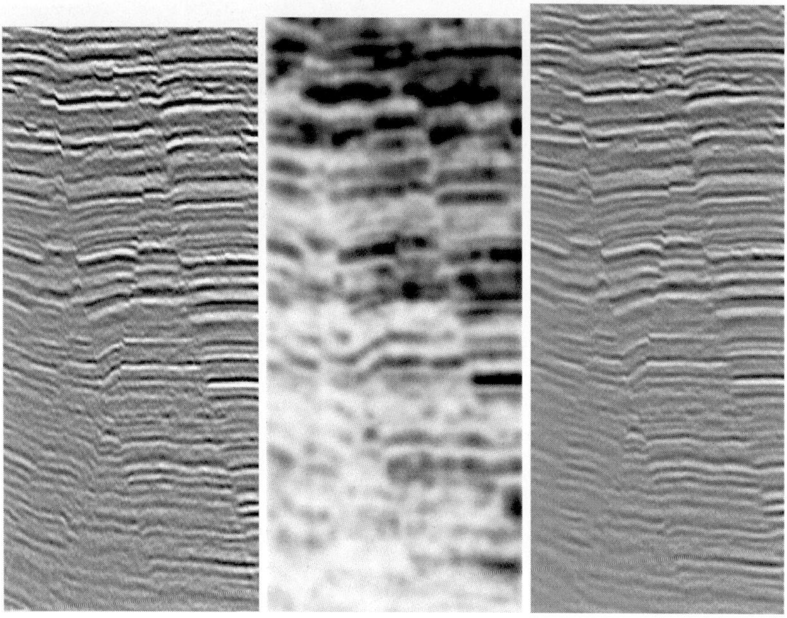

(a) (b) (c)

Fig. 18. (b) Illustration of the use of a Gaussian low-pass filter for extracting the local signal magnitude of (a). Here the Gaussian filter with $\sigma = 3.0$ in all three dimensions was preceded by an absolute value pre-processor. (c) Gaussian dip-guided layer-parallel smoothing result. Observe how the continuity of the seismic signal has been enhanced in the filtered image compared to the original (a).

produces a cube with high values at positions where the local seismic magnitude is high (irrespective of whether the seismic signal has a zero crossing, etc. at that voxel). The illustration in Figure 18 (b) is an example for this.

6.2 Gaussian Dip-Guided Filter – Layer-Parallel Smoothing

Key Principle: Smooths a data set by local averaging with a Gaussian weighted averaging along flow surfaces following the local dip/azimuth (see Section 2), thus obtaining increased layer continuity without sacrificing vertical resolution.

Main Uses:

- Increased reflector continuity.
- Increased signal to noise ratio.

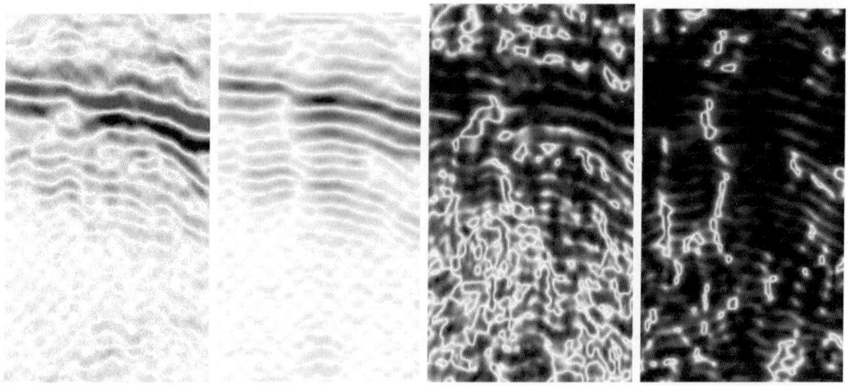

(a) (b) (c) (d)

Fig. 19. Gaussian dip-guided layer-parallel smoothing result (b). Observe how the continuity of the seismic signal has been enhanced. Fault attribute cubes corresponding to the data in (a) and (b) can be seen in figures (c) and (d), respectively. The improved fault mapping is clearly observed.

Key Features:

- Structurally consistent smoothing along reflections.
- No smoothing vertically.

By knowing the locally dominating strata orientation in the seismic data (the dip and azimuth), it is possible to do any kind of filtering parallel or orthogonal to the strata. One incarnation of this general concept is smoothing parallel to the layers, while not applying smoothing orthogonally. With such a filter, the *vertical* resolution is preserved, but lateral continuity is enhanced. A Gaussian filter applied *parallel to* the dominating orientation is used. This module first estimates the dip and azimuth (see Section 2) of the seismic data, then extracts surfaces guided by the dip and azimuth and finally smooths the signal along these surfaces. See the examples in Figures 18 (c) and 19(c).

6.3 Gaussian Dip-Guided Flow Line Filter – Layer-Orthogonal Smoothing

Key Principle: Smooths a data set by local averaging with a Gaussian weighted averaging along flow lines orthogonal to the local dip/azimuth (see Section 2), thus obtaining increased continuity of vertical events without sacrificing horizontal resolution.

Main Uses: Increasing fault attribute continuity.

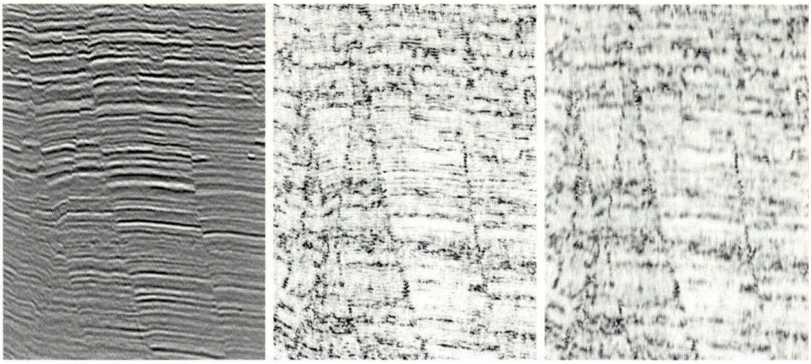

(a) (b) (c)

Fig. 20. Gaussian dip-guided layer-orthogonal smoothing result. The seismic image (left) is used for computing the dip and azimuth, and then the projected principal gradient fault attribute (Subsection 5.2) (middle) is smoothed along short layer-orthogonal traces by a Gaussian filter to yield the result to the right.

Key Features:

- Structurally consistent smoothing perpendicular to reflections.
- No smoothing horizontally.

When enhancing vertically aligned events, for example faults, a filter smoothing orthogonal to the layer structure may be desirable. Such a filter may be realized by applying a Gaussian filter to traces extracted orthogonal to the locally dominating dip and azimuth (see Section 2). A result of this approach is illustrated in Figure 20.

Acknowledgement

The work was partly supported by the European Union through the TriTex project, contract no. IST-1999-20500, and the Norwegian Research Council through the Demo2000 program, contract no. 136706/298. We also wish to thank Dave Hunt, Kai Hogstad, Geir Elvebakk, Daniel Stoddart, and license owners Norsk Hydro, Statoil, AGIP, and Fortum and for collaboration and permission to publish data.

Hessian-based (otherwise known as DET) corner detector are reviewed. In Section 4, we present the 3D first and second order structure tensors and their properties. Section 5 demonstrates the use of these tensors in the analysis of 2D and 3D seismic data, and in Section 6 we draw our conclusions.

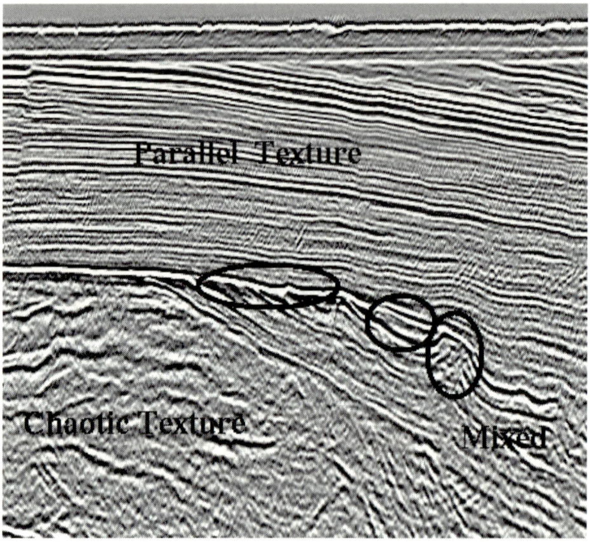

Fig. 1. A seismic section showing three different textures. The ovals mark regions of interest.

2 Literature Review

2.1 First Order Tensor

The first order structure tensor has been extensively used in image analysis for locating distinctive regions in images. This tensor is known by many different names: gradient structure tensor, second-moment matrix, scatter matrix, interest operator and windowed covariance matrix. It is defined in terms of the first derivative of the image and has been introduced for the detection of lines, edges and corners. One of the first attempts to use it was that of Moravec [29] who simply calculated the average variation (un-normalised autocorrelation function) of the image inside a rectangular window, by shifting the window in several directions and measuring the minimum value of the autocorrelation function. This minimum value was a measure of curvature of the point under consideration [30]. Constant image intensity would produce a small variation in all shifts, and this would imply absence of significant local curvature. An edge would produce a small variation along one direction and a large variation perpendicular to the edge, and this would indicate the presence of a small curvature along the direction of the edge. Finally, if a corner point existed in the windowed area, then a large variation would have been observed in all shifts, indicating the presence of high local curvature. As a result, corner points were detected in images as the points where the variation in all shifts was above a certain thereshold.

Harris and Stephens [16] identified the problem created by using rectangular windows, which enhance noise due to their sharp edges, and proposed the use of smoothly decaying windows such as Gaussian ones. Another shortcoming pointed out by Harris and Stephens was the anisotropy introduced due to the use of four directions for the estimation of the variation, which was dealt with by using first order Taylor expansion about the origin to include all possible small shifts. They defined the first order image structure tensor, and introduced a measure of *cornerness* and *edginess* based on its eigenvalues. The eigenvalues of the tensor were shown to be proportional to the principal curvatures of the autocorrelation function seen as a surface. A mathematical proof that explains how the Harris corner detector estimates the image surface curvature is given by Noble [31].

The gradient structure tensor has been used extensively in the estimation of local structure and local orientation. This is known as the shape-from-texture problem where the distortion of texture patterns can provide information about surface shape or orientation. Some of the first works based on the gradient tensor was that of Knutsson [23]. The tensor was used for the estimation of the orientation of 3D surfaces so that the ambiguity introduced by the use of the normal vector was overcome. Even before that, the structure tensor constructed from the first derivatives had been adopted by Kass and Witkin [21] for the calculation of the local orientation of texture patterns in 2D images. Vliet and colleagues produced a considerable amount of work extending Knutsson's approach. They used the average gradient-square tensor for orientation estimation of anisotropic patterns in 2D. Its advantage was the removal of the ambiguity introduced by averaging the gradient orientation vectors in windows which normally cancel out in areas of linear structures [45]. The curvature of 2D curvilinear structures [46] and 3D ones [2] was also calculated by a modified version of the gradient structure tensor. Local curvature of iso-grey-level surfaces in images was estimated with the help of the gradient structure tensor used to estimate local orientation [36, 35].

Based on Knutsson's work, Westin et al., detected effectively thin bone structures in 3D Computer Tomography (CT) data [51] and enhanced blood vessels in 3D Magnetic Resonance Angiography (MRA) images [52], by performing adaptive filtering using the gradient tensor for the estimation of the local orientation.

Scale space theory is another tool of image analysis. The idea is to produce a one-parameter family of versions of the original image which are gradually smoothed with the details being progressively blurred. The terms diffusion tensor and scale-space are highly related [47, 48]. Diffusion is the physical process of equalising concentration differences between two areas, creating a flux. The diffusion equation relates the flux to the gradient of the concentration. In image processing the concentration represents the grey values of the image. A diffusion tensor was introduced to describe the diffusion mechanism. This diffusion tensor is either constant over the image, which is the case of isotropic diffusion, or it is applied in specific directions, which is the case of anisotropic diffusion. In addition, the diffusion may be linear or nonlinear depending on the linearity of the diffusion equation. Diffusion is used to smooth images in all directions when linear isotropic diffusion filtering is applied, or along specific structures as in anisotropic diffusion filtering. The first case yields the theory of scale-space and is equivalent to a simple smoothing with a Gaussian filter. The second case has been shown to process better line-like or other structures. A significant amount of work that uses the diffusion tensor is based on the idea of applying anisotropic diffusion filtering on the image which results in smoothing the image along preferred directions, in different scales. The preferred orientation or coherence orientation is estimated from the eigenvectors of the structure tensor of the image. The diffusion tensor is also constructed from these eigenvectors so that diffusion is applied in one direction. Thus, flow-like structures [11, 49] and creases (ridges and valleys) [39] can be detected in images.

Creases in images may also be detected by use of the gradient structure tensor directly [28]. Ridges in medical images were detected with the use of both the gradient and Hessian tensors in different scales of smoothing, with the gradient tensor outperforming the Hessian tensor [41].

Diffusion Tensor Magnetic Resonance Imaging, or DT-MRI, is a technology that measures the random motion of hydrogen atoms within water molecules referred to as diffusion. Water in tissues containing a large number of fibres, like skeletal muscle, cardiac muscle, and brain white matter, diffuse fastest along the direction of the fibres, and slowest along the direction orthogonal to it. In contrast, in tissues which contain few fibres, water diffuses in a spherical pattern. DT-MRI includes such information about how water diffuses in tissues, and it adds to conventional MRI the capability of measuring the random motion of water molecules in three dimensions. Basser et al. [6, 7], were the first to describe the explicit relationship between a diffusion tensor and the magnetic resonance signal. Since then, there has been a large amount of work on the diffusion tensor used in medical images [50].

Finally, the structure tensors found also application in image and video analysis. Hladuvka et al. [17] performed direction driven grey level interpolation along the directions identified by the analysis of the structure tensor in 3D images. More specifically, they calculated the eigensystem at grid points and employed interpolation in the directions defined by the eigenvectors of the structure tensor for the estimation of the grey values at off-grid localities.

Image sequences have been regarded as functions of space and time, thus they define hyper-surfaces. The curvature of the hyper-surfaces gives important information, so the structure tensor has been used in relation to motion parameter estimation [3, 4, 5] and optical flow estimation [20, 40].

2.2 Second Order Tensor

The second order tensor or Hessian matrix of the image function is defined in terms of the second partial derivatives of the image.

It was initially used by Baudet [8], who calculated the image Gaussian curvature and searched for saddle points on the image brightness surface. More specifically, he found the extrema of the product of the eigenvalues of the Hessian matrix which actually correspond to the Gaussian curvature. Dreschler [12] detected corners as points between two extrema of the Gaussian curvature.

The second order tensor in 2D images was also used by Noble [31], who pointed out that the local image surface can be classified according to the Hessian matrix determinant as a planar point (zero determinant), parabolic point (zero determinant), hyperbolic point (negative determinant) and elliptic point (positive determinant). Points of interest are those which contain strong intensity variation such as the hyperbolic and elliptic points.

The Hessian matrix appears also in scale-space theory. The connection of the diffusion equation with the image structure expressed in terms of the Hessian matrix in different scales was initially made by Koenderink [24]. Frangi et al. [14] exploited the three eigenvalues of the second order Hessian matrix of 3D medical images with the purpose of developing vessel enhancement filters. A *vesselness* function was introduced, which combined the three eigenvalues with the scale of the Gaussian filter used for differentiation. The final result was obtained by integrating the vesselness measure for different scales of the filter. In this way, both thin and wide vessels were detected. Furthermore, Sato et al. used the Hessian matrix first to detect lines by multi-scale integration [37], and then to detect sheets and blobs in 3D images [38]. Lin [26] introduced an alternative way of describing the shapes in an image by using the spherical harmonics in combination with the Hessian matrix, and performed enhancement and detection of string structures and stenosis in 3D MRA medical images.

Hladuvka et al. [18] thresholded two out of the three eigenvalues of the Hessian matrix of 3D brain images to reduce the volume data for visualisation purposes. Blood vessels were identified in 3D medical images by Bhalerao and Wilson [9, 10]. Initially, the local structures were modelled with Gaussian profiles. For the estimation of the parameters of the profiles and their classification into lines, planes, or spheres, the Multiresolution Fourier Transform was used, which takes into consideration different scales enabling the local structure to be fixed in a specific scale.

Finally, Medioni and colleagues [25, 42, 43, 44] used the tensor voting technique in different applications of image analysis. Tensor voting is an approach of extracting salient structures by encoding data and corresponding uncertainties in the Hessian matrix.

2.3 Structure Tensors in Seismic Image Analysis

Although the image structure tensor has recently been introduced in the medical imaging field, where clear structures exist and need to be identified, there is limited amount of work done in the field of seismic images. This may be explained by the fact that structures present in the Earth's subsurface are not of the same nature one can find in other images depicting specific objects that need to be identified and segmented.

Segmentation of chaotic patterns in 3D seismic images, based on the eigenvalues of the covariance matrix of the gradient vectors, was performed in [33]. Curvilinear structures called channels were also detected in seismic volumes, [2], by use of the eigenvalues of the gradient tensor.

Just recently, Bakker et al. [2] and Bakker [1] detected channels and faults in 2D and 3D seismic images, by using the gradient structure tensor for detecting the position of these structures. More specifically, they estimated their orientation by using the eigenvectors of the tensor and used steered adaptive anisotropic filters, elongated according to the shape of the structure under examination along the estimated orientation. These filters enhanced the structure by noise removal without degrading it. Furthermore, the gradient structure tensor was modified into a curvature corrected structure tensor, to account for plane-like and line-like curvilinear structures.

In this chapter, we shall demonstrate the use of both the first and second order tensors in 2D and 3D seismic images for the detection and segmentation of areas of interest for geophysicists. The identification of the chaotic area can be based on the fact that the small line segments that make up the texture have very different orientations, creating junctions at the points they intersect, called corners. Thus, a simple corner detection algorithm could identify such areas. The Harris corner detector (otherwise known as the Plessey algorithm) [16], is an improvement of the Moravec algorithm and has been proven to work well in detecting corners in the past. Its use is demonstrated here on seismic cross-sections for the isolation of chaotic regions.

3 Definitions of the 2D Structure Tensors

3.1 First Order Tensor

The image structure tensor has often been used especially for corner detection [8, 16, 29]. The idea is based on the fact that the change of image intensity or *squared difference* of the image will dramatically vary in areas of points of interest, e.g. corners. The squared difference of an image is defined as

$$E(x,y) \equiv \sum_{x',y'} w(x',y')[I(x'+x,y'+y) - I(x',y')]^2$$
$$= \langle [I(x'+x,y'+y) - I(x',y')]^2 \rangle, \tag{1}$$

where (x,y) is the position of a pixel in the image, $w(x,y)$ is a window function which can be a rectangle or a smooth Gaussian filter, and $I(x,y)$ is the intensity value at the specific position of the image. The symbol $\langle\cdot\rangle$ represents the average value, produced by the smoothing procedure. The squared difference is related to the autocorrelation function $C(x,y)$ by the expression

$$E(x,y) = \langle I(x'+x,y'+y)^2 \rangle + \langle I(x',y')^2 \rangle - 2\langle I(x'+x,y'+y)I(x',y') \rangle$$
$$= \langle I(x'+x,y'+y)^2 \rangle + \langle I(x',y')^2 \rangle - 2C(x,y).$$

Note that if the local energy of the image is constant, i.e.,

$$\langle I(x'+x,y'+y)^2 \rangle = \langle I(x',y')^2 \rangle = \text{const},$$

then the minima of the square difference and the maxima of the autocorrelation function will coincide.

For *small* shifts (x,y), the image function at $(x'+x,y'+y)$ can be approximated by Taylor series expansion at point (x',y') as

$$I(x'+x,y'+y) \approx I(x',y') + xI_x + yI_y + \frac{x^2}{2}I_{xx} + \frac{y^2}{2}I_{yy} + xyI_{xy} \tag{2}$$

and therefore,

$$I(x'+x,y'+y) - I(x',y') \approx xI_x + yI_y, \tag{3}$$

where I_x and I_y are the first partial derivatives of the image in direction x and y, respectively. Moreover, I_{xx} and I_{yy} are the second order partial derivatives in these two directions and I_{xy} is the mixed second order partial derivative, all computed at point (x',y'). In the above Taylor expansion the left hand side of equation (2) can be approximated by the first three terms only of the right hand side. In this case, the difference of the image in two positions can be represented by the equation of a line, as it can be seen from equation (3). The square which appears in equation (1) is then given by

$$[I(x'+x,y'+y) - I(x',y')]^2 \approx [xI_x + yI_y]^2$$
$$= x^2I_x^2 + y^2I_y^2 + 2xyI_xI_y$$
$$= (x\ y)\begin{pmatrix} I_x^2 & I_xI_y \\ I_xI_y & I_y^2 \end{pmatrix}\begin{pmatrix} x \\ y \end{pmatrix}.$$

Since only the values of the derivatives depend on (x', y'), $E(x, y)$ can be rewritten as

$$E(x,y) = \begin{pmatrix} x & y \end{pmatrix} \begin{pmatrix} \langle I_x^2 \rangle & \langle I_x I_y \rangle \\ \langle I_x I_y \rangle & \langle I_y^2 \rangle \end{pmatrix} \begin{pmatrix} x \\ y \end{pmatrix} \equiv \begin{pmatrix} x & y \end{pmatrix} M \begin{pmatrix} x & y \end{pmatrix}^T, \qquad (4)$$

where T denotes the transpose of a matrix and M is called the *image structure tensor*. It has been proven that this matrix is invariant under translation and its eigenvectors are rotationally invariant.

It is immediately obvious that two scales were introduced in the definition of the tensor, the scale involved in differentiation (local scale) and the scale involved in the smoothing window (integration scale) [27]. The first scale is the local smoothing done when the derivatives are determined, since derivatives estimated from un-smoothed data are very noisy. The integration scale refers to the size of window $w(x, y)$ which defines the area around which the average is taken, and is related to the size of the structure that needs to be detected. The two scales are closely related since the integration scale is generally an upper limit of the local scale. More specifically, the integration scale is three to ten times the local scale [1]. In general, a Gaussian function is used for the smoothing since it is rotationally invariant and separable.

The squared difference will have a specific shape according to the local feature present in the image. It will be bowl-shaped if pixel (x, y) is different from its surroundings, implying a corner in the region; it will be ridge shaped parallel to a linear feature, if (x, y) is a point of a line, and it will be constant in the case (x, y) is in a constant grey value region.

The information extracted from the structure tensor is based on the eigenvectors and the eigenvalues of it. Due to the fact that the tensor is symmetric and positive semi-definite, the structure tensor can be graphically represented by an ellipse shown in Figure 2. Also it is expected that the eigenvalues are real and positive and they represent the gradient magnitude. The largest eigenvalue corresponds to the dominant orientation of the largest variation. The eigenvectors define a gauge-coordinate system and the eigenvector corresponding to the largest eigenvalue shows the direction along which the first derivative is maximal.

The eigenvalues λ of the image structure tensor can be used to detect lines, corners or constant grey value regions. The characteristic equation of matrix M is

$$\lambda^2 - (\langle I_x^2 \rangle + \langle I_y^2 \rangle)\lambda + \langle I_x^2 \rangle \langle I_y^2 \rangle - \langle I_x I_y \rangle^2 = 0.$$

The solutions of this characteristic equation are the eigenvalues of tensor M:

$$\lambda = \frac{1}{2} \left[\langle I_x^2 \rangle + \langle I_y^2 \rangle \pm \sqrt{(\langle I_x^2 \rangle + \langle I_y^2 \rangle)^2 - 4[\langle I_x^2 \rangle \langle I_y^2 \rangle - \langle I_x I_y \rangle^2]} \right].$$

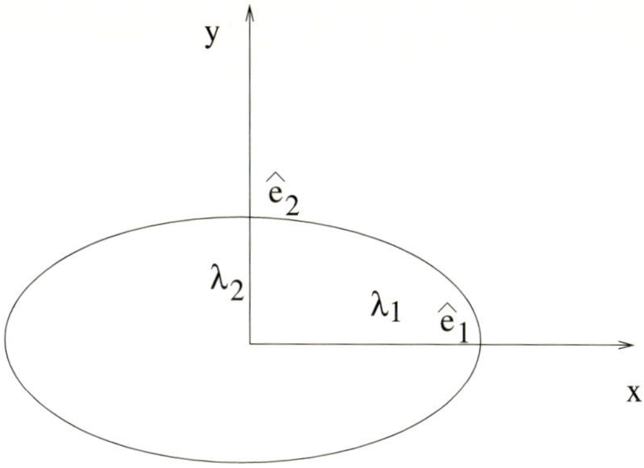

Fig. 2. Ellipse corresponding to the 2D tensor.

The orientation ϕ of each eigenvector is given by

$$\tan\phi = \frac{\lambda - \langle I_x^2 \rangle}{\langle I_x I_y \rangle},$$

and the determinant $|M|$ of the tensor is $|M| = \langle I_x^2\rangle\langle I_y^2\rangle - \langle I_x I_y\rangle^2$.

It is easy to show how information extracted from the structure tensor is related to the structure present in an image. Let us consider that image I depicts a linear feature which can be represented by a function $f(ax + by)$, where a and b are some constants. The direction of the linear structure is $-a/b$. The first partial derivatives I_x and I_y are

$$I_x = \frac{\partial f}{\partial x} = af' \quad \text{and} \quad I_y = \frac{\partial f}{\partial y} = bf'. \tag{5}$$

The ratio of I_x over I_y is a constant, equal to $c \equiv a/b$. Then $I_x = c \times I_y$ and the determinant $|M|$ becomes

$$\begin{aligned} |M| &= \langle I_x^2\rangle\langle I_y^2\rangle - \langle I_x I_y\rangle^2 \\ &= \langle c^2 I_y^2\rangle\langle I_y^2\rangle - \langle c I_y I_y\rangle^2 \\ &= c^2\langle I_y^2\rangle\langle I_y^2\rangle - c^2\langle I_y^2\rangle^2 = 0. \end{aligned}$$

The solutions of the characteristic polynomial are in this case

$$\lambda_1 = \langle I_x^2\rangle + \langle I_y^2\rangle \quad \text{and} \quad \lambda_2 = 0.$$

Note that one eigenvalue is zero, corresponding to the direction along the line where there is no grey value variation, and the other eigenvalue corresponds to the direction of the greatest variation. The orientation of the eigenvector corresponding to the non-zero eigenvalue, is given by

differentiated using second derivatives of Gaussian filters, along the other diagonal direction. The determinant of tensor B is expected to become zero in regions of diagonal linear features and non-zero in regions of linear horizontal or vertical features.

The simultaneous consideration of the largest determinant values of both tensors A and B should isolate areas which are not linear in the vertical, horizontal or diagonal directions. This would be a useful approach in detecting the chaotic regions, as well as regions of interest in the seismic images which do not contain linear features.

4 Definitions of the 3D Structure Tensors

4.1 First Order Tensor

The *squared difference* of a 3D image is defined as

$$\begin{aligned} E(x,y,z) &\equiv \sum_{x',y',z'} w(x',y',z')[I(x'+x,y'+y,z'+z) - I(x',y',z')]^2 \\ &= \langle [I(x'+x,y'+y,z'+z) - I(x',y',z')]^2 \rangle, \end{aligned} \tag{9}$$

where (x,y,z) is the position of a voxel in the image, $w(x,y,z)$ is a window function which can be a rectangle or a smooth Gaussian filter, and $I(x,y,z)$ is the intensity value at the specific position of the data.

For *small* shifts (x,y,z), the image function at $(x'+x,y'+y,z'+z)$ can be expanded in Taylor series as

$$\begin{aligned} I(x'+x,y'+y,z'+z) \approx\ & I(x',y',z') + xI_x + yI_y + zI_z \\ & + \frac{x^2}{2}I_{xx} + \frac{y^2}{2}I_{yy} + \frac{z^2}{2}I_{zz} \\ & + xyI_{xy} + xzI_{xz} + yzI_{yz}. \end{aligned} \tag{10}$$

Then

$$I(x'+x,y'+y,z'+z) - I(x',y',z') \approx xI_x + yI_y + zI_z, \tag{11}$$

where I_x, I_y and I_z are the first order partial derivatives of the image in the three directions x, y and z respectively, I_{xx}, I_{yy} and I_{zz} are the second order partial derivatives in the three directions, and I_{xy}, I_{xz} and I_{yz} are the mixed second order partial derivatives, all computed at point (x',y',z'). In the above Taylor expansion, the left hand side of equation (10) can be approximated by the first four terms only of the right hand side. In this case, the difference of the image in two positions can be represented by the equation of a plane, as it can be seen from equation (11). The square which appears in equation (9) is then given by

$$
\begin{aligned}
&[I(x'+x, y'+y, z'+z) - I(x', y', z')]^2 \\
&\approx [xI_x + yI_y + zI_z]^2 \\
&= x^2 I_x^2 + y^2 I_y^2 + z^2 I_z^2 + 2xyI_xI_y + 2xzI_xI_z + 2yzI_yI_z \\
&= \begin{pmatrix} x & y & z \end{pmatrix} \begin{pmatrix} I_x^2 & I_xI_y & I_xI_z \\ I_xI_y & I_y^2 & I_yI_z \\ I_xI_z & I_yI_z & I_z^2 \end{pmatrix} \begin{pmatrix} x \\ y \\ z \end{pmatrix}.
\end{aligned}
$$

Since only the values of the derivatives depend on (x', y', z'), $E(x, y, z)$ can be rewritten as

$$
\begin{aligned}
E(x,y,z) &= \begin{pmatrix} x & y & z \end{pmatrix} \begin{pmatrix} \langle I_x^2\rangle & \langle I_xI_y\rangle & \langle I_xI_z\rangle \\ \langle I_xI_y\rangle & \langle I_y^2\rangle & \langle I_yI_z\rangle \\ \langle I_xI_z\rangle & \langle I_yI_z\rangle & \langle I_z^2\rangle \end{pmatrix} \begin{pmatrix} x \\ y \\ z \end{pmatrix} \\
&\equiv \begin{pmatrix} x & y & z \end{pmatrix} N \begin{pmatrix} x & y & z \end{pmatrix}^T,
\end{aligned}
$$

where N is the image structure tensor. The squared difference will have a specific shape according to the feature present in the image.

The characteristic equation of N is

$$
\begin{aligned}
&\lambda^3 - \left(\langle I_x^2\rangle + \langle I_y^2\rangle + \langle I_z^2\rangle\right)\lambda^2 \\
&- \left(\langle I_x^2\rangle\langle I_y^2\rangle + \langle I_x^2\rangle\langle I_z^2\rangle + \langle I_y^2\rangle\langle I_z^2\rangle - \langle I_xI_y\rangle^2 - \langle I_xI_z\rangle^2 - \langle I_yI_z\rangle^2\right)\lambda \\
&+ \langle I_x^2\rangle\langle I_y^2\rangle\langle I_z^2\rangle - \langle I_x^2\rangle\langle I_yI_z\rangle^2 - \langle I_z^2\rangle\langle I_xI_y\rangle^2 - \langle I_y^2\rangle\langle I_xI_z\rangle^2 \\
&+ 2\langle I_xI_y\rangle\langle I_xI_z\rangle\langle I_yI_z\rangle = 0.
\end{aligned}
$$

This matrix is positive semi-definite and so it is expected to have three positive eigenvalues. The solution of this cubic equation is not a trivial matter. Instead of estimating the roots of the characteristic equation in this case, it is preferable to use a more general method and estimate the eigenvalues of matrix A, defined by equation (8), by use of the Jacobi transformation of the matrix. The Jacobi transformation (or orthogonalisation) is an iterative application of rotations to a matrix until all its off-diagonal values are zero at machine precision.

The orientation of each eigenvector can be described by two angles, the zenith angle θ and the azimuth angle ϕ:

$$
\cos\phi = \frac{x}{\sqrt{x^2+y^2}}, \quad \sin\phi = \frac{y}{\sqrt{x^2+y^2}}, \quad \cos\theta = \frac{z}{\sqrt{x^2+y^2+z^2}}, \tag{12}
$$

where the two angles have values in the ranges

$$
0^o < \phi < 360^o \quad \text{and} \quad 0^o < \theta < 180^o.
$$

The system of equations that has to be solved for estimating the orientation of the eigenvectors is the following one.

$$\begin{pmatrix} \langle I_x^2 \rangle & \langle I_x I_y \rangle & \langle I_x I_z \rangle \\ \langle I_x I_y \rangle & \langle I_y^2 \rangle & \langle I_y I_z \rangle \\ \langle I_x I_z \rangle & \langle I_y I_z \rangle & \langle I_z^2 \rangle \end{pmatrix} \begin{pmatrix} x \\ y \\ z \end{pmatrix} = \lambda \begin{pmatrix} x \\ y \\ z \end{pmatrix}$$

and therefore,

$$(\langle I_x^2 \rangle - \lambda)x + \langle I_x I_y \rangle y + \langle I_x I_z \rangle z = 0,$$

$$\langle I_x I_y \rangle x + (\langle I_y^2 \rangle - \lambda)y + \langle I_y I_z \rangle z = 0,$$

$$\langle I_x I_z \rangle x + \langle I_y I_z \rangle y + (\langle I_z^2 \rangle - \lambda)z = 0.$$

After some algebraic manipulations of the above system of equations, we obtain the angle values

$$\cos\phi = \frac{\langle I_x I_z \rangle(\langle I_y^2 \rangle - \lambda) - \langle I_x I_y \rangle \langle I_y I_z \rangle}{\sqrt{\left[(\langle I_y^2 \rangle - \lambda)\langle I_x I_z \rangle - \langle I_x I_y \rangle \langle I_y I_z \rangle\right]^2 + \left[(\langle I_x^2 \rangle - \lambda)\langle I_y I_z \rangle - \langle I_x I_y \rangle \langle I_x I_z \rangle\right]^2}}$$

$$\sin\phi = \frac{\langle I_y I_z \rangle(\langle I_x^2 \rangle - \lambda) - \langle I_x I_y \rangle \langle I_x I_z \rangle}{\sqrt{\left[(\langle I_y^2 \rangle - \lambda)\langle I_x I_z \rangle - \langle I_x I_y \rangle \langle I_y I_z \rangle\right]^2 + \left[(\langle I_x^2 \rangle - \lambda)\langle I_y I_z \rangle - \langle I_x I_y \rangle \langle I_x I_z \rangle\right]^2}}$$

$$\begin{aligned} \cos\theta = {} & \left\{\langle I_x I_y \rangle - (\langle I_x^2 \rangle - \lambda)(\langle I_y^2 \rangle - \lambda)\right\} \times \\ & \Big\{\left[\langle I_x I_y \rangle^2 - (\langle I_x^2 \rangle - \lambda)(\langle I_y^2 \rangle - \lambda)\right]^2 + \left[(\langle I_y^2 \rangle - \lambda)\langle I_x I_z \rangle - \langle I_y I_z \rangle \langle I_x I_y \rangle\right]^2 \\ & + \left[(\langle I_x^2 \rangle - \lambda)\langle I_y I_z \rangle - \langle I_x I_z \rangle \langle I_x I_y \rangle\right]^2\Big\}^{-1/2}. \end{aligned}$$

Let us consider that image I depicts a linear feature which can be represented by a function $f(ax + by + cz)$, where a, b and c are some constants. The first partial image derivatives I_x, I_y and I_z are in this case given by

$$I_x = \frac{\partial f}{\partial x} = af', \quad I_y = \frac{\partial f}{\partial y} = bf', \quad I_z = \frac{\partial f}{\partial z} = cf'.$$

The ratio of I_x over I_y is a constant, equal to $d \equiv a/b$, i.e., $I_x = d \times I_y$. In the same way, the ratio of I_z and I_y is a constant value $e \equiv c/b$, i.e., $I_z = e \times I_y$.

Then, the determinant $|N|$ becomes

$$\begin{aligned} |N| &= \langle I_x^2 \rangle \langle I_y^2 \rangle \langle I_z^2 \rangle - \langle I_x^2 \rangle \langle I_y I_z \rangle^2 - \langle I_z^2 \rangle \langle I_x I_y \rangle^2 - \langle I_y^2 \rangle \langle I_x I_z \rangle^2 + 2\langle I_x I_y \rangle \langle I_x I_z \rangle \langle I_y I_z \rangle \\ &= d^2 \langle I_y^2 \rangle \langle I_y^2 \rangle e^2 \langle I_y^2 \rangle - d^2 \langle I_y^2 \rangle e^2 \langle I_y^2 \rangle^2 - e^2 \langle I_y^2 \rangle d^2 \langle I_y^2 \rangle^2 - \langle I_y^2 \rangle e^2 d^2 \langle I_y^2 \rangle^2 \\ &\quad + 2c \langle I_y^2 \rangle cd \langle I_y^2 \rangle d \langle I_y^2 \rangle = 0. \end{aligned}$$

In this case, the characteristic equation of N becomes

$$\lambda^3 - \left(\langle I_x^2 \rangle + \langle I_y^2 \rangle + \langle I_z^2 \rangle\right) \lambda^2 = 0.$$

It is obvious that this equation has three real solutions, two of which are zero and one different from zero ($\lambda_1 = \langle I_x^2 \rangle + \langle I_y^2 \rangle + \langle I_z^2 \rangle$, $\lambda_2 = \lambda_3 = 0$). This proves that when the local structure is stratified, the structure tensor has one non-zero eigenvalue and two eigenvalues approximately zero. The orientation of the eigenvector corresponding to the non-zero eigenvalue is the same as the normal to the strata.

In summary, the shape of the 3D tensor describes locally the structure of the image, in terms of how much it resembles a plane, a line or a sphere. The tensor can be visualised in 3D by an ellipsoid, as shown in Figure 3, the axes of which are along the directions of the eigenvectors of the tensor and their semi-lengths are equal to the magnitudes of the corresponding eigenvalues.

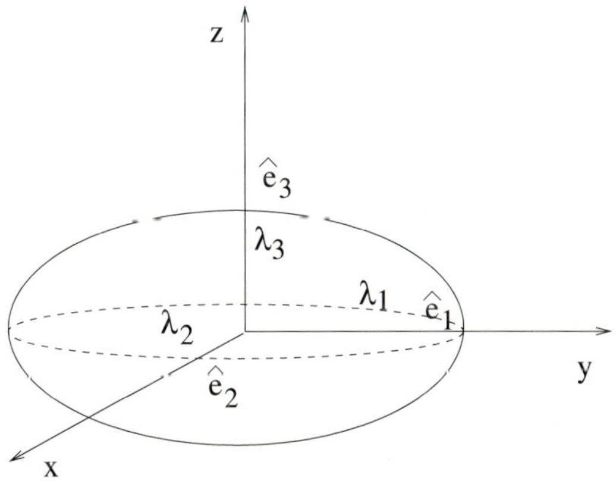

Fig. 3. Ellipsoid corresponding to the 3D structure tensor.

For linear features, the eigenvector associated with the smallest eigenvalue is the direction of the line. For planes, the eigenvector associated with the largest eigenvalue is the normal to the plane. In the isotropic case, such as for the sphere, there is no dominant orientation of change. These cases are summarised in Table 2.

The assumption that the variation of a 3D image can be represented by an ellipsoid was mainly based on the fact that all eigenvalues of the matrix are positive since the gradient matrix is usually positive definite or positive semi-definite.

Local Structure	Eigenvalues
Plane	$\lambda_1 \gg \lambda_2 \approx \lambda_3 \approx 0$
Line	$\lambda_1 \approx \lambda_2 \gg \lambda_3 \approx 0$
Uniform	$\lambda_1 \approx \lambda_2 \approx \lambda_3$

Table 2. Local structure conditions.

4.2 Second Order Tensor

In the previous subsection we kept only the first order terms in the expansion of equation (10). Another structure tensor can be defined as

$$\begin{aligned} &I(x'+x, y'+y, z+z') \\ &\approx I(x',y',z') + xI_x + yI_y + zI_z \\ &+ \frac{x^2}{2}I_{xx} + \frac{y^2}{2}I_{yy} + \frac{z^2}{2}I_{zz} + xyI_{xy} + yzI_{yz} + xzI_{xz}, \end{aligned}$$

provided that the second order approximation of the image is used or if ten terms of the Taylor expansion are kept, and so

$$\begin{aligned} &I(x'+x, y'+y, z+z') - I(x',y',z') - xI_x - yI_y - zI_z \\ &\approx \frac{x^2}{2}I_{xx} + \frac{y^2}{2}I_{yy} + \frac{z^2}{2}I_{zz} + xyI_{xy} + yzI_{yz} + xzI_{xz}. \end{aligned} \tag{13}$$

The left hand side of (13) represents the departure of the local structure from stratification. Following the previous steps, the average deviation from stratified structure can be written as

$$\begin{aligned} &\langle I(x'+x, y'+y, z+z') - I(x',y',z') - xI_x - yI_y - zI_z \rangle \\ &\approx \frac{1}{2}\langle [x^2 I_{xx} + y^2 I_{yy} + z^2 I_{zz} + 2xyI_{xy} + 2yzI_{yz} + 2xzI_{xz}] \rangle \\ &= \frac{1}{2}\begin{pmatrix} x \; y \; z \end{pmatrix} \begin{pmatrix} \langle I_{xx}\rangle & \langle I_{xy}\rangle & \langle I_{xz}\rangle \\ \langle I_{xy}\rangle & \langle I_{yy}\rangle & \langle I_{yz}\rangle \\ \langle I_{xy}\rangle & \langle I_{yz}\rangle & \langle I_{zz}\rangle \end{pmatrix} \begin{pmatrix} x \\ y \\ z \end{pmatrix} \\ &\equiv \frac{1}{2}\begin{pmatrix} x \; y \; z \end{pmatrix} C \begin{pmatrix} x \; y \; z \end{pmatrix}^T. \end{aligned}$$

This new structure tensor C is the Hessian matrix of the data function, and it is defined in terms of the second partial derivatives of the volume data. The second order local structures that can be classified using the eigenvalues of the tensor are sheet-like, line-like and blob-like.

The characteristic equation of the above matrix is the cubic polynomial

$$\begin{aligned}
&\lambda^3 - \left(\langle I_{xx}\rangle + \langle I_{yy}\rangle + \langle I_{zz}\rangle\right)\lambda^2 \\
&\quad - \left(\langle I_{xy}^2\rangle + \langle I_{yz}^2\rangle + \langle I_{xz}^2\rangle - \langle I_{xx}\rangle\langle I_{yy}\rangle - \langle I_{yy}\rangle\langle I_{zz}\rangle - \langle I_{xx}\rangle\langle I_{zz}\rangle\right)\lambda \\
&\quad + \langle I_{xx}\rangle\langle I_{yy}\rangle\langle I_{zz}\rangle - \langle I_{xx}\rangle\langle I_{yz}^2\rangle - \langle I_{zz}\rangle\langle I_{xy}^2\rangle - \langle I_{yy}\rangle\langle I_{xz}^2\rangle \\
&\quad + 2\langle I_{xy}\rangle\langle I_{xz}\rangle\langle I_{yz}\rangle = 0.
\end{aligned}$$

The orientation of each eigenvector may be described by the two angles mentioned earlier, the zenith angle θ and the azimuth angle ϕ defined by equation (12). After some algebraic manipulations we deduce that

$$\cos\phi = \frac{\langle I_{xz}\rangle(\langle I_{yy}\rangle - \lambda) - \langle I_{xy}\rangle\langle I_{yz}\rangle}{\sqrt{[(\langle I_{yy}\rangle - \lambda)\langle I_{xz}\rangle - \langle I_{xy}\rangle\langle I_{yz}\rangle]^2 + [(\langle I_{xx}\rangle - \lambda)\langle I_{yz}\rangle - \langle I_{xy}\rangle\langle I_{xz}\rangle]^2}},$$

$$\sin\phi = \frac{\langle I_{yz}\rangle(\langle I_{xx}\rangle - \lambda) - \langle I_{xy}\rangle\langle I_{xz}\rangle}{\sqrt{[(\langle I_{yy}\rangle - \lambda)\langle I_{xz}\rangle - \langle I_{xy}\rangle\langle I_{yz}\rangle]^2 + [(\langle I_{xx}\rangle - \lambda)\langle I_{yz}\rangle - \langle I_{xy}\rangle\langle I_{xz}\rangle]^2}},$$

$$\begin{aligned}
\cos\theta = &\left\{\lambda(\langle I_{xx}\rangle + \langle I_{yy}\rangle + \langle I_{zz}\rangle - \lambda) + \langle I_{xy}\rangle^2 - \langle I_{xx}\rangle\langle I_{yy}\rangle\right\} \\
&\Big\{\left[\lambda(\langle I_{xx}\rangle + \langle I_{yy}\rangle + \langle I_{zz}\rangle - \lambda) + \langle I_{xy}\rangle^2 - \langle I_{xx}\rangle\langle I_{yy}\rangle\right]^2 \\
&\quad + [(\langle I_{yy}\rangle - \lambda)\langle I_{xz}\rangle - \langle I_{yz}\rangle\langle I_{xy}\rangle]^2 \\
&\quad + [(\langle I_{xx}\rangle - \lambda)\langle I_{yz}\rangle - \langle I_{xz}\rangle\langle I_{xy}\rangle]^2\Big\}^{-1/2}.
\end{aligned}$$

Let us consider the form of I for the case of stratified local structure, which can be represented by a function $f(ax + by + cz)$, where a, b and c are some constants. The second order partial derivatives I_{xx}, I_{yy}, I_{zz}, I_{xy}, I_{yz}, and I_{xz} are in this case given by

$$\begin{aligned}
I_{xx} &= \frac{\partial^2 f}{\partial x^2} = a^2 f'', \quad & I_{yy} &= \frac{\partial^2 f}{\partial y^2} = b^2 f'', \quad & I_{zz} &= \frac{\partial^2 f}{\partial z^2} = c^2 f'', \\
I_{xy} &= \frac{\partial^2 f}{\partial x \partial y} = abf'', \quad & I_{yz} &= \frac{\partial^2 f}{\partial y \partial z} = bcf'', \quad & I_{xz} &= \frac{\partial^2 f}{\partial x \partial z} = acf''.
\end{aligned}$$

Next we form the ratios:

$$I_{xx} = \frac{a^2}{b^2} I_{yy}, \; I_{zz} = \frac{c^2}{b^2} I_{yy}, \; I_{xy} = \frac{b}{c} I_{xz}, \; I_{yz} = \frac{b}{a} I_{xz}, \; I_{xz} = \frac{ac}{b^2} I_{yy}.$$

Then, determinant $|N|$ becomes $|N| = 0$. In this case the characteristic equation is

$$\lambda^3 - \left(\langle I_{xx}\rangle + \langle I_{yy}\rangle + \langle I_{zz}\rangle\right)\lambda^2 = 0.$$

It is obvious that the above equation has two zero solutions and yields only one non-zero eigenvalue ($\lambda_1 = \langle I_{xx}\rangle + \langle I_{yy}\rangle + \langle I_{zz}\rangle$, $\lambda_2 = \lambda_3 = 0$). The eigenvector that corresponds to the non-zero eigenvalue is orthogonal to the direction of the strata.

$0.05 < \lambda_1/\lambda_2 < 1$. Considering this, t should be in the range $(1, 20)$. We chose to experiment with two values of t, $t = 2$ and $t = 20$ which, according to equation (15), yield $k = 0.2$ and the default value $k = 0.04$, respectively. The image on which we perform the corner detection is shown in Figure 4 (d).

The Harris corner pixels[1] are shown in Figure 5 in red. This result was produced with $k = 0.2$, standard deviation of the Gaussian smoothing filter $\sigma = 2$ and without any non-maxima suppression. The same procedure was followed for the case of $k = 0.04$, $\sigma = 2$ but too many candidate corner pixel were detected. For the moment we are only studying the areas of candidate corner pixels. These areas actually correspond to positive R values since as it was mentioned earlier, $R > 0$ represents corners, $R < 0$ represents edges and $R \approx 0$ represents constant grey value areas. A threshold value T_R other than 0 could be used for R, for isolating the corner pixels. In general the positive R values are very large, with very few pixels having small positive R values. After experimenting, it was concluded that any threshold T_R in the range $(0, 10^5)$ would provide very similar results, so we adopted for threshold T_R the value 0. As it can be observed from Figure 5, the chaotic area, where most of the corners lie, is very well isolated, but many small regions dense in corners have also been identified all over the image.

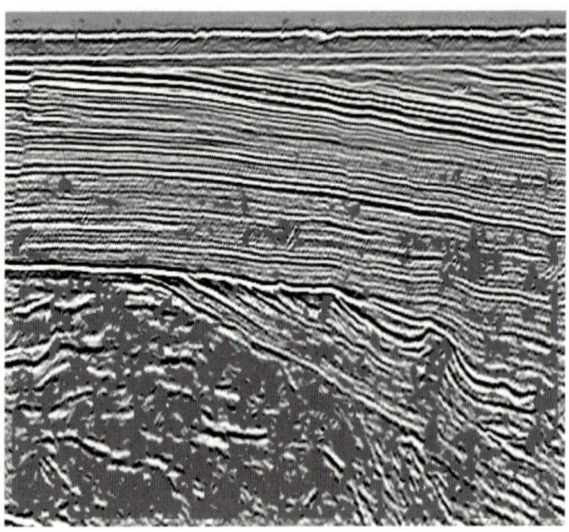

Fig. 5. [Reproduced in colour in Plate 6 on page 424.] Corner pixels detected by the Harris corner detector for $k = 0.2$, $\sigma = 2$.

This result may be postprocessed in various ways. The first post-processing is that of Harris and Stephens [16], i.e., to apply non-maxima suppression for identifying the real corner points out of the candidate corner points. A candidate corner pixel is compared with its eight neighbours and if it is the local

[1]We use the terminology *Harris corner pixels* for abbreviation of pixels that are candidate corner pixels.

maximum in the neighbourhood, then it is accepted as a corner pixel. Due to the large number of points detected in this case, only some of them are kept, the most important ones, namely those which exhibit the highest positive value of the cornerness measure R. The most significant 100 corner points, as detected by the Harris corner detector for the two cases of $k = 0.2$, $\sigma = 5$ and $k = 0.04$, $\sigma = 4$, are shown in Figure 6. We can still see too many corner points scattered all over the image, without any significant difference in the small regions we are interested in.

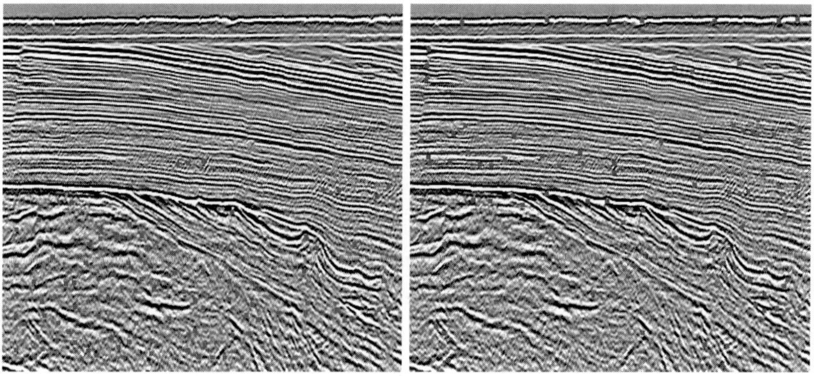

(a) (b)

Fig. 6. [Reproduced in colour in Plate 7 on page 424.] (a) The most significant 100 corner points for $k = 0.2$, $\sigma = 5$ and (b) for $k = 0.04$, $\sigma = 4$.

Another possible post-processing of the results shown in Figure 5 is to use connected component analysis, in order to remove all agglomerations of corner pixels that consist of fewer than a certain number of pixels. The image is first transformed into a binary image, with black marking the corner areas and white marking the background pixels as shown in Figure 7 (a). Then, all black pixels in the image are grouped, based on pixel connectivity and are assigned a label so that the image is divided into patches of different labels. Figure 7 (b) shows the different labels by different grey values. Then, connected components with fewer than $T_2 = 1000$ pixels are marked as background pixels. The result is shown in Figure 7 (c) and superimposed on the original image in Figure 7 (d). This is a very promising result, which in addition to the large chaotic region has also isolated the small chaotic patches of interest.

The points of interest can be separated by performing a further connected component analysis on Figure 7 (c). First, we are interested in isolating the very small insignificant areas. The image in Figure 7 (c) is divided into patches of different labels and connected components with fewer than a threshold $T_3 = 200$ pixels are marked as background. The resulting image is shown in Figure 7 (e). Then, connected component analysis is performed again but this time we are specifically interested in eliminating the large chaotic area and keeping only small areas of interest. Thus, connected components with more than a threshold of $T_4 = 5000$ pixels are marked as background and connected components with fewer than $T_4 = 5000$ pixels are marked as of interest. The resulting image is shown in Figure 7 (f).

Finally, a third post-processing technique which can be applied on the output of the Harris corner detector of Figure 7 (a) is based on mathematical morphology. Figure 8 (a) shows the result of eroding the binary image of Figure 7 (a) by a 3×3 structuring element: any detail of the black regions smaller than 3×3 is removed. Figure 8 (b) shows the image of Figure 7 (a) dilated by a 3×3 structuring element: any *hole* in it smaller than 3×3 has been filled in. Applying erosion again to the result in Figure 8 (a) with the same structuring element is equivalent to having eroded Figure 7 (a) by a 5×5 structuring element. This result is shown in Figure 8 (c). The result of double dilation is shown in Figure 8 (d). Applying dilation in the eroded image or erosion to the dilated image produces the result of opening (Figure 8 (e)) and closing (Figure 8 (f)) of the original image. The opening and closing are important morphological transformations. Opening is used to eliminate small image details while closing fills up small holes and connects objects close to each other. These two transformations do not distort the original shape or change the size of the depicted object. An interactive user may choose to apply these operations with whatever size structuring element they wish by repeating the erosions and dilations applied. These results, however, are not satisfactory enough, as we either loose the regions of interest, or have too many of them.

5.3 The Enhanced Hessian-Based Corner Detector Applied to Seismic Sections

First, it is necessary to determine the second order derivatives of the image differentiating it horizontally and vertically to create matrix A for each voxel.

Smoothing the image $I(x, y)$ is done by convolving it with Gaussian filter $w(u)$ in the two directions x and y. The second derivative I_{xx} is estimated by convolving the image with the Gaussian filter $w(u)$ in the vertical direction and convolving the result with the second derivative of the Gaussian filter $w''(u)$ in the horizontal direction. The second derivative I_{yy} is estimated by convolving the image with $w(u)$ in the horizontal direction and the result with $w''(u)$ in the vertical direction. Finally, the second derivative I_{xy} is estimated by convolving the image with $w'(u)$ in the horizontal direction and the result with $w'(u)$ in the vertical direction. The derivatives of the smoothing filter (14) are given by

$$w'(u) = -\frac{u}{\sigma^2} e^{-\frac{u^2}{2\sigma^2}},$$
$$w''(u) = -\frac{1}{\sigma^2} e^{-\frac{u^2}{2\sigma^2}} + \frac{u^2}{\sigma^4} e^{-\frac{u^2}{2\sigma^2}}. \tag{16}$$

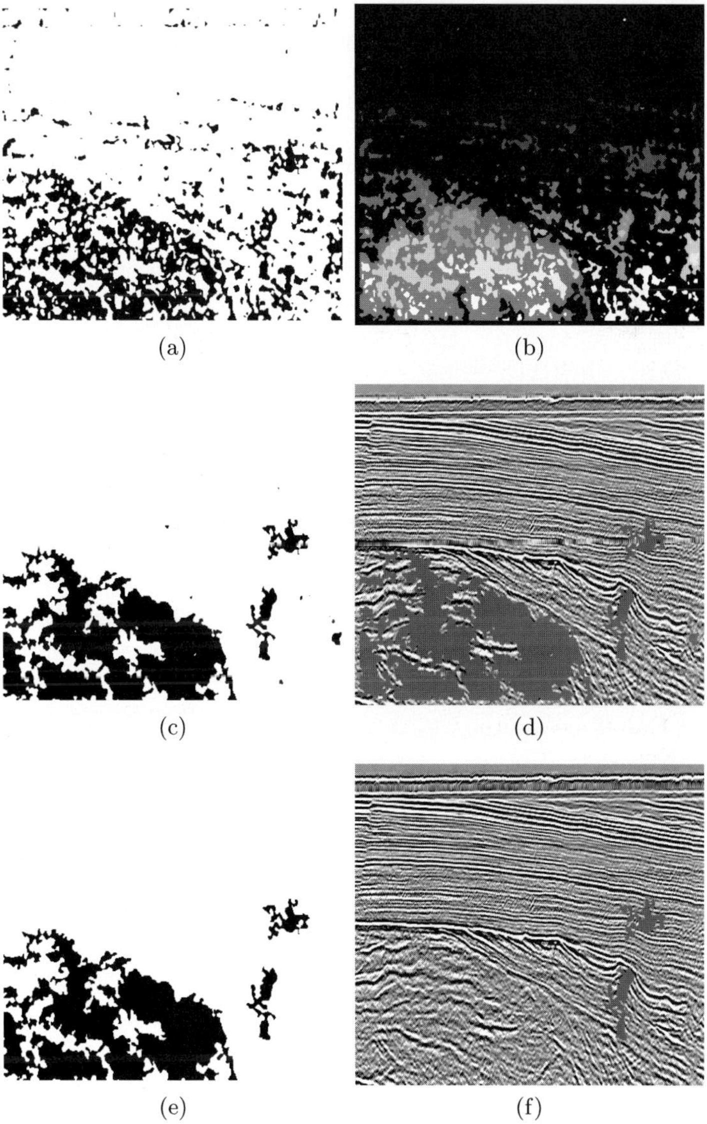

Fig. 7. [Reproduced in colour in Plate 8 on page 425.] (a) Binary image with corner points, (b) different labels of each connected component, (c) binary image after small regions have been thresholded away, (d) final result superimposed on the original image, (e) binary image after very small regions have been thresholded away, (f) areas of interest.

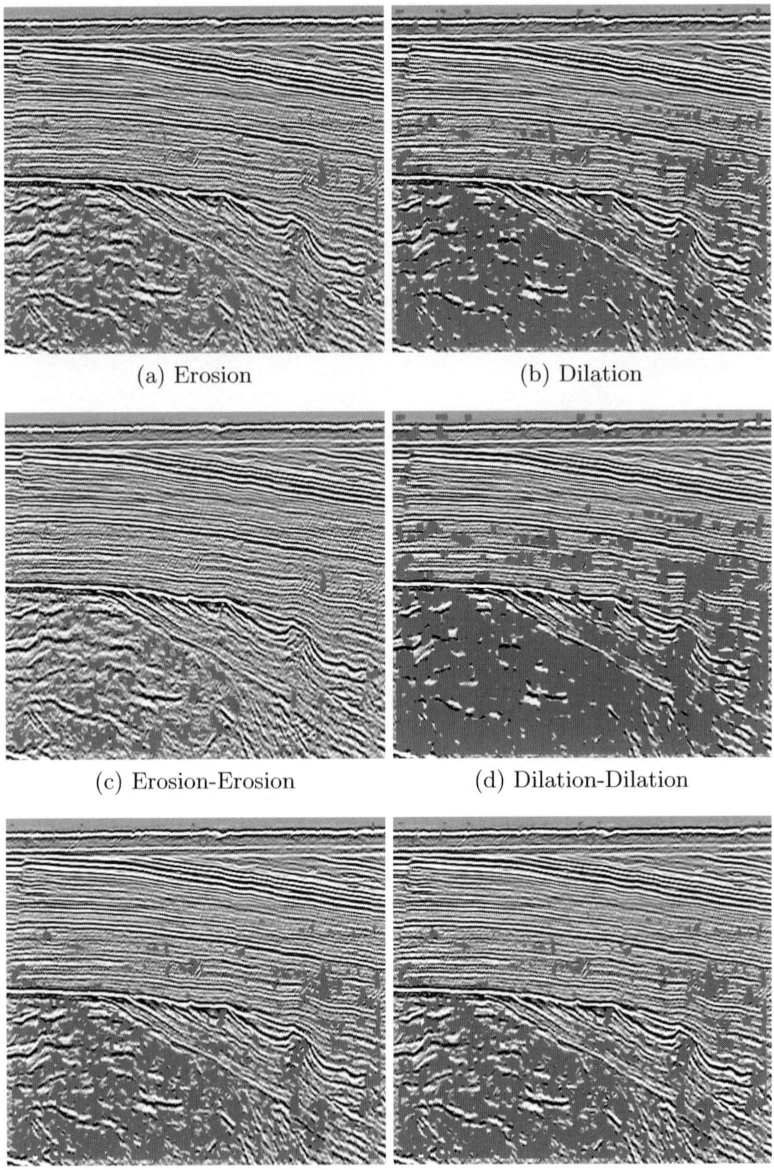

(a) Erosion (b) Dilation

(c) Erosion-Erosion (d) Dilation-Dilation

(e) Opening (f) Closing

Fig. 8. [See Colour Plate 9 on page 426.] Several morphological transformations.

The standard deviation of the Gaussian filter was chosen to be $\sigma = 2$ corresponding to a filter of size 13. These functions are used to compute the filter weights at sampling points which correspond to the integer values of their arguments. Filters $w'(u)$ and $w''(u)$ should have zero direct component as they are high pass filters. Therefore, in order to normalise them, we subtract from each filter weight the mean value of the originally computed weights using equations (16). With such a normalisation, the three filters used in the convolution procedure in both directions x and y are given in Table 3.

filter	-6	-5	-4	-3	-2	-1	0
w	0.002	0.009	0.027	0.065	0.121	0.176	0.199
w'	0.016	0.055	0.135	0.243	0.303	0.220	0.000
w''	-0.023	-0.059	-0.103	-0.103	-0.001	0.164	0.249

Table 3. Filter weights. Due to symmetries, only the left half of each filter is shown. The values for sampling points 1–6 are symmetric about sampling point 0 for the top and bottom filters, while for the middle filter they are equal to the weights in their symmetric positions on the left of 0, but with opposite sign, due to the antisymmetry of this filter.

The image convolved with these filters is the image of Figure 4 (d). The derivatives of the image in both directions are shown in Figure 9.

It is also necessary to determine the diagonal second order derivatives of the image for creating determinant $|B|$. For that reason, the Gaussian filter is convolved with the image along the diagonal directions. The filter used for the diagonal convolution is the Gaussian filter, of length 13 and $\sigma = 2$. In order for the filter length to remain equal to 13 interpixel distances as for the horizontal and vertical convolutions, the number of pixels the filter occupies in the diagonal direction becomes $13/\sqrt{2} \approx 9$, since $\sqrt{2}$ is the distance of two pixels along the diagonal direction. The filter values are also normalised and we end up with the filter weights for the diagonal convolution shown in Table 4.

The second derivative I_{d11} is estimated by convolving first with the Gaussian filter $w(u)$ along the diagonal direction from top right to bottom left, and the result with the second derivative of the Gaussian filter $w''(u)$ along the diagonal direction from top left to bottom right. The second derivative I_{d22} is estimated by convolving the image with $w(u)$ along the top left to bottom right direction and the result with $w''(u)$ along the top right to bottom left direction. Finally, the second derivative I_{d12} is estimated by convolving the image with $w'(u)$ along one diagonal direction and the result again with $w'(u)$ along the other diagonal direction.

The results of the differentiation are shown in Figure 10.

After tensors A and B are formed for each pixel, the absolute values of their determinants are estimated. The histogram of the absolute values of the determinant of A is shown in Figure 11 (a). We threshold the determinant values, choosing a threshold $T_A = 4000$. The pixels, for which the determinant

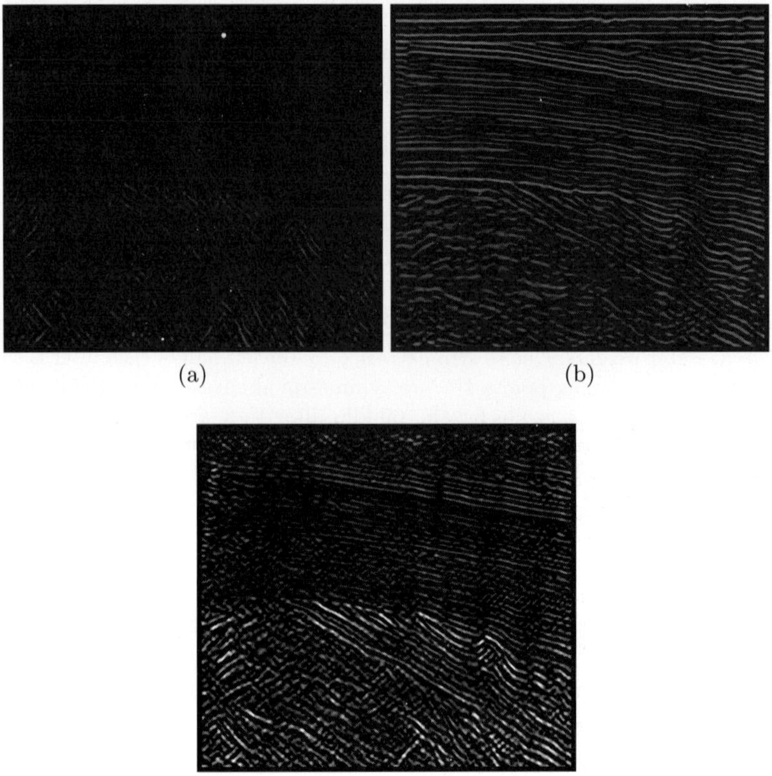

(a) (b)

(c)

Fig. 9. (a) Horizontal second derivative I_{xx} of Figure 4 (d), (b) vertical second derivative I_{yy} of Figure 4 (d), and (c) mixed second derivative I_{xy} of Figure 4 (d).

filter	-4$\sqrt{2}$	-3$\sqrt{2}$	-2$\sqrt{2}$	-$\sqrt{2}$	0	$\sqrt{2}$	2$\sqrt{2}$	3$\sqrt{2}$	4$\sqrt{2}$
w	0.0026	0.0149	0.0520	0.1100	0.141	0.1100	0.0520	0.0149	0.0026
w'	0.0259	0.1120	0.2600	0.2750	0.000	-0.2750	-0.2600	-0.1120	-0.0259
w''	-0.0350	-0.0960	-0.0950	0.0940	0.246	0.0940	-0.0950	-0.0960	-0.0350

Table 4. Filter weights for diagonal convolution.

exceeds the threshold, are marked with red colour in Figure 12 (a). As it was expected, the diagonal features are also detected.

In a similar way, the histogram of the absolute values of the determinant of B is shown in Figure 11 (b). We threshold the determinant values choosing a threshold $T_B = 3000$. The pixels, for which the determinant exceeds the threshold, are marked again with red colour in Figure 12 (b). As it was expected, the linear horizontal features are also detected.

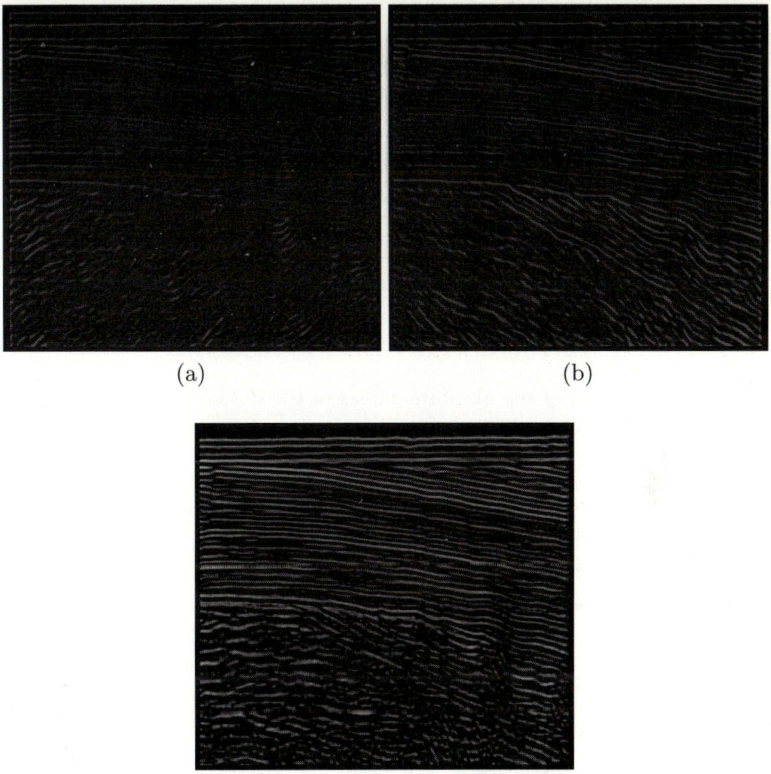

Fig. 10. (a) Diagonal top left-bottom right second derivative of image I_{d11}, (b) diagonal top right-bottom left second derivative of image I_{d22}, and (c) mixed diagonal second derivative of image in both diagonal directions I_{d12}.

A combination of both determinants by using the above threshold values produces a map of points of interest as shown in Figure 12 (c) where the red points represent points, whose determinants $|A|$ and $|B|$ exceed the chosen thresholds. Several points of interest in the chaotic and mixed regions are identified.

In order to improve the results by reducing these unwanted features and obtain better segmentation, connected component analysis is applied to the output image. The binary image which contains the points under examination is shown in Figure 13 (a) where the points of interest are black and the background is white. After pixels that belong to patches with fewer pixels than a certain threshold T_C are removed, we obtain Figure 13 (b). The result of segmentation after applying connected component analysis of 4-connectivity is shown in Figure 14 (a) where the identified points of interest are superimposed on the initial image.

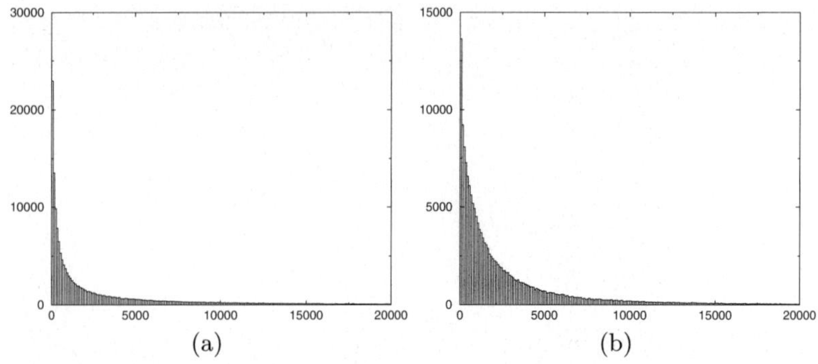

Fig. 11. (a) Histogram of the absolute values of the determinant of matrix A, (b) histogram of the absolute values of the determinant of matrix B.

In the final result, the regions of interest in the mixed and chaotic areas are differentiated from the stratified areas although some wrongly detected lines still persist.

More experiments can be performed, by varying the three threshold values, T_A, T_B and T_C. Another example is shown in Figure 14 (b) which shows the combination of the two determinants A and B where $T_A = 6000$, $T_B = 2000$ and $T_C = 20$. In this case, the unwanted features in the stratified area are fewer but so are the points detected in the chaotic area.

As it will be shown next, it is possible to isolate the chaotic from the mixed area and extract regions of interest by doing some further analysis.

5.4 Detection of Chaotic Areas and Points of Interest

The gradient operator is generally used to detect rapid changes in an image. The gradient of an image has extensively been used for edge detection [32] where the magnitude of the gradient represents the edge magnitude and the gradient orientation is rotated with respect to the edge orientation by 90°. The gradient direction has also been used in the past for corner detection. At the position of a corner point, the direction of the gradient changes significantly [22, 54].

The gradient has been used in texture segmentation before. More specifically, the dominant local orientation of the gradient vector has been employed to characterise oriented textures [34].

In this subsection, the local gradient orientation is employed for segmenting the seismic image. Especially the chaotic area is isolated due to its randomness in orientation. Also, the points of interest are detected by combining the gradient results with the Hessian results, described in the previous subsection.

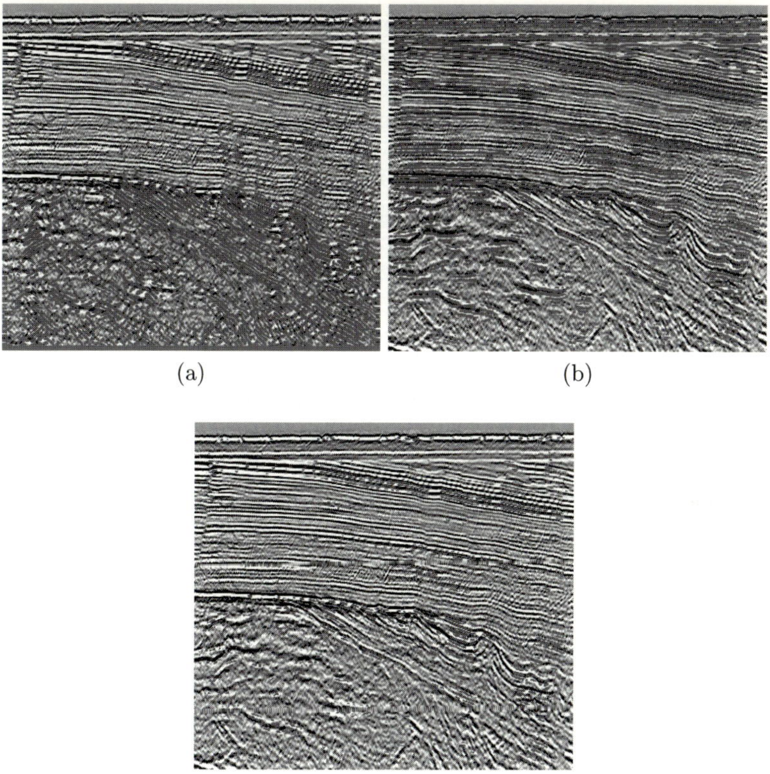

(a) (b) (c)

Fig. 12. [Reproduced in colour in Plate 10 on page 427.] (a) Image after thresholding the determinant of tensor A with $T_A = 4000$, (b) image after thresholding the determinant of tensor B with $T_B = 3000$, and (c) combination of results (a) and (b) with an AND operator.

Local Image Orientation. The magnitude of the local gradient of the image is defined as $|\nabla I(x,y)| = (I_x^2 + I_y^2)^{1/2}$, and the orientation of the gradient vector as $\tilde{\theta} = \tan^{-1}(I_y/I_x)$.

Special care should be taken so that the estimated orientation angles θ fall in the range 0° to 360°. This can be achieved by considering the sign of the derivatives I_x and I_y,

- $I_x > 0$ and $I_y > 0$, $\theta = \tilde{\theta}$,
- $I_x > 0$ and $I_y < 0$, $\theta = \tilde{\theta} + 2\pi$,
- $I_x < 0$ and $I_y > 0$, $\theta = \tilde{\theta} + \pi$,
- $I_x < 0$ and $I_y < 0$, $\theta = \tilde{\theta} + \pi$,
- $I_x = 0$ and $I_y \neq 0$, $\theta = \pi/2$.

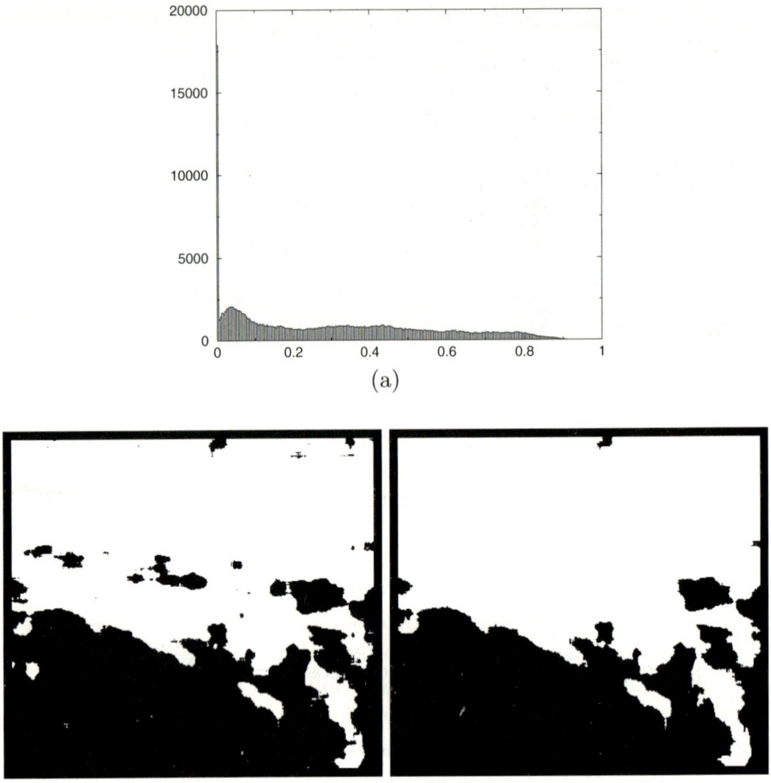

(a)

(b) (c)

Fig. 15. (a) Histogram of the average gradient magnitude image, (b) magnitude image after histogram thresholding, (c) magnitude image after connected component analysis and removal of small regions.

be observed, some points of interest are quite well-separated, and application of connected component analysis and removal of small areas of less than $T_F = 20$ pixels, improves the results as shown in Figure 16 (b).

In Figure 16 (c) the red points are actually red points in Figure 14 (b) which also belong to the white area of Figure 15 (c). Application of connected component analysis and removal of small areas of fewer than $T_F = 40$ pixels, improves the results as shown in Figure 16 (d), with fewer pixels from the linear regions appearing. Points of horizon termination and other geologically interesting areas are roughly detected.

Local Average Image Orientation: Separating the Linear from the Mixed Regions. If we exclude the chaotic region from our analysis, since it was isolated quite accurately in Figure 15 (c), then we expect that the

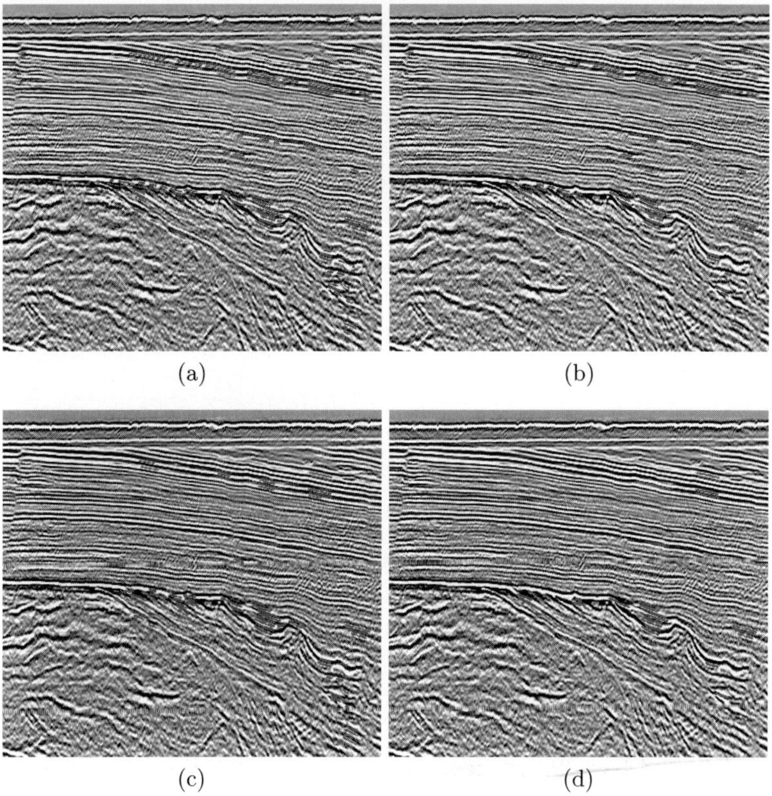

Fig. 16. [Reproduced in colour in Plate 12 on page 429.] (a),(c) Combination of both determinant and magnitude methods, (b),(d) combination of both determinant and magnitude methods after connected component analysis and region thresholding.

average orientation angle inside local windows will be near zero in the linear areas and non-zero in the mixed areas.

In order to separate the parallel from the mixed areas, first the orientation angle of each pixel in the image is doubled as before, and the angle of 360° is subtracted from the new doubled angle, if this angle is larger than 360°. In this way, we ensure that the new angle falls in the range 0 to 360°. Then the average value of the new angles is estimated within sliding windows of 21×21 and also divided by 2. The histogram of these values is shown in Figure 17 (a). The histogram is bimodal, and by picking a threshold value $T_G = 1.87$ it is possible to separate two texture areas. Pixels with values larger than the threshold are shown as grey in Figure 17 (b), while pixels in the stratified region are shown as black. The pixels previously identified with the chaotic region are flagged white in the same image. It is clear that the grey area represents the mixed texture and the black area is the parallel texture area. Thus, in this way, we managed to approximately isolate the areas of interest lying in the mixed texture area, as shown in Figure 17 (c) where the grey areas are superimposed on the initial cross-section as red points. Finally, by removing regions with fewer than 70 pixels, we end up with Figure 17 (d), which highlights quite clearly the areas of interest.

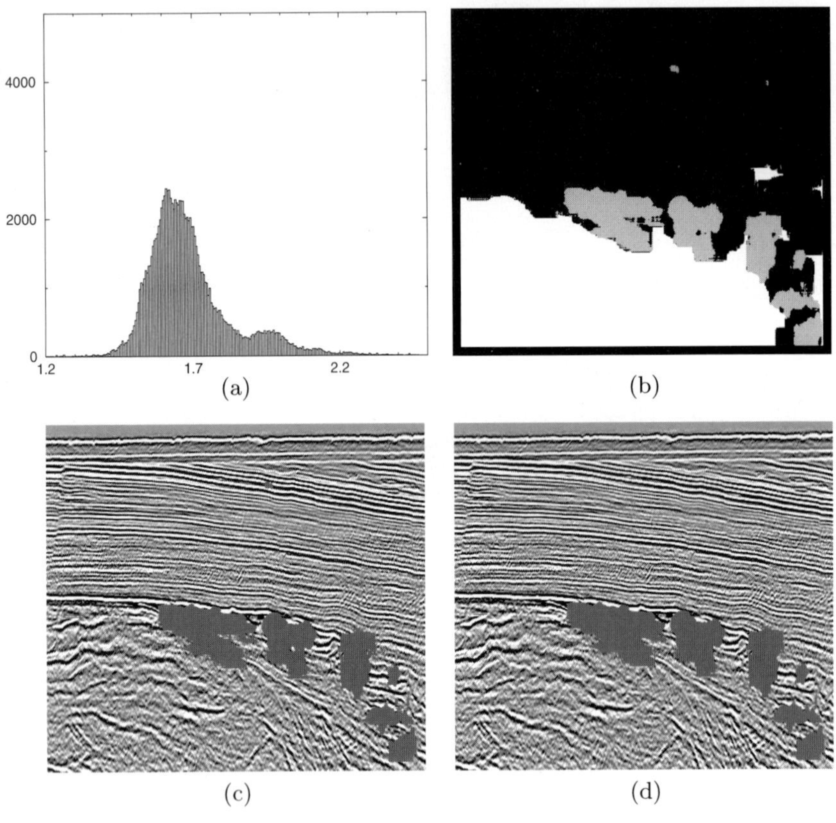

Fig. 17. [Reproduced in colour in Plate 13 on page 430.] (a) Histogram of the orientation image averaged in windows, (b) final thresholded image, (c) areas of interest on the original cross-section, and (d) areas of interest after connected component analysis.

5.5 Experiments in 3D

Figure 18 shows the histograms of the eigenvalues of the first order structure tensor in 3D. If we accept that the chaotic region is characterised by all three eigenvalues being large, then the use of a simple threshold, $T = 2800$ allowed us to produce Figure 19 (a) where we show a cross section of the volume data inside which all voxels with three eigenvalues above the chosen threshold are highlighted.

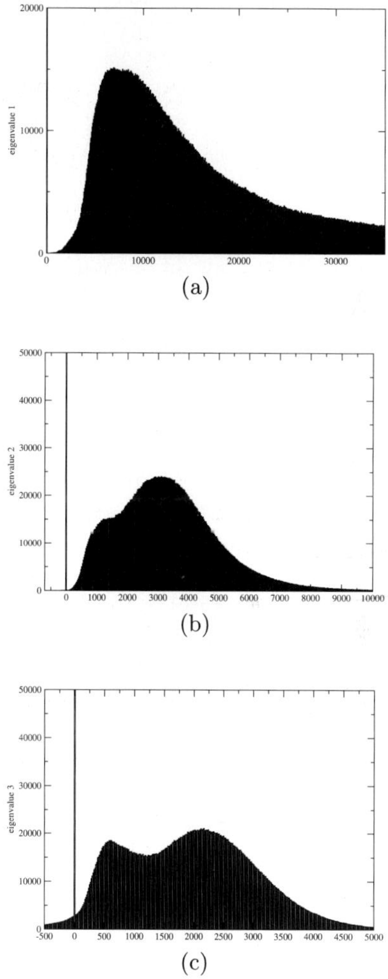

Fig. 18. (a) Histogram of the eigenvalue λ_1, (b) histogram of the eigenvalue λ_2, and (c) histogram of the eigenvalue λ_3.

In a similar way, by playing with the choice of such thresholds, we were able to produce Figure 19 (b), where some of the faults present in the volume are highlighted. This result was produced by highlighting all voxels with $\lambda_1, \lambda_2 > 6000$ and $\lambda_3 > 1300$.

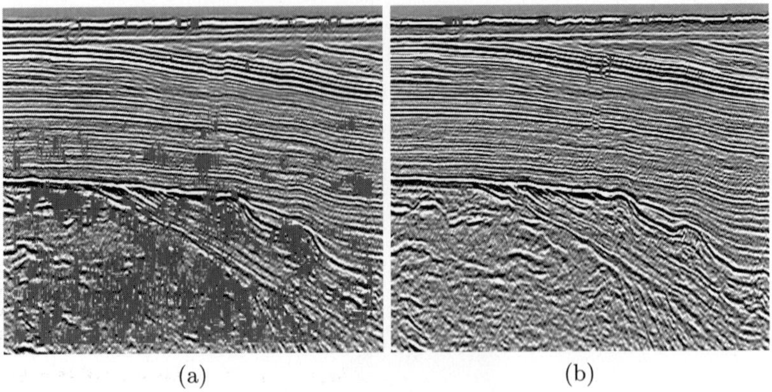

Fig. 19. [Reproduced in colour in Plate 14 on page 431.] (a) Chaotic region and mounds, (b) faults detected.

6 Conclusions

The segmentation of seismic images is a crucial step in seismic analysis. We demonstrated here the use of image structure tensors and features extracted from them, for segmenting seismic sections and highlighting some regions of interest. Image tensors have been very popular in the past due to their simplicity, speed and easiness of implementation, since only convolutions of the image with certain filters is necessary.

The structure tensor consisting of the first image derivatives was used for corner detection. It was shown to be useful in isolating chaotic regions. The chaotic area was also successfully detected by the use of a simple average of the local orientation of the image.

A tensor which consists of the second order image derivatives was also used, and in combination with a tensor consisting of the diagonal second order derivatives of the image, it proved useful in detecting corner points. This corner detector proved quite useful in separating the chaotic and mixed regions from the stratified regions.

Finally, combining the results of chaotic region detection and corner detection or average orientation estimation, allowed the isolation of geologically distinct regions in the seismic image.

The field of 2D seismic data analysis is quite mature now, but the field of 3D data analysis is still under intensive research. In particular, the potential of the 3D structure tensors to isolate structures of interest in the data has not been fully investigated yet. We presented here some results using the 3D structure tensors. They appear to be promising in enhancing various structures, including faults.

Acknowledgement

The work was partly supported by the European Union through the TriTex project, contract no. IST-1999-20500. We also wish to thank Dave Hunt, Kai Hogstad, and Geir Elvebakk and the Barents Sea Seismic Area C partners (Norsk-Hydro, Statoil, AGIP, and Fortum) for collaboration and permission to publish data.

References

1. P. Bakker (2002) *Image structure analysis for seismic interpretation.* PhD thesis, Pattern Recognition Group, Faculty of Applied Physics, Delft University of Technology, Delft, The Netherlands, June 2002.
2. P. Bakker, L.J. van Vliet, and P.W. Verbeek (2001) Confidence and curvature of curvilinear structures in 3D. Proceedings of the Eighth IEEE International Conference on Computer Vision, Vancouver, Canada, July 2001.
3. E. Barth (2000) The minors of the structure tensor. Mustererkennung, 221–228.
4. E. Barth and M. Ferraro (2000) On the geometric structure of spatio-temporal patterns. Proc. Second International Workshop Algebraic Frames for the Perception-Action Cycle (AFPAC), Springer-Verlag, Heidelberg, 134–143.
5. E. Barth, I. Stuke, and C. Mota (2002) Analysis of motion and curvature in image sequences. Proc. Fifth IEEE Southwest Symposium on Image Analysis and Interpretation (SSIAI), Santa Fe, New Mexico, April 2002, 206–210.
6. P.J. Basser, J. Mattiello, and D. Le Bihan (1994) Estimation of the effective self-diffusion tensor from the NMR spin echo. Journal of Magnetic Resonance **103**, 247–254.
7. P.J. Basser, J. Mattiello, and D. Le Bihan (1994) MR diffusion tensor spectroscopy and imaging. Biophys J. **66**, 259–267.
8. P.R. Baudet (1978) Rotationally invariant image operators. 4th International Conference on Pattern Recognition, 579–583.
9. A. Bhalerao and R. Wilson (2001) Estimating local and global structure using a gaussian intensity model. Proceedings of Medical Image Understanding and Analysis (MIUA).
10. A. Bhalerao and R. Wilson (2001) A Fourier approach to 3D local feature estimation from volume data. Proceedings of British Machine Vision Conference (BMVC), volume 2, 461–470.
11. K. Deguchi, T. Izumitani, and H. Hontani (2002) Detection and enhancement of line structures in an image by anisotropic diffusion. Pattern Recognition Letters **23**, 1399–1405.
12. L. Dreschler and H.H. Nagel (1981) Volumetric model and 3D trajectory of a moving car derived from monocular TV frame sequences of a street scene. International Joint Conference on Artificial Intelligence (IJCAI), 692–697.
13. L. Florack (1997) *Image Structure.* Computational Imaging and Vision, volume 10, Kluwer Academic Publishers.

14. A.F. Frangi, W.J. Niessen, K.L. Vincken, and M.A. Viergever (1998) Multiscale vessel enhancement filtering. Medical Image Computing and Computer-Assisted Intervention (MICCAI), Lecture Notes in Computer Science, volume 1496, Springer-Verlag, 130–137.
15. G.H. Granlund and H. Knutsson (1995) *Signal Processing for Computer Vision.* Kluwer Academic Publisher.
16. C. Harris and M. Stephens (1988) A combined corner and edge detector. Proc. 4th Alvey Vision Conf., 189–192.
17. J. Hladuvka and E. Gröller (2001) Direction-driven shape-based interpolation of volume data. Proceedings of Vision, Modeling, and Visualization, Stuttgart, Germany, November 2001.
18. J. Hladuvka, A. Konig, and E. Gröller (2001) Exploiting eigenvalues of the Hessian matrix for volume decimation. Proceedings of the 9th International Conference in Central Europe on Computer Graphics, Visualisation and Computer Vision(WSCG).
19. B. Jähne (1993) *Digital Image Processing.* Springer-Verlag.
20. B. Jähne, H. Haussecker, H. Scharr, H. Spies, D. Schmundt, and U. Schurr (1998) Study of dynamical processes with tensor-based spatiotemporal image processing techniques. Proc. European Conference on Computer Vision (ECCV), 323–336.
21. M. Kass and A. Witkin (1987) Analyzing oriented patterns. Computer Vision, Graphics and Image Processing **37**, 362–385.
22. L. Kitchen and A. Rosenfeld (1988) Grey-level corner detection. Pattern Recognition letters **8**, 311–318.
23. H. Knutsson (1989) Representing local structure using tensors. Proceedings of the 6th Scandinavian Conference Image Analysis, Oulu, Finland, June 1989.
24. J.J. Koenderink (1984) The structure of images. Biological Cybernetics **50**, 363–370.
25. P. Kornprobst and G. Medioni (2000) Tracking segmented objects using tensor voting. Proc. of Conputer Vision and Pattern Recognition, 118–125.
26. Q. Lin (2001) *Enhancement, Detection, and Visualization of 3D Volume Data.* PhD thesis, Linköping University, Department of Electrical Engineering, Computer Vision Laboratory, Linköping, Sweden, October 2001.
27. T. Lindeberg (1994) *Scale-Space Theory in Computer Vision.* Kluwer Academic Publishers.
28. A.M. López, F. Lumbreras, J. Serrat, and J.J. Villanueva (1999) Evaluation of methods for ridge and valley detection. IEEE Pattern Analysis and Machine Intelligence **21**(4), 327–335.
29. H.P. Moravec (1977) Towards automatic visual obstacle avoidance. Proc. Int. joint. Conf. Artificial Intelligence, page 584.
30. M.S. Nixon and A.S. Aguado (2002) *Feature Extraction and Image Processing.* Butterworth-Heinemann.
31. J.A. Noble (1988) Finding corners. Image and Vision Computing Journal **6**(2), 121–128.
32. M. Petrou (1994) The differentiating filter approach to edge detection. Advances in Electronics and Electron Physics **88**,297–345.
33. T. Randen, E. Monsen, C. Signer, A. Abrahamsen, J. Hansen, T. Saeter, and J. Schlaf (2000) Three-dimensional texture attributes for seismic data analysis. Society of Exploration Geophysicists(SEG), Annual Meeting, Expanded Abstracts, 668–671.

34. A.R. Rao and B.G. Schunck (1989) Computing oriented texture fields. Computer Vision and Pattern Recognition **32**, 61–68.
35. B. Rieger, F.J. Timmermans, and L.J. van Vliet (2002) Estimation of curvature on surfaces in 3D grey-value images. Proc. 8th Annual Conference of the Advanced School for Computing and Imaging (ASCI), Delft, The Netherlands, June 2002, 170–177.
36. B. Rieger and L.J. van Vliet (2002) Curvature of n-dimensional space curves in grey-value images. IEEE Transactions on Image Processing **11**(7), 738–745.
37. Y. Sato, S. Nakajima, and N. Shiraga (1998) 3D multi-scale line filter for segmentation and visualization of curvilinear structures in medical images. Medical Image Analysis **2**(2), 143–168.
38. Y. Sato, C.F. Westin, A. Bhalerao, S. Nakajima, N. Shiraga, S. Tamura, and R. Kikinis (2000) Tissue classification based on 3D local intensity structures for volume rendering. IEEE Transactions on Visualization and Computer Graphics **6**(2), 160–179.
39. A.F. Solé, A. López, and G. Sapiro (2001) Crease enhancement diffusion. Computer Vision and Image Understanding **84**, 241–248.
40. H. Spies and B. Jähne (2001) A general framework for image sequence analysis. Fachtagung Informationstechnik, Magdeburg, Germany, March 2001, 125–132.
41. J. Stoeckel, F.M. Vos, P.H. Vos, and A.M. Vossepoel (2000) An evaluation of ridge extraction methods for portal imaging. Proc. of International Conference on Pattern Recognition (ICPR), Barcelona, Spain, September 2000, 3433–3436.
42. C.K. Tang and G. Medioni (2002) Curvature-augmented tensor voting for shape inference from noisy 3D data. IEEE Transactions on Pattern Recognition and Machine Intelligence, 858–864.
43. C.K. Tang, G. Medioni, and M.S. Lee (2001) N-dimensional tensor voting, with application to epipolar geometry estimation. IEEE Transactions on Pattern Recognition and Machine Intelligence **23**(8), 829–844.
44. W.S. Tong, C.K. Tang, and G. Medioni (2001) First order tensor voting, and application to 3-D scale analysis. Proc. of Computer Vision and Pattern Recognition, 175–182.
45. L.J. van Vliet and P.W. Verbeek (1995) Estimators for orientation and anisotropy in digitized images. Proceedings of the first Conference of the Advanced School for Computing and Imaging (ASCI), 442–450.
46. P.W. Verbeek, L.J. van Vliet, and J. van de Weijer (1998) Improved curvature and anisotropy estimation for curved line bundles. Proc. 14th Int. Conf. on Pattern Recognition, Brisbane, Australia, August 1998.
47. J. Weickert (1997) A review of nonlinear diffusion filtering. Scale-Space Theory in Computer Vision, Lecture Notes in Comp. Science, volume 1252, Springer-Verlag, Berlin, 3–28.
48. J. Weickert (1998) *Anisotropic Diffusion in Image Processing*. ECMI Series, Teubner-Verlag, Stuttgart.
49. J. Weickert (1999) Coherence-enhancing diffusion filtering. Int. J. Computer Vision **31**, 111–127.
50. C.F. Westin, S.E. Maier, B. Khidhir, P. Everett, F.A. Jolesz, and R. Kikinis (1999) Image processing for diffusion tensor magnetic resonance imaging. Second International Conference on Medical Image Computing and Computer-Assisted Intervention (MICCAI), Springer-Verlag, 441–452.

51. C.F. Westin, A. Bhalerao, H. Knutsson, and R. Kikinis (1997) Using local 3D structure for segmentation of bone from computer tomography images. Proceedings of IEEE Conf on Conputer Vision and Pattern Recognition (CVPR), 794–800.
52. C.F. Westin, L. Wigstrom, T. Loock, L. Sjokvist, R. Kikinis, and H. Knutsson (2001) Three-dimensional adaptive filtering in magnetic resonance angiography. Journal of Magnetic Resonance Imaging **14**, 63–71.
53. Z. Zheng, H. Wang, and E.K. Teoh (1999) Analysis of gray level corner detection. Pattern Recognition Letters **20**, 149–162.
54. O.A. Zuniga and R.M. Haralick (1983) Corner detection using the facet model. Proc. Conf. Pattern Recognition Image Processing, 30–37.

Automated Structural Interpretation Through Classification of Seismic Horizons

Hilde G. Borgos, Thorleif Skov, and Lars Sønneland

Schlumberger Stavanger Research, P.O. Box 8013, N-4068 Stavanger, Norway

Summary. A novel method for extracting geometry primitives from seismic data is presented. All events in the 3D seismic cube will be detected and can be combined into geometric primitives based on similarities in the local wave form. No assumptions of continuity in the geometric primitives are required. The geometric primitives can therefore represent faulted horizons, which furthermore facilitates quantification of the fault displacement.

The accuracy with which the local waveform can be represented is defined as a user input, implying that subtle lateral changes in the reflectivity can be detected and exploited. This characteristic enables analysis of stratigraphic variations along horizons.

1 Introduction

Interpretations of seismic horizons play an important role in structural characterization of 3D seismic data. Interpretations can be obtained manually, by signal consistent automatic extraction of minimum, maximum or zero crossing surfaces, or by combining automated surface extraction with manual interpretation. Structurally complex regions are a challenge to the existing interpretation procedures, since a continuity requirement is often imposed by automatic tracker algorithms. When the seismic signal along the horizon being tracked changes abruptly, or discontinuities occur in the horizon, the algorithm may fail to provide a good estimate. The classification approach presented in this paper addresses this problem, by attempting to group horizon segments according to similarities in the shape of the seismic signal around the horizons. The grouping is then performed without continuity requirement, but only on the basis of the seismic wave form. The shape of each seismic horizon can be represented through a set of attributes with one-point support, selected along minimum or maximum positions. This reduces the number of data points to be classified, allowing the 3D classification to be run on a sparse 3D volume.

Classification tools are widely used in both 3D and 4D reservoir characterization, for example in mapping 3D structures, lithological properties and production effects [8]. In this work we extend the area of application of the classification methodology into automated interpretation of seismic horizons

[1]. By classifying the seismic signal along horizons we gain an improved automated interpretation that performs well also in structurally complex regions, e.g., across faults. The classification may not provide a complete automated interpretation, but instead a set of surface segments is obtained. These surface segments serve as primitives for constructing a horizon interpretation, where a final manual interpretation is performed by grouping and combining these surface primitives. Thus, the classification procedure constitutes a tool for automating parts of the traditionally manual horizon interpretation. In particular, the new methodology has its strength as a guide when the complexity of the seismic data may result in ambiguities in the manual interpretation. It is also demonstrated how the surface primitives are applicable for automated fault displacement estimation.

2 Seismic Extrema Representation

A 3D seismic volume is represented by a cube $\mathbf{S} = \{S(x, y, z)\}$ containing seismic amplitudes at each voxel position (x, y, z), where x and y are positions in the two horizontal directions, measured in distance, and z is the position in the vertical dimension, measured in depth or two-way travel-time. The column $S(x_0, y_0, z)$ at position (x_0, y_0) is denoted a seismic trace. A seismic horizon is a piecewise continuous surface in 3D, consisting of positions $z = g(x, y)$ within the seismic volume. Throughout this work, seismic horizon segments are used as primitives for structural interpretation. Horizon interpretations are often placed on minimum, maximum or zero crossing events in the seismic cube, and are defined as piecewise continuous surfaces in 3D falling along these types of events. Minimum and maximum events in a seismic cube are commonly referred to as seismic extrema, and are chosen as the basis for the automated horizon interpretation presented in this paper. A set of seismic extrema can be described as a contour surface

$$g(x, y) = \left\{ z : \frac{\mathrm{d}S(x, y, z)}{\mathrm{d}z} = 0 \right\} .$$

All extrema within a 3D seismic volume can be represented by two sparse 3D cubes, containing only information about the minimum and maximum events in the seismic data. A vertical trace of the first cube contains the actual maximum or minimum seismic amplitude values along this trace, stored in the vertically closest voxel along the trace. This cube is referred to as the extrema value cube. The second cube, denoted the extrema position cube, contains sub-sample information about the exact location of the extrema, i.e., the vertical correction to the seismic sampling resolution. Both extrema cubes are sparse cubes, with value zero at voxel positions not falling on an extremum. The set of voxels containing extrema data is the same for the two cubes, but contains amplitudes and sub-sample positions respectively.

The extrema cubes are generated using *volume reflection spectral analysis* technology [9, 10], where each seismic trace is locally reconstructed using the orthogonal basis defined by the Chebyshev polynomials. Thus, locally the seismic trace is represented as

$$S(x_0, y_0, z) = b_0 p_0(z) + b_1 p_1(z) + \cdots + b_n p_n(z),$$

where $p_i(z)$ is a polynomial of order i and b_i, $i = 1, \ldots, n$, are coefficients estimated from the seismic trace in a window around z. The analytical form of the local polynomial representation of the seismic trace enables sub-sample precision calculations of all the extrema positions $\mathbf{E} = \{E(x, y, z)\}$ within the seismic cube, defined as

$$E(x, y, z) = \begin{cases} S(x, y, z), & \text{if } \mathrm{d}S(x, y, z)/\mathrm{d}z = 0, \\ 0, & \text{otherwise.} \end{cases} \quad (1)$$

Figure 1 shows an example of a sparse extrema representation of a seismic section, illustrating the cube containing amplitude values. All events from the original seismic section are preserved in the extrema value cube, but are represented only through the position of its minimum or maximum.

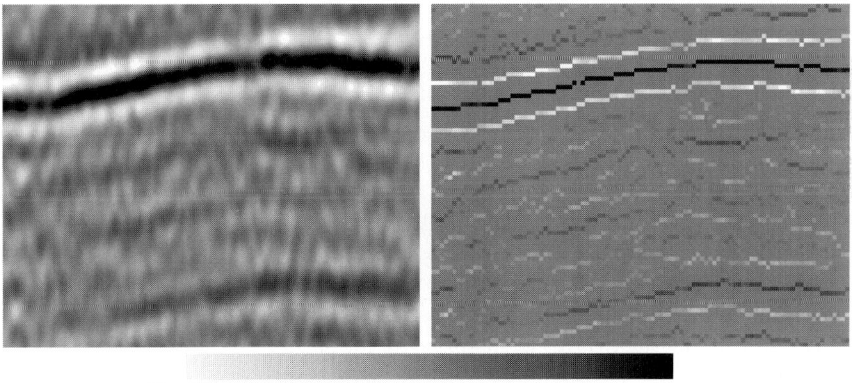

Fig. 1. The left image shows a vertical section through a seismic cube. The right image shows the corresponding section of the sparse extrema value cube, where only minimum or maximum positions have non-zero values.

3 Seismic Wave Form Attributes

The seismic wave form around a seismic extremum can be described through a set of seismic attributes. The seismic attributes are obtained using reconstruction techniques of the seismic trace, such that the seismic signal is locally

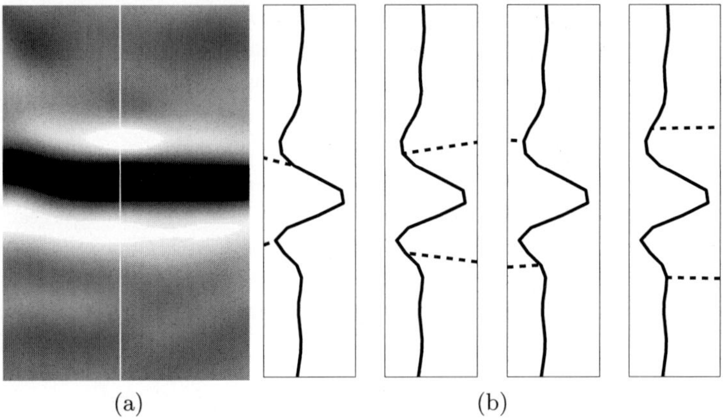

Fig. 2. Reconstruction of a single trace from a seismic section. Image (a) shows the seismic section, with the seismic trace marked with white. The plots in (b) show the seismic trace (solid curve) and corresponding reconstructions (dashed curves). The number of attributes applied in the reconstructions increases from left to right.

described through a limited number of attributes $a_1, a_2, \ldots, a_{n_A}$, being the coefficients from the reconstruction of the seismic trace. These attributes can then be used as input to classification, to enable grouping of seismic horizon segments based on their local wave form.

Figure 2 (a) gives an illustration of an observed seismic trace, and Figure 2 (b) shows corresponding reconstructions of the trace obtained from the attribute values at an extremum point along the trace. Several reconstructions using an increasing number of attributes are displayed. It is observed that the error term is insignificant within a vertical window around the extremum point, covering the distance to the next shallower and deeper extrema points when using a sufficiently large number of attributes. Thus, from one single extremum point the overall local shape of the seismic signal is well described through the chosen set of attributes. The quality of the seismic reconstruction depends on the number of attributes applied, and by increasing the number of attributes (Figure 2 (b)), a minor error is observed in an increasing vertical window around the extremum. In this way, the choice of number of attributes can control how much of the seismic wave form should be taken into account in the automated interpretation of seismic horizons.

4 Extrema Classification

The shape of the seismic signal along a specific seismic horizon is assumed to inhabit similar characteristics laterally. This results in similar values of the attributes along the extrema surface representing the horizon. Classification is applied to group the attributes into classes of similar responses. The use of clustering algorithms to detect natural clusters of attribute responses is referred to as *unsupervised classification*. Each resulting class may then represent a single horizon, or a group of horizon segments with similar seismic wave form. Classification based on a set of training data, which describes the attribute responses and thus the wave form of the seismic horizons of interest, is referred to as *supervised classification*. The resulting classes then represent horizon patches of similar seismic wave form as the chosen training regions, which are likely to belong to the same seismic horizon.

4.1 Statistical Model of Input Attributes

The classification of the seismic wave form attributes is performed in a statistical framework, by defining a probability density function (pdf) of the input attributes. Only the set of voxels representing extrema positions in (1), denoted $\mathcal{E} = \{(x, y, z); E(x, y, z) \neq 0\}$, are included in the classification, and the notation $k \in \mathcal{E}$ is used below to represent an element (x_k, y_k, z_k) in $\mathcal{E}$. The set of n_A input attributes is denoted $\mathbf{A} = \{\mathbf{A}(x, y, z)\} = \{\mathbf{A}_k\}$, where each element $k \in \mathcal{E}$ has an attribute vector $\mathbf{A}_k$ of length n_A representing the seismic wave form around this extremum position. The set of attribute responses is treated as a stochastic variable, and a statistical notation is applied, where capital letter $\mathbf{A}$ represents the stochastic variable and lower case letter $\mathbf{a}$ a realized value of the variable. In both the unsupervised and the supervised classification scheme, the underlying model assumption is that attribute vectors within each class follows a multivariate Gaussian distribution of dimension n_A. Correlation between input attributes within a class is accounted for through the covariance matrix of the Gaussian distribution, while all attribute vectors are assumed to be spatially independent. Let $\mathbf{c} = \{c_k\}$, where $c_k \in \{1, 2, \ldots, n_C\}$, represent the class indexes in a classification into n_C classes, and denote the mean and variance of the Gaussian distributions $\mu = \{\mu_1, \mu_2, \ldots, \mu_{n_C}\}$ and $\Sigma = \{\Sigma_1, \Sigma_2, \ldots, \Sigma_{n_C}\}$ respectively. The mean vector of one class represents the center of attribute values from this class in the n_A–dimensional attribute space, and the covariance matrix describes the shape and orientation of the class in the attribute space. The pdf of the independent attribute vectors is then given as

$$
\begin{aligned}
f(\mathbf{a}|\mathbf{c}, \mu, \Sigma) &= \prod_{k \in \mathcal{E}} f(\mathbf{a}_k | c_k, \mu_{c_k}, \Sigma_{c_k}) \\
&= \prod_{k \in \mathcal{E}} (2\pi)^{-n_A/2} |\Sigma_{c_k}|^{-1/2} \exp\left\{-(\mathbf{a}_k - \mu_{c_k})' \Sigma_{c_k}^{-1} (\mathbf{a}_k - \mu_{c_k})/2\right\}, \quad (2)
\end{aligned}
$$

where $f(\mathbf{a}_k | c_k, \mu_{c_k}, \Sigma_{c_k})$ is the marginal pdf of attribute vector $\mathbf{A}_k$. The aim of the classification procedure is to estimate the unknown parameters of the pdf (2), i.e., the set of Gaussian parameters μ and Σ, and the classification indexes $\mathbf{c}$. The estimation is performed based on the likelihood function of

c, μ and Σ, which is derived by considering the right hand side of (2) as a function of these parameters, given the observed attribute vectors **a**.

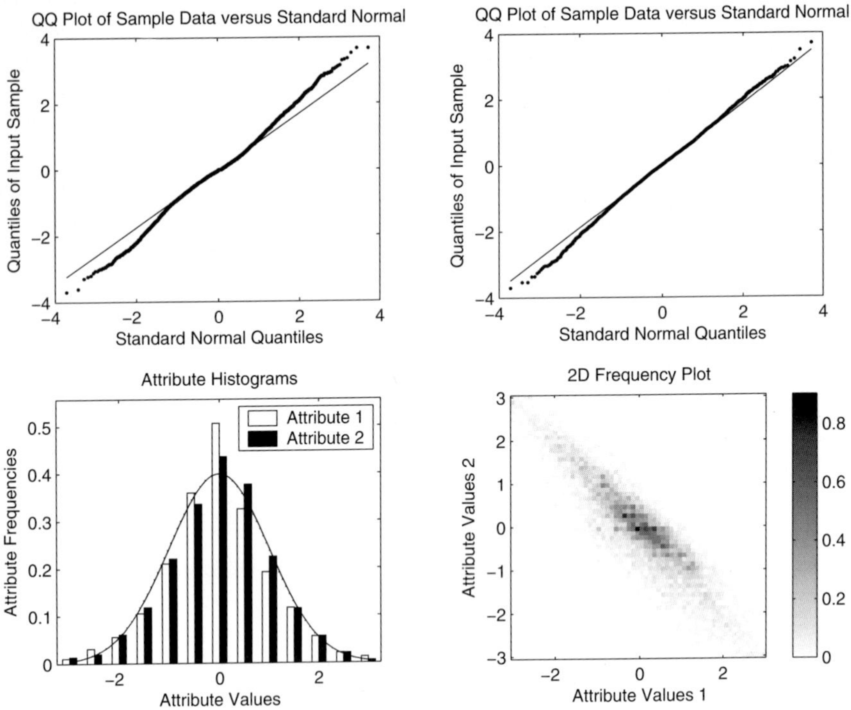

Fig. 3. The figure examines graphically the Gaussianity of 5000 attribute vectors $\mathbf{a}_k = (a_{k1}, a_{k2})$, at randomly selected locations k within a seismic cube. The upper plots show quantile-quantile plots of a_{k1} and a_{k2}, comparing their empirical quantiles to the theoretical quantiles of a Gaussian distribution (dots). The lines join the first and third quartiles of the distributions. The lower left figure shows histograms of a_{k1} and a_{k2}, with fitted Gaussian distributions superimposed for a_{k1} (dotted line) and a_{k2} (solid line). The lines are almost identical due to normalization. The lower right figure illustrates a joint frequency plot of a_{k1} and a_{k2}.

The choice of an underlying Gaussian distribution is validated graphically in Figure 3, through quantile-quantile plots, histograms and joint frequency plots of the two first attributes a_{k1} and a_{k2}. The plots illustrate the Gaussianity of normalized seismic attributes, prior to classification, and at any location within the seismic cube, not only extrema. This is done to illustrate that these types of seismic attributes can in general be fitted fairly well by Gaussian distributions. For a perfect Gaussian distribution, quantile quantile plots should produce straight lines. The plots for a_{k1} and a_{k2} show a proper linearity in the center of the distribution, with slightly heavier tails for the seismic attributes than a theoretical Gaussian distribution. Similar plots for higher order attributes, not shown here, produce an ever more pronounced linearity in quantile-quantile plots. The histogram of a_{k1} and a_{k2} show a good fit with the Gaussian pdfs obtained by estimating the mean and variance of the two attributes. The joint 2D frequency plot of a_{k1} and a_{k2} shows that the two attributes are highly correlated, supporting the choice of a full covariance matrix in the statistical classification model.

A joint Gaussian behavior of attribute vectors $\mathbf{A}_k$ with $n_A > 2$ cannot be examined graphically, due to the dimensionality. However, if the attribute vector $\mathbf{A}_k$ follows a Gaussian distribution with parameters μ and Σ, the univariate variable

$$W_k = (\mathbf{A}_k - \mu)'\Sigma^{-1}(\mathbf{A}_k - \mu) \tag{3}$$

follows a χ^2 distribution with n_A degrees of freedom. Figure 4 illustrates a histogram of w_k, based on 5000 vectors $\mathbf{a}_k$ of length $n_A = 8$, showing a good fit between the histogram and the theoretical pdf of the χ^2 distribution.

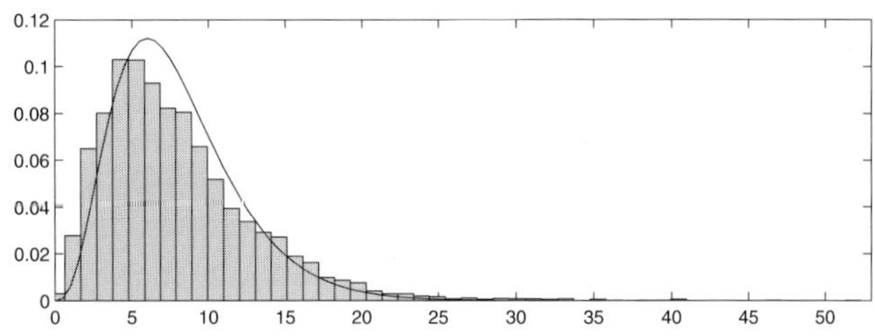

Fig. 4. A histogram of the variables w_k in (3), with $n_A = 8$ attributes and estimated parameters μ and Σ, is compared to the pdf of the corresponding χ^2 distribution with 8 degrees of freedom (solid line).

4.2 Supervised Classification

In supervised classification, a set of training data $\mathbf{T}_c = \{\mathbf{A}_k; k \in \mathcal{E}_{T_c}\}$ is provided for each class $c = 1, 2, \ldots, n_C$. The training set consists of observed attribute values $\mathbf{A}_k$ from a set of extrema positions $k \in \mathcal{E}_{T_c}$ belonging to class c. The training set $\mathcal{E}_{T_c}$ should thus be chosen as a set of extrema positions along one specific seismic horizon. The training data can be picked manually, or they can be generated automatically as described in Subsection 5.3.

Numerous criteria for supervised classifications exist, see for example [5]. In this work a maximum likelihood criterion is applied for classification. Maximum likelihood estimators of the Gaussian parameters μ_c and Σ_c within each class $c = 1, 2, \ldots, n_C$ are first obtained based on the training data

$$\hat{\mu}_c = \frac{1}{|\mathcal{E}_{T_c}|} \sum_{k \in \mathcal{E}_{T_c}} \mathbf{A}_k, \tag{4}$$

$$\hat{\Sigma}_c = \frac{1}{|\mathcal{E}_{T_c}|} \sum_{k \in \mathcal{E}_{T_c}} (\mathbf{A}_k - \hat{\mu}_c)^2, \tag{5}$$

and the classification is performed based on these estimated values. The likelihood of class c_k in point (x_k, y_k, z_k), given the attribute vector $\mathbf{A}_k$ and the estimated class parameters $\hat{\mu}_c$ and $\hat{\Sigma}_c$, is given as

$$f(c_k|\mathbf{a}_k, \hat{\mu}_c, \hat{\Sigma}_c) = \frac{f(\mathbf{a}_k|c_k, \hat{\mu}_{c_k}, \hat{\Sigma}_{c_k})}{\sum_{c=1}^{nC} f(\mathbf{a}_k|c, \hat{\mu}_c, \hat{\Sigma}_c)}. \tag{6}$$

The corresponding maximum likelihood classification rule is

$$\hat{c}_k = \underset{c_k}{\operatorname{argmax}}\, f(c_k|\mathbf{a}_k, \hat{\mu}_c, \hat{\Sigma}_c), \tag{7}$$

classifying the point (x_k, y_k, z_k) into the class giving the highest probability to the observed attribute values $\mathbf{A}_k$.

4.3 Unsupervised Classification

In the case of unsupervised classification, the classification is aiming at revealing natural clusters in the attribute set $\mathbf{A}$. This approach is also referred to as *clustering analysis*. When running unsupervised classification on the extrema attribute set, each resulting class may represent a single seismic horizon, or a group of horizon segments with similar seismic wave form. In unsupervised classification, both the class indexes $\mathbf{c} = \{c_k\}$, and the Gaussian parameters $\mu = \{\mu_1, \mu_2, \ldots, \mu_{nC}\}$ and $\Sigma = \{\Sigma_1, \Sigma_2, \ldots, \Sigma_{nC}\}$ are unknown, and must be estimated in the classification algorithm. This can for example be done by performing maximum likelihood estimation of the unknowns by applying an expectation–maximization (EM) algorithm [2] or an agglomerative hierarchical clustering method [3]. Another approach is to define prior distributions on $\mathbf{c}$, μ and Σ in a Bayesian framework and solve the inversion using a Markov chain Monte Carlo algorithm, see [4] and references therein.

In this chapter a maximum likelihood approach is chosen, and an EM-algorithm is applied to estimate the classification indexes and the Gaussian parameters describing each class. The maximum likelihood estimates are derived by maximizing $f(\mathbf{a}|\mathbf{c}, \mu, \Sigma)$ in (2) with respect to $\mathbf{c}$, μ and Σ. The estimates of each μ_c and Σ_c are obtained through differentiation of (2) with respect to these parameters, and setting the derivatives equal to zero. It can be shown that the solution to these equations are the expressions in (4) and (5), with $\mathcal{E}_{T_c}$ replaced by $\mathcal{E}_c$, defined as $\mathcal{E}_c = \{k; c_k = c\}$, i.e., the set of all voxel positions k with classification index c. Thus, the estimators $\hat{\mu}_c$ and $\hat{\Sigma}_c$ depends on the unknown classification indexes of the complete set of voxels $\mathcal{E}$. The maximum likelihood estimator for c_k is given as

$$\hat{c}_k = \underset{c_k}{\operatorname{argmax}} f(\mathbf{a}_k | c_k, \mu_c, \Sigma_c) \tag{8}$$

which, except for the normalizing constant, is equal to (6) and (7). The maximum likelihood estimators for $\mathbf{c}$, μ and Σ are mutually dependent, forming a nonlinear system of equations that cannot be solved analytically. Instead an EM-algorithm is applied to solve the set of equations iteratively. In each iteration of the EM-algorithm, $\hat{\mathbf{c}}$, $\hat{\mu}$ and $\hat{\Sigma}$ are updated by plugging in the current values of the parameters in (4), (5) and (8) successively. Alternatively, the estimate $\hat{\mu}_c$ in (4) could be replaced by

$$\hat{\mu}_c = \frac{1}{|\mathcal{E}|} \sum_{k \in \mathcal{E}} f(c|\mathbf{A}_k, \hat{\mu}_c, \hat{\Sigma}_c)\mathbf{A}_k$$

and correspondingly for $\hat{\Sigma}_c$ in (5), where the estimates of the Gaussian parameters of class c are derived from weighted sums, with weights given by the class likelihood in each data point k.

4.4 Number of Attributes

The number of attributes applied in the extrema classification, n_A, controls the length of the segment of the seismic trace being represented in the classification (Figure 2). Figure 5 illustrates the effect of increasing the number of attributes. The figure displays three images of the class index extracted along an existing horizon, where the white regions are fault zones with lacking horizon interpretation. Ideally, if the class index is constant along the horizon, the horizon is solely described by this class. Random variations in class index along the horizon are interpreted as noise, while systematic variations can be caused by changes in the seismic wave form along the horizon. In Figure 5 the number of attributes increases from left to right, and it is observed that the noise level in the class index along the horizon decreases accordingly. Thus,

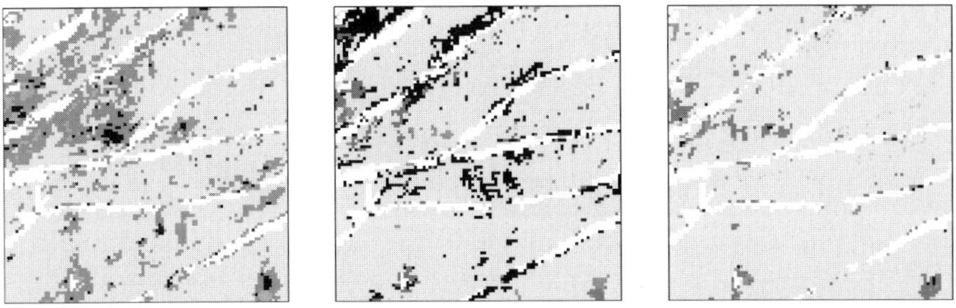

Fig. 5. Extrema class indexes are extracted along a horizon interpretation, where the extrema classification was run with two (left), four (center) and seven (right) attributes. Each color represents one class, and the white regions are fault zones void of horizon interpretation.

a higher number of attributes implies a better continuity in the classification result along a horizon, due to an improved description of the seismic wave form around the extrema. However, if the number of attributes is chosen too high, dependencies in the attributes (Figure 3) may lead to covariance matrices Σ_c that are close to non-singular, resulting in numerical instabilities in the classification algorithm. The choice of number of attributes should thus be a compromise between continuity and numerical stability.

5 Geometry Extraction

The output of the extrema classification is a sparse class cube with values only at extrema locations, so all classified data points are positioned on a seismic horizon. Each extremum point is assigned a class index, and neighbouring extrema points are likely to be assigned the same class index, since the seismic wave form typically is similar from one trace to its neighbour. Thus, starting in an extremum point with a specific class index, it is possible to track a connected region of extrema points with equal class index. This tracked region is referred to as a *horizon patch*, and serves as a primitive for geometry extraction from the extrema class cube.

Fig. 6. A set of horizon patches is extracted from a seismic cube using extrema classification. The patches belong to three different classes, assigned three different gray scale colors, and are drawn as 3D surfaces expanding out from a vertical seismic cross section.

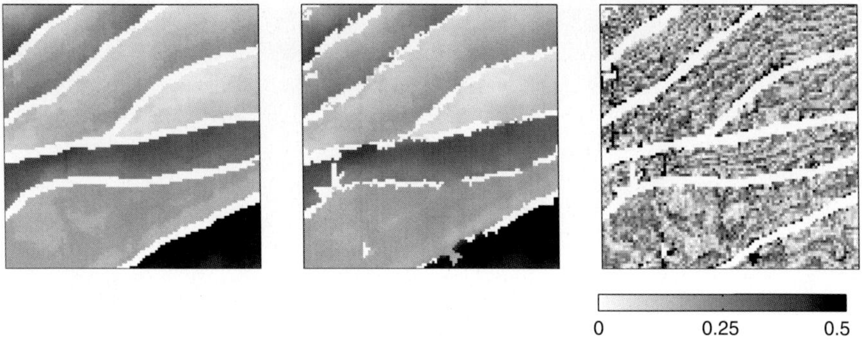

Fig. 7. The figure compares an existing seismic horizon interpretation (left) with the corresponding interpretation obtained using extrema classification (center). White regions are fault zones void of horizon interpretation. The absolute value of the difference between the two interpretations (right) is mainly within 0.5 sample.

Figure 6 illustrates a set of horizon patches extracted from a faulted seismic volume using extrema classification. The patches belong to three different classes, and represent a vertical sequence of three different seismic horizons. The vertical sequence is repeated in each fault block. If each class represents one seismic horizon, then this repetition makes sense geologically, since the geological layer sequence typically would be the same within each fault block. Thus, the constant vertical sequence of classes laterally across faults is a strong indicator of consistency in the extrema classification results. Figure 7 gives a further validation of extrema classification results, where a horizon extracted with this new technology is compared to an existing interpretation of the same seismic horizon. The difference between the two is mainly within 0.5 sample in the vertical direction.

5.1 Seismic Horizons

Continuous patches of class consistent horizon interpretations extracted from an extrema class cube can be combined laterally to form a complete horizon interpretation covering the full area of interest. A class consistent horizon interpretation then consists of extrema points with a similar shape of the seismic wave form in a neighborhood around the extrema, which are likely to belong to the same horizon. Extraction of class consistent interpretations provides a way to automatically interpret horizons in structurally complex regions, for example across faults, since lateral combination of horizon patches is performed based on class index, and not only spatial connectivity.

The unsupervised extrema classification method (Subsection 4.3) has been applied to automatically extract an overburden horizon from a North Sea field, offshore Norway. The extrema classification was applied within a sub-

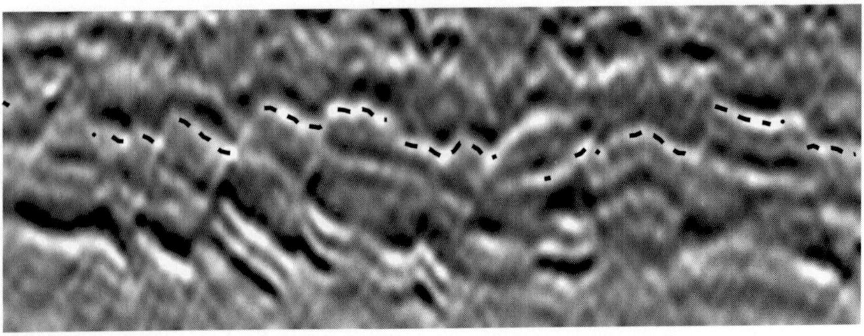

Fig. 8. A seismic horizon extracted using extrema classification is marked (dashed line) on a vertical cross section through the seismic cube.

volume of size 191 × 461 samples horizontally and 38 samples vertically. The horizontal dimension is meters, with sample rates of 25m and 12.5m in the two directions. The vertical dimension is two-way travel-time, with a sample rate of 4ms. The number of attributes in this classification run was eight, and the extrema points were classified into eight classes. The number of classes was chosen in relation to the number of seismic horizons observed within this vertical interval.

Figure 8 shows the horizon interpretation plotted on a vertical section through the 3D seismic volume and Figure 9 shows the horizon as a surface in 3D. The horizon is obtained by combining patches belonging to one class, this class being consistent throughout the section. Thus, the corresponding

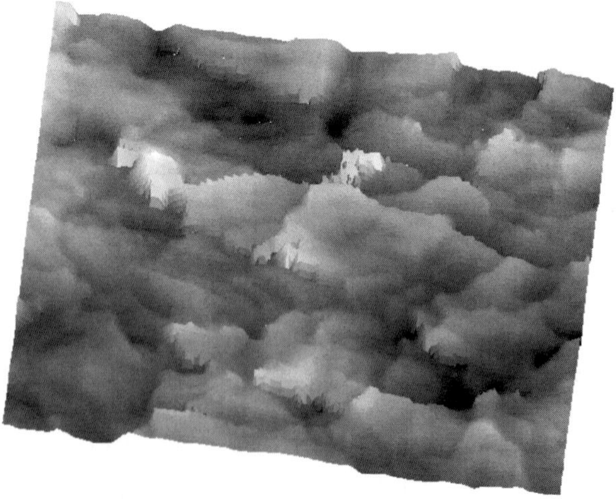

Fig. 9. The extracted horizon from Figure 8 is plotted as a surface in 3D. The fault network crossing the horizon is visible as discontinuities in the surface.

horizon has a characteristic seismic wave form that is recognized laterally through the classification procedure. The mapped horizon has a polynomial structure, and is interpreted automatically across a number of faults. In the left part of the vertical cross section in Figure 8, it is observed how the horizon extraction using extrema classification has been able to complete the interpretation within a number of relatively small, tilted fault blocks. Furthermore, both the positions and offsets of the faults are indirectly mapped by the extrema classification interpretation. The original horizon interpretation contained a number of void regions in the interior of the fault blocks, corresponding to minor classification errors, as well as void regions along the faults. These regions have been interpolated in Figure 9, in order to display a full coverage horizon interpretation. The fault network intersecting the horizon is then visible as discontinuities in the surface.

5.2 Seismic Bodies

In the following example unsupervised extrema classification (Subsection 4.3) is applied to map the extent of high amplitude anomalies observed within an overburden formation in a North Sea field, offshore Norway. The anomalies are interpreted to represent volumes of ancient gas stored within a channel and moraine system in the formation. The top and base of the formation is interpreted manually in advance, and the extrema classification is run in a seismic sub-volume confined by these two interpretations. The sub-volume covers a horizontal region of 711×446 samples, with a sample rate of 12.5m in both directions. The vertical extent of the classification volume is up to 50 samples, with an average of 29 samples, and the vertical two-way travel-time sample rate is 4ms. The extrema points within the formation was classified into six classes, using eight attributes.

In the example presented in Subsection 5.1, the class of interest defined a horizon with full lateral coverage. In this example, however, several classes are of interest, and the classes are occurring as a vertical grouping of horizon

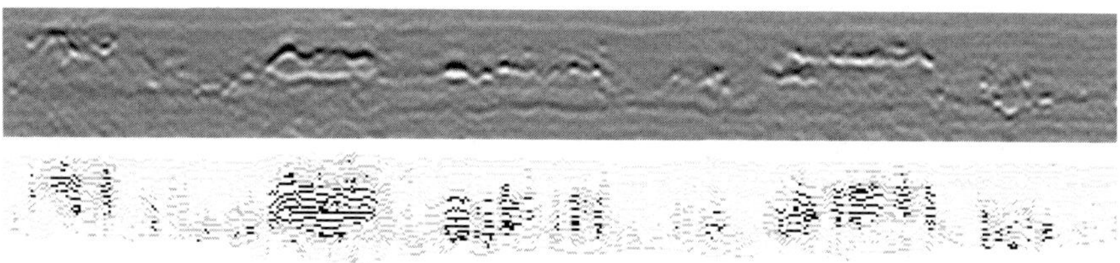

Fig. 10. The figure shows a vertical cross section through a seismic cube (top) and the resulting extrema class cube (bottom), where classification has been performed within one geological formation. The darkest colored classes correspond to the high amplitude anomalies observed in the seismic cube, interpreted as channel and moraine systems containing ancient gas.

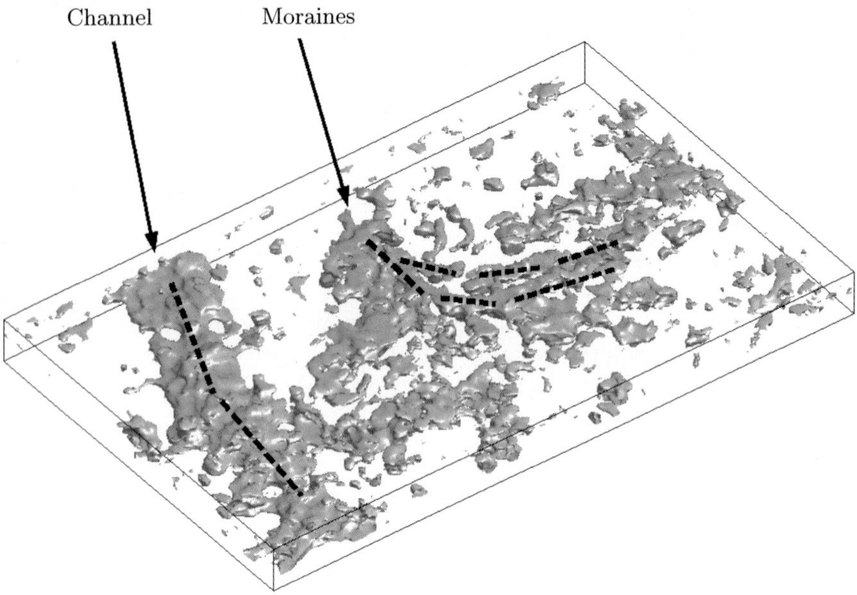

Fig. 11. Three dimensional bodies are generated by defining the closure of the darkest colored classes in Figure 10. A channel and a system of end moraines are interpreted from these 3D bodies, as marked with arrows and dashed lines.

segments belonging to the same class, as illustrated in Figure 10. Thus, each class does not represent one unique, consistent horizon in 3D, but collectively they constitute a 3D body representation of the channels and moraines. The outlines of these bodies in 3D are defined in Figure 11 as the volumetric closure of the classes of interest. By examining these bodies in 3D, a clear channel structure and crescent shaped features resembling end moraines are observed.

5.3 Fault Displacement Estimation

The examples of horizon patches in Figure 6 and the extracted horizon in Figure 8 illustrate how automatic interpretations from extrema classification contain discontinuities along faults. The actual magnitude of the discontinuity, i.e., the fault displacement, can be quantified based on the horizon interpretation on either side. The extrema classification methodology is now adapted to fault displacement estimation, by focusing only on horizon segments close to fault surfaces, instead of extracting the full coverage interpretations of the seismic horizons.

Faults are represented as surfaces, described through a set of coordinates in 3D, and are extracted from the seismic data in advance ([6, 7]). Horizons on one side of the fault surface are first identified automatically as continuous surfaces in the extrema value cube, and attributes extracted along these horizon segments are used as training data. A minimum size of the extracted horizons is required, to ensure sufficiently large training sets. Next, extrema points on the opposite side of the fault surface are classified, and continuous, class consistent horizon patches are extracted on this side of the fault. The classification approach is thus supervised with training data automatically picked, hence no manual effort is needed. The extrema classification results are applied to match pairs of corresponding horizon patches on opposite sides of the fault surface. For consistency, the same vertical sequence of horizon patch classes is required on either side of the fault, and horizon patches out of sequence are not included in the displacement estimations. All pairs of horizon patches are extrapolated from each side of the fault until the intersection with the fault surface is reached, and the displacement is calculated as the distance between the intersection points in the direction orthogonal to the fault plane. Figure 12 shows a sketch of the fault displacement estimation in a vertical cross section orthogonal to the fault plane. Figure 13 gives a 3D illustration of pairs of horizon patches on opposite sides of a fault plane, and their intersection points with the fault plane.

The displacement distance is measured along the fault plane, thus it provides an estimate of the dip-slip component of the displacement, not only a vertical throw estimate. However, since the vertical dimension is often two-way travel-time and the horizontal dimension is meter, the displacement should be decomposed into a vertical throw component measured in milliseconds and a horizontal heave component measured in meters. Since several horizon pairs are involved in the displacement estimation, and all horizon pairs contribute with displacement measurements along an intersection line,

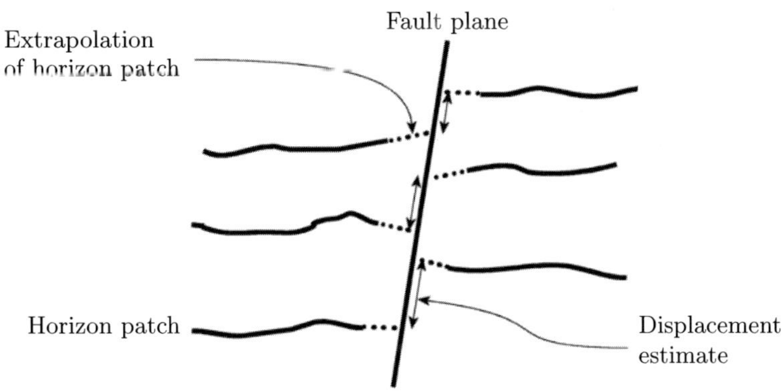

Fig. 12. In a vertical cross section through a fault plane, horizon patches on either side of the fault are extrapolated into the fault plane (dotted lines). The displacement is measured as the distance between pairs of intersection points belonging to a pair of matching horizon patches.

numerous point estimates of displacement are provided for each fault surface. The displacement estimates may vary spatially along the fault surface, and thus describe both the vertical and lateral change in displacement along the fault surface. The described procedure can be applied for all estimated fault surfaces, and constitutes an automatic fault displacement estimation of a complete fault population.

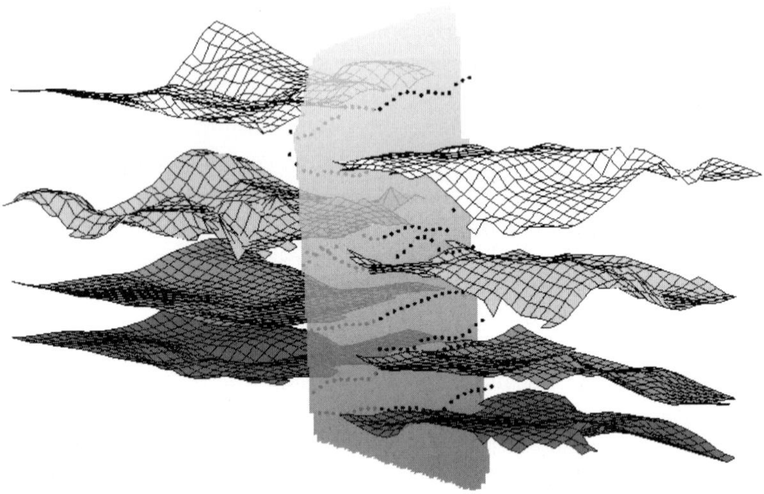

Fig. 13. Pairs of horizon patches on opposite sides of the fault plane (transparent), belonging to the same class, are assigned equal gray scale color. Estimated intersection points between the extrapolation of the horizon patches and the fault plane are plotted (dots), and the displacement is measured as the distance between pairs of points from opposite horizon patches.

6 Discussion

Classification of seismic horizons, denoted extrema classification, has been described, and has proved valuable for mapping automatically seismic horizons in structurally complex regions. Furthermore, a new procedure for automatic fault displacement estimation across pre-interpreted fault surfaces is designed with the extrema classification as the core methodology. The performance of the extrema classification method has been illustrated through a set of real data examples.

The main assumption of the presented methodology is that seismic horizons inhabit lateral similarities in seismic wave form. Similarities in signal wave form along horizons are revealed through classification of a set of wave form attributes, enabling grouping of horizon segments. This assumption is not always fulfilled, for example when the acoustic impedance contrast changes due to changing lithology above or below a reflector, or in seismic tuning regions where the wave form changes due to interference between closely located reflectors. In these cases one should expect a possible change of extrema class index along the seismic horizon. However, such a change of class index may also provide information about the underlying physical conditions leading to the change in the seismic wave form.

The extrema classification helps detect the most possible lateral continuation of a horizon or structure in complex regions, in particular across faults. Structural interpretations can be obtained by focusing on a single class, and extracting automatically class consistent surfaces from the sparse extrema classification cube. The automatic extraction provides a set of horizon patches, where the structural interpreter is afterwards able to manually combine the segments into a full coverage horizon interpretation. Horizon patches obtained through extrema classification also form the basis for automated fault displacement assessment.

The choice of attributes and the statistical classification algorithms are not crucial to the extrema classification methodology. In the extrema classification the seismic wave form attributes are defined as coefficients from a reconstruction of the seismic trace. These attributes can in general be replaced by other set of attributes describing the shape of the seismic trace, as the classification algorithm itself is valid for any type of input attribute vectors. Furthermore, the statistical framework can be generalized, and the Gaussian model in the statistical classification model can be replaced by other appropriate statistical models. Alternatively, the unsupervised classification algorithm can be replaced by other clustering algorithms, for example neural-net approaches. The core of the extrema classification methodology is thus the choice of the extrema as the primitives for geometry interpretation, and the classification approach itself for lateral recognition of seismic horizons.

Acknowledgement

The work was partly supported by the European Union through the TriTex project, contract no. IST-1999-20500, and the SACS2 project, contract no. ENK6-CT-1999-00014. We wish to thank Anna-Lena Skjerdingstad, Jan Petter Fjellanger, and the Grane license owners Norsk Hydro, ConocoPhillips, ExxonMobil and Petoro AS for collaboration and permission to publish data. We would also like to thank the Sleipner partners Statoil, Esso Norge, Norsk Hydro, Total and the SACS2 partners BP, Norsk Hydro, Mobil, Vattenfall, British Geological Survey, Bureau de Recherches Géologiques et Minières, Geological Survey of Denmark and Greenland, Institut Français du Pétrole, Netherlands Institute of Applied Geoscience/National Geological Survey and Sintef Petroleum Research for permission to publish data.

References

1. H.G. Borgos, T. Skov, T. Randen, and L. Sønneland (2003) Automated geometry extraction from 3D seismic data. Expanded Abstr., Int. Mtg., Soc. Exploration Geophys., 1541–1544.
2. G. Celeux and G. Govaert (1995) Gaussian parsimonious clustering models. Pattern Recognition **28**(5), 781–793.
3. C. Fraley (1998) Algorithms for model-based Gaussian hierarchical clustering. SIAM Journal of Scientific Computing **20**, 270–281.
4. C. Fraley and A.E. Raftery (2002) Model-based clustering, discriminant analysis, and density estimation. Journal of the American Statistical Association **97**(458), 611–631.
5. R.A. Johnson and D.W. Wichern (1992) *Applied Multivariate Statistical Analysis.* Prentice Hall International, Inc., 3rd edition.
6. S.I. Pedersen, T. Skov, T. Randen, and L. Sønneland (2004) Automatic fault extraction using artificial ants. This volume.
7. T. Randen, S.I. Pedersen, and L. Sønneland (2001) Automatic extraction of fault surfaces from three-dimensional seismic data. Expanded Abstr., Int. Mtg., Soc. Exploration Geophys., 551–554.
8. L. Sønneland, P. Tennebo, T. Gehrmann, and Ø. Yrke (1994) 3D model-based Bayesian classification. Extended Abstr., Eur. Assoc. Expl. Geophys.
9. L. Sønneland, P. Tennebø, T. Gehrmann, Ø. Yrke, K. Boge, and G. Berge (1994) Volume reflection spectral analysis. Technical report, Schlumberger, Research & Development, July 1994, 1–17.
10. L. Sønneland, P. Tennebø, T. Gehrmann, Ø. Yrke, K. Boge and G. Berge (1998) Orthogonal polynomial spectral decomposition. PCT Patent No. WO/9837437.

Automatic Fault Extraction Using Artificial Ants

Stein Inge Pedersen, Thorleif Skov, Trygve Randen, and Lars Sønneland

Schlumberger Stavanger Research, P.O. Box 8013, N-4068 Stavanger, Norway

Summary. A high-level fault interpretation workflow using automatically extracted surfaces is presented. The first step of the workflow is to generate a fault attribute that enhances the discontinuities in the seismic data. Fault-like surfaces are then extracted using an algorithm called *ant tracking*. The surfaces are then loaded into an analysis tool where the interpreter, by interactively working with the surfaces, decides on the final interpretation. The interpreter works on two levels in the analysis tool. Firstly, on the system level, where the fault surfaces are split into separate systems according to their strikes. Faults that are created at the same time period typically form a fault system. This separation is geologically meaningful and gives the interpreter an overview of the structural history of the area. Secondly, the interpreter groups and modifies individual surfaces within each fault system to form the final interpretation. The workflow is demonstrated as a case study of two fields offshore mid Norway.

1 Introduction

In geomechanics and structural geology there is vast evidence from both laboratory experiments and field observations - that faults developed when the differential stress has reached a critical value, are not randomly oriented, but their orientation is controlled by the direction of the principal stresses [3]. A striking feature is the occurrence of conjugate sets of parallel faults that often cooperate in accommodating imposed deformations. This feature exists on any scale, ranging from microscopical size to fault systems of continental dimensions. Structural geologists use rose diagrams and polar diagrams to help analyzing these fault systems. We present a procedure for extracting faults automatically from 3D seismic data [4] and propose a change of paradigm for seismic fault interpretation by allowing the geologist to analyze fault systems instead of manually mapping individual faults. Analyzing the fault systems in this manner will help reconstructing the paleo-stress orientations. This type of information can be valuable e.g. when fracture orientations of reservoir rocks are being mapped.

in the volume, and hence be strongly marked by pheromone. Noise and surfaces unlikely to be faults should be unmarked or weakly marked, and will be removed by thresholding.

Figure 3 and Figure 4 show some results from applying ant tracking to fault attributes. Note how noisy responses from chaotic layers in the attributes have been removed and how the fault surfaces have been made sharper and more continuous.

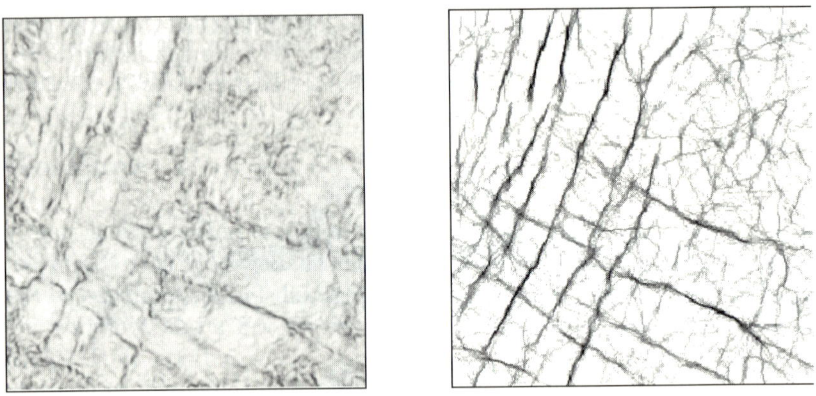

(a) (b)

Fig. 3. (a) Time slice of a fault attribute (variance) with (b) corresponding ant tracking results.

The agents make an orientation estimate of the surfaces they are tracking. This estimate is stored along with the extracted surface as the surface normal, and it is hence possible to make selections of surfaces on their orientations. Other surface properties, like the input attribute's value along the surface, are also stored, allowing separation of faults on how strongly they are supported in the input data. The surface properties play an important role in the structural analysis, and their use will be explained in the next section.

4 Fault Analysis

In order to obtain a consistent fault interpretation, human interaction is required. This interaction is provided through a fault analysis tool. The most important role of this tool is to provide the interpreter with functionality to validate the extracted surfaces. Since the fault attributes do not only enhance events due to faulting, surfaces that are not faults will be extracted and have to be deleted. Second, the analysis tool must provide the interpreter with

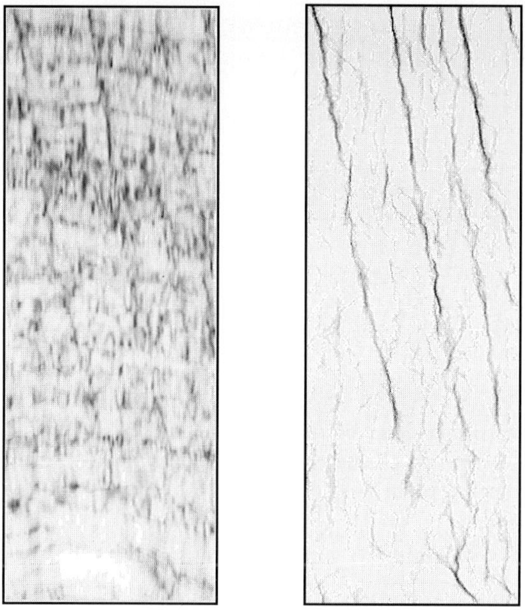

(a) (b)

Fig. 4. (a) Inline section of fault attribute (variance) with (b) ant tracking results.

some surface manipulation functionality. The surface extraction algorithm may not have been successful in combining surface patches that are clearly parts of the same surface, and the interpreter may want to group and re-triangulate these into larger surfaces.

3D visualization is imperative to the analysis tool. Figure 5 shows an example of 3D visualization of extracted surfaces from a field offshore mid Norway. The horizontal slice is an *enhanced attribute* produced by the ant tracking, mapping all extracted surfaces, and a transparent seismic section. More about applying the fault interpretation workflow on this data can be found in [8].

Along with the extracted surfaces, orientation, size and some confidence measures of the surfaces are stored. These properties are used to make selections among the surfaces. If less detail is desired, the interpreter can use the size property to remove all small surfaces. By thresholding on confidence measures, the low confidence surfaces can be removed. Confidence can be measured as the amplitude of the fault attribute along the surface, or as the amplitude of the *ant tracks* [4]. Plotting the surface properties as functions of orientation may also give valuable information. Averaging the value of the property in each cell of the polar plot, provides an indication of the property of the fault system. For example, plotting a confidence measure will indicate

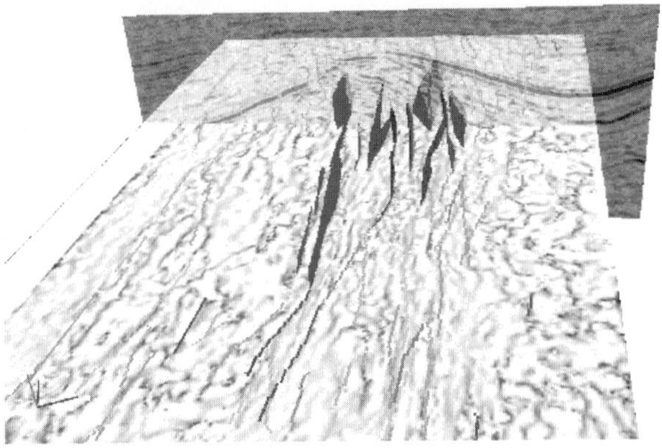

Fig. 5. The fault surfaces can be displayed as 2D lines on seismic or attributes, or as 3D patches by double clicking on the 2D lines. A seismic section and a fault attribute slice from the main field example are displayed together with the fault surfaces.

which fault systems are strongly supported in the data. A method to estimate displacement along a fault surface has recently been developed by Borgos [2], and applied on the extracted surfaces. A displacement estimate will then be attached to each surface and can be used for surface filtering like the other properties.

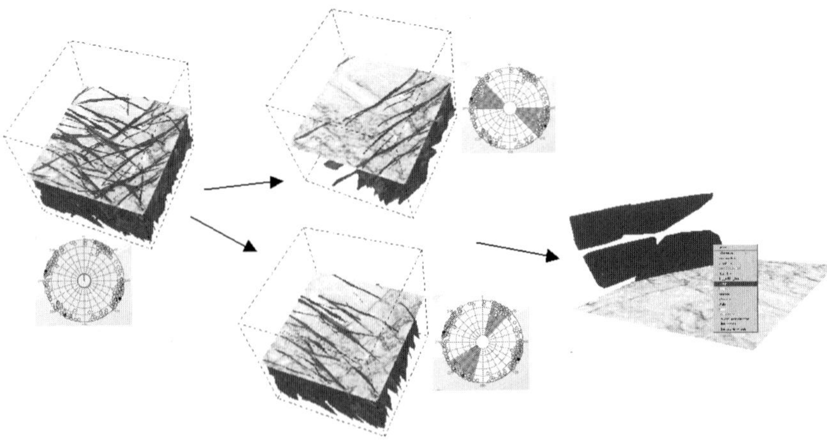

(a) (b) (c)

Fig. 6. (a) Fault systems can be seen as clusters being symmetrical around the center of the polar plot. (b) The surfaces can be split into fault systems by marking regions on the polar plot. (c) Surface patches can be combined and retriangulated into larger surfaces.

The orientation is represented by the surface normal given in spherical coordinates, and can be plotted in a *polar plot*. It is known that faults, due to the direction of the stress field that once created them, typically appear in systems, i.e., having the same strike and conjugate dips. In a polar plot, such a system will appear as clusters symmetrical around the center, as can be seen from the polar plot in Figure 6. The interpreter can make surface selections directly from the polar plot (Figure 6 (b)). By selecting the clusters, the surfaces can be split into separate systems. Without the interference of other surfaces, the interpreter can then validate each system and consider if they are likely fault systems, or if they are due to e.g. coherent noise or processing artifacts in the data. Having studied the various systems, the interpreter can keep the systems that are likely to be due to faulting, and remove those that are not. The interpreter completes the interpretation by working (group and triangulate, smooth, delete) on individual surfaces within each fault system (Figure 6 (c)). By working with the surfaces both individually and in terms of systems, the interpreter has an overall impression on the fault network and the flexibility to change individual surfaces.

5 Data Examples

The workflow has been applied to two fields offshore mid Norway, where the motivation was to obtain an objective fault interpretation and to enhance small-scale heterogeneities produced by small faults. The two fields are located close to each other and are covered by the same data set (Figure 7). The first field is defined by a *horst structure* with relatively horizontal layers. The simple geometry makes it easy to verify the automatically extracted internal faults with manual interpretation in the seismic data. The second field, which is the main focus in this work, is defined by a large and strongly rotated fault block. This field is structurally much more complex with locally intensive internal deformation.

From the horst structure (the first field), the overall impression is that the automatic method is able to track reasonable continuous surfaces of the internal faults. Still, some structural editing is needed. From the seismic section displayed in Figure 8, several of the surfaces (displayed in 2D) clearly represent offsets and/or flexures of the seismic reflections. Other surfaces represent only subtle flexures or dimming of the amplitude, and are consequently more uncertain. The result is considered to give a good qualitatively reference for the rotated block, regarding tracking of faults of the same size and the data quality.

The rotated block is more deformed, and a high number of surfaces were tracked. To simplify the editing, the surfaces were organized after the following properties: (1) The surfaces were subdivided into several work-areas based on inline intervals of the seismic cube. (2) The surfaces were subdivided into five orientation populations. (3) The surfaces were filtered by a top and

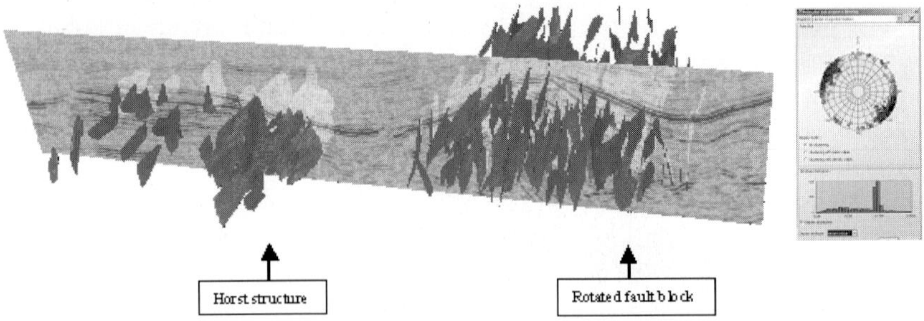

Fig. 7. 3D visualization of automatically extracted fault surfaces from the two reservoir fields, offshore mid Norway (left). A polar plot shows the orientation distribution of the extracted surfaces (right). Properties of the surfaces are displayed in the histogram below the polar plot.

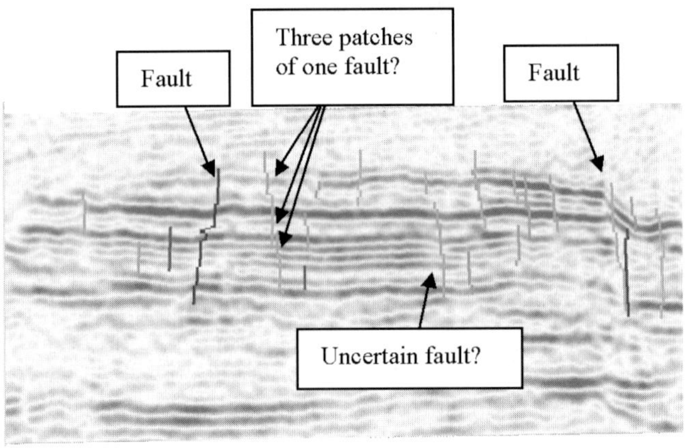

Fig. 8. Edited faults in the horst structure, displayed in 2D. The interpreter has to decide which faults are real, and which fault segments need to be combined.

a base horizon (Figure 10), meaning all potential faults cutting the reservoir zones. (4) The surfaces were filtered by size, meaning that the smallest surfaces were considered to have too low confidence.

To preserve the objectiveness from the ant tracking as good as possible, the editing was performed in the following way. Surfaces that were obviously not representing faults, such as surfaces following a layer boundary or surfaces with unreal geometry, were deleted. Surfaces that intersected and had the same orientation were anticipated to represent the same fault and were thereby grouped and retriangulated. After analyzing each of the subsets, they were added back together. This implied that the structural editing was reconsidered with the same criteria (surfaces that cross or merge). After the last editing iteration, the final result was considered to represent a detailed network of internal faults.

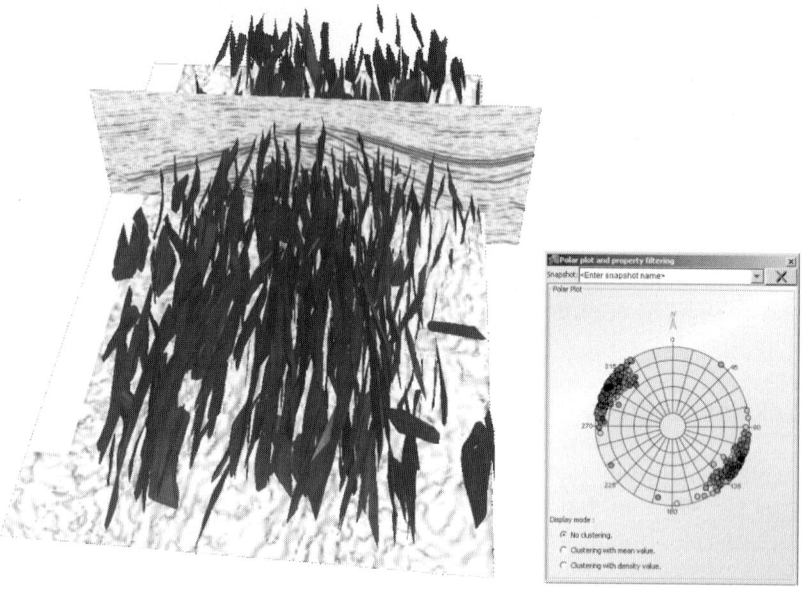

Fig. 9. Edited faults from the rotated fault block displayed with a seismic inline and a fault-attribute time slice (left). Pole plot of the edited internal faults in the rotated fault block (right).

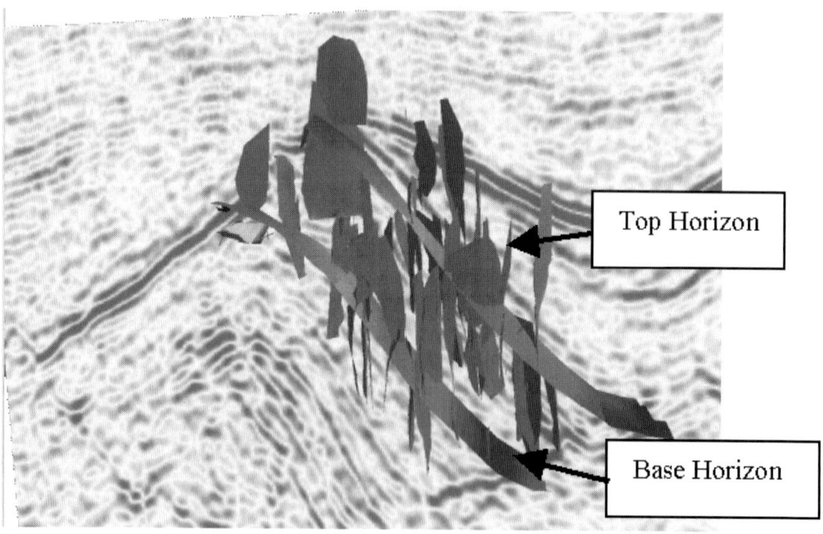

Fig. 10. A top and a base horizon can be used to focus the structural editing to a rock volume, e.g. reservoir zone. This means that every fault surface that cuts this volume is kept while others are filtered out.

6 Conclusion

The workflow outlined represents a paradigm shift in seismic fault interpretation. Automatic surface extraction results in a highly detailed mapping of discontinuities. First principles from structural geology and geomechanics state that faults occur in conjugated sets of parallel faults. The workflow enables the interpreter to operate on such fault systems, as well as on individual fault surfaces. The extracted surfaces needs to be edited in order to obtain the final interpretation. The workflow reduces interpretation time and increases the level of detail.

Acknowledgement

Thanks to Statoil for permission to show results, and to Øyvind Steen, Arne Grønlie and Jan Ove Hansen for their support and contribution. The work was partly supported by the European Union, grant no. IST-1999-20500.

References

1. E. Bonbcau and G. Theraulaz (2000) Swarm smarts. Scientific American, 51–64.
2. H.G. Borgos, T. Skov, T. Randen, and L. Sønneland (2003) Automated geometry extraction from 3d seismic data. Expanded Abstr., Int. Mtg., Soc. Exploration Geophys., 1541–1544.
3. G. Mandl (1998) *Mechanics of techtonic faulting*. Elsevier.
4. S.I. Pedersen, T. Randen, L. Sønneland, and Ø. Steen (2002) Automatic fault extraction using artificial ants. Expanded Abstr., Int. Mtg., Soc. Exploration Geophys., 512–515.
5. T. Randen, E. Monsen, C. Signer, A. Abrahamsen, J.O. Hansen, T. Sater, J. Schlaf, and L. Sønneland (2000) Three-dimensional texture attributes for seismic data analysis. Expanded Abstr., Int. Mtg., Soc. Exploration Geophys.
6. T. Randen, S.I. Pedersen, and L. Sønneland (2001) Automatic extraction of fault surfaces from three-dimensional seismic data. Expanded Abstr., Int. Mtg., Soc. Exploration Geophys., 551–554.
7. T. Randen, L. Sønneland, A. Carrillat, T.S. Valen, T. Skov, S.I. Pedersen, B. Rafaelsen, and G. Elvebakk (2003) Preconditioning for optimal 3d stratigraphical and structural inversion. Extended Abstr., Eur. Assoc. Expl. Geophys.
8. T. Skov, S.I. Pedersen, T.S. Valen, P. Fayemendy, A. Grønlie, J.O. Hansen, A. Hetlelid, T. Iversen, T. Randen, and L. Sønneland (2003) Fault system analysis using a new interpretation paradigm. Extended Abstr., Eur. Assoc. Expl. Geophys.

Part II

Geological Model Building

Geological Modelling and Reservoir Simulation

Chris L. Farmer[1,2]

[1] Schlumberger, Abingdon Technology Centre, Abingdon, UK
[2] University of Oxford, Oxford Centre for Industrial and Applied Mathematics

Summary. The main mathematical techniques used in building geological models for input to fluid flow simulation are reviewed. The subject matter concerns the entire geological and reservoir simulation modelling workflow relating to the subsurface. To provide a realistic illustration of a complete fluid flow model, a short outline of two-phase incompressible flow through porous media is given. The mathematics of model building is discussed in a context of seismic acquisition, processing and interpretation, well logging and geology. Grid generation, geometric modelling and spatial statistics are covered in considerable detail. A few new results in the area of geostatistics are proved. In particular the equivalence of radial basis functions, general forms of kriging and minimum curvature methods is shown. A Bayesian formulation of uncertainty assessment is outlined. The theory of inverse problems is discussed in a general way, from both deterministic and statistical points of view. There is a brief discussion of upscaling. A case for multiscale geological modelling is made and the outstanding research problems to be solved in building multiscale models from many types of data are discussed.

1 Introduction

1.1 Model Based Decision Making under Uncertainty

Hydrocarbons are found in rocks at depths of up to five or more kilometers below the surface of the Earth. Temperatures can be higher than 130 ^{o}C and pressures can reach the order of 1000 atmospheres. The rocks may be more than 100 million years old.

Finding and recovering hydrocarbons uses knowledge from most of the geosciences, physics and engineering. The focus of this article is upon the contribution of mathematics to these disciplines in building, analysing and applying models of fluid flow in the subsurface.

Hydrocarbon recovery involves drilling wells into the rock. Usually these wells are cased with metal tubing. The tubing is perforated at various places, called completions, to allow flow in and out of the well. Some wells are injectors in which fluid, usually water and sometimes gas, is pushed into the rocks to force other fluid toward the producing wells. Designing and operating the system of wells, and the associated surface facilities is an expensive and risky activity.

Reservoir engineers, who decide where to drill, how to complete and how to operate the wells, work in conditions of considerable uncertainty. Although the dominating uncertainty is the price of oil, uncertainties concerning subsurface structure and properties are under some control. This control involves assimilating various measurements into models. Sometimes only one model will be constructed, but usually several models, all incorporating the available data, are made. Some models will be optimistic, some pessimistic, but all are designed to characterise the reservoir and the uncertainty about the reservoir. There is a definite trend toward ensemble reservoir forecasting, where a wide range of models are developed that sample probability distributions of reservoir parameters. Such stochastic models represent, in some sense, an infinite number of deterministic models.

The models may be simple conceptual models in the mind of an engineer, sophisticated but relatively simple mathematical models that can be understood with analytical methods or more complicated models that involve numerical approximations and computer simulation. The idea is that an optimal decision over the models will be optimal in the real world.

The following explains how geological and flow simulation model building takes place.

1.2 About this Chapter

The following chapter is mainly a review. However, there are a few new mathematical results in Section 4, and some personal opinions of the author, about the current state of the art and possible ways to progress, are expressed throughout.

At the centre of the decision making workflow is the fluid flow model. This model is normally solved using numerical methods. The review describes a simplified version of such an approach so that the focus - building the flow model input - makes sense to a general reader.

Fluid flow models require quantitative description of all input parameters. These include functions of spatial position such as porosity and permeability. Techniques for building quantitative descriptions are reviewed in the following.

The following material is a mixture of general background, mathematical detail and a guide to the literature. It is hoped the review adds some extra ingredients helping relate the referenced articles and books to one another to develop a coherent story.

Structure of the Chapter: This chapter is written with a view to helping engineers, mathematicians, geoscientists and others, gain an appreciation of the more mathematical parts of an interdisciplinary subject.

The main stages in the workflow are to gather seismic observations, take cores from wells, to log wells, to study outcrops chosen via examination of cores, to perform well tests, and to incorporate production history into the models. All stages involve direct interaction with data. To use the data there are intermediate stages involving geometric modelling, property interpolation and assimilation of measurements into the models.

It is impossible to devise a structure, following the workflow used in oil companies and oil service companies, that is comprehensible to all readers without excessive use of footnotes and digressions. The structure of the following is thus designed to enable a linear development of the subject.

The structure follows an order where, at the beginning, essentially isolated subjects are reviewed. In the later sections these are brought together in a description of geological and simulation modelling. The first major topic, in Section 2, is flow through porous media and reservoir simulation. Then, in Section 3 the topic of grid generation is covered. Following this, in Section 4, the subject of spatial statistics, also known as geostatistics, scattered data interpolation or property modelling is reviewed in some depth. Section 4 is at the centre of the chapter and places the most demands upon the reader. This section contains some new results that link different methods together.

In Section 5 a general discussion of approaches to inverse problems is provided. The geoscience aspects of the chapter start in Section 6 where the various types of field measurements are described. With all the main concepts in place, Section 7 describes how geological models are constructed. This section on geological modelling also contains additional material on grid generation and spatial statistics that is special to geological modelling. Section 8 covers the topics of upscaling and upgridding - the theory of approximating detailed models with less detailed models. Upscaling and upgridding are currently necessary, in general, if a flow simulation is to be performed with the geological model as input. Of course, upscaling is always needed when it is necessary to relate data on one scale to data on a different scale.

The topic of history matching, where the inverse problem for reservoir simulation is tackled is the subject of Section 9. This builds on the general discussion of inverse problems given in Section 5. In Section 10 an outline analysis of possible workflows is given. The concluding discussion, Section 11, lists some of the suggested open problems and fruitful areas for further research and finishes with a summary of the main conclusion.

This structure, it is hoped, will help readers, depending on their backgrounds and their objectives, to read this chapter in a selective and perhaps nonlinear manner.

Summary of the Argument: With a view to building models suitable for flow simulation, the current approach is explained and criticised. The current approach involves building as detailed model as possible of the *whole* system. The detailed model is then *coarsened* by averaging out detail that cannot be resolved in flow models.

The difficulties with this approach are that (i) the initial level of detail is insufficient to resolve all that is important and (ii) the averaging process leads to inaccurate coarse models even if the fine model is, itself, accurate. This is a dilemma; on the one hand the model is not fine enough and on the other hand it is too fine.

A way around this dilemma, at least for stable flow processes, is suggested in which exceptionally detailed models, but only in the vicinity of the wells are subjected to careful analysis of their scaling behaviour. It is suggested that the coarse model is then constructed by interpolation (stochastic or deterministic) of the scaled local models.

2 Flow Through Porous Media and Reservoir Simulation

This section, after some general remarks, outlines the theory of two-phase flow through porous media. The intention is to introduce readers to a concrete example for later discussion on inverse modelling, scale dependence and heterogeneity. This example also motivates ideas and activities involved in geological and reservoir simulation modelling. Detailed reviews of fluid flow can be found in [11, 37, 50, 128]. For information on reservoir engineering in general and more about reservoir simulation see [9].

Some Remarks about Continuum Mechanics: Continuum mechanics studies deformation and flow of material bodies defined as continua of material points. This idealisation is analogous to idealisations in geometry where properties of shapes and surfaces, defined by simple rules or equations, are analysed. Indeed some scientists view continuum mechanics as a body of theoretical knowledge of the same status as other branches of mathematics such as differential geometry. If one takes this view, then it is necessary that the axioms of the theory are not regarded as *a priori* truths but rather as assumptions characterising a *possible* material that might be found to approximate the behaviour of some physically real material.

Continuum theories characterise the state of a physical system using scalar, vector and tensor valued functions of position x, a three dimensional point, and time t, a real number. These functions sometimes correspond to observable characteristics, such as velocity or position, and other times to more indirectly observable characteristics such as stress. Indirectly observable fields are theoretical constructs of a theory, and need the theory to determine their values. Note that in the following, the symbol x is used to represent a 3D point, an N-D point or the x-component of a 3D vector, $x = (x, y, z)$ or a 2D vector, $x = (x, y)$; the particular meaning of the symbol is always clear from the context.

The principles of continuum theory fall into two classes: balance laws, pertaining to wide classes of materials, and constitutive laws defining particular materials. Wide classes of system are subject to balances of energy, momentum, angular momentum, and mass, but only special systems are subject to constitutive laws such as Fourier's law (heat conduction) or Darcy's law (flow through porous media). The use of the word *law* does not imply a universal truth; it merely distinguishes an important component of a mathematical model.

The Balance Laws of Porous Medium Flow: Porous medium flow theories only use balance laws of mass and energy. This introduction will only define objects using a *mass balance.*

The following considers flow of a fluid with two phases in which there is no mass transfer between the phases. The phases are called phase-A and phase-B. One might picture them, for example, as water and oil. The fluids are characterised by pressures $p_A = p_A(x,t)$ and $p_B = p_B(x,t)$, two scalar functions of position and time. The phase densities, ρ_A and ρ_B are generally functions of the phase pressures, so that $\rho_A = \rho_A(p_A)$ and $\rho_B = \rho_B(p_B)$. The rock is characterised by the *porosity* $\phi = \phi(x)$, a function of position such that the integral

$$V_f = \int_V \phi(x) d^3x$$

is interpreted as the volume of space within the volume V accessible to fluid flow. In practice one must use sufficiently large volumes V, so that the granularity of the porous medium does not significantly influence volume values obtained in measurements. A more general model might make the porosity a function of the *average pressure* $p = \frac{1}{2}(p_A + p_B)$. Yet more general theories use a complete elastic model.

Introducing the function $S = S(x,t)$, called the *saturation of phase-A*, the integral

$$M_A(t) = \int_V \rho_A \phi(x) S(x,t) d^3x$$

is interpreted as the mass of phase-A in the volume V in the pore space V_f.

By definition the phase-B saturation is $1 - S(x,t)$.

Volumetric flux vectors, u_A and u_B, are defined so that the surface integrals, over the surface ∂V of V, where n is the outward pointing unit normal vector,

$$\int_{\partial V} \rho_A u_A \cdot n ds$$

and

$$\int_{\partial V} \rho_B u_B \cdot n ds$$

are interpreted as net mass rates of flow through the surface of the volume V. Finally introducing the *source terms* q_A and q_B, the *mass balance* laws are,

$$\frac{d}{dt}\int_V \rho_A \phi(x) S(x,t) d^3x + \int_{\partial V} \rho_A u_A \cdot n ds = \int_V q_A d^3x,$$

$$\frac{d}{dt}\int_V \rho_B \phi(x)(1 - S(x,t)) d^3x + \int_{\partial V} \rho_B u_B \cdot n ds = \int_V q_B d^3x.$$

Darcy's Law and Generalisations: Balance laws provide a framework requiring further assumptions before interesting or useful mathematical structures, to be used as models of real systems, can be finalised.

Now assume, for each phase, the generalised Darcy's law, due to Muskat and his associates (see the book by Bear [16] for the Muskat and other references of historical interest),

$$u_A = -\frac{k k_A}{\mu_A}(\nabla p_A - \rho_A g \nabla h),$$

and

$$u_B = -\frac{k k_B}{\mu_B}(\nabla p_B - \rho_B g \nabla h).$$

In these equations k is a tensor valued function of position called the *permeability*, $k_A = k_A(S, x)$ is the scalar valued *relative permeability* of phase-A and μ_A its *viscosity*. Similar definitions apply to phase-B. The function $h = h(x)$ is the *depth* of the point x *below* some horizontal datum plane and g is the gravitational acceleration. In general k is a second order symmetric tensor. However, it is very convenient when writing simulators or upscaling packages, to approximate k as diagonal in the selected coordinate system. This assumption is made here, as the use of tensors is not central to our topic. (Note, however, that the wider use of tensors by geologists when building geological models, could reduce the eventual resolution requirements.) The relative permeability of phase-A is a monotonic increasing function of S with values in $[0, 1]$ that must include zero. The relative permeability of phase-B is a monotonic decreasing function of S with values in $[0, 1]$ that must also include zero.

Capillary Pressure: The system of equations is closed by the capillary pressure function $p_c = p_c(S, x)$ such that

$$p_B - p_A = p_c(S, x).$$

The capillary pressure function is a monotonic function of S, often with regions of near zero or near infinite gradient with respect to S. Capillary pressure differences have their origin in surface tension effects, and in some explicit forms of the capillary pressure function the surface tension appears as a parameter. In most applications of continuum theory to porous medium flow there is no need to consider a microscopic explanation and the capillary function is measured in the laboratory or inferred from field data.

Phase Behaviour: In more general models there can be three or even more phases with mass transfer between phases. For example the *black oil model* is a three component system of oil, gas and water components separating into three phases; oleic, gaseous and aqueous. Gas can move between the oleic, gaseous and aqueous phases. The oleic phase is a mixture of oil and gas, the aqueous phase a mixture of water and gas, and the gaseous phase is just gas.

Compressible Single Phase Flow: Single phase flow is of theoretical and practical importance. In this simpler model there is only one phase so that a single balance law, for phase-A say, and a single constitutive equation, the classical Darcy's law

$$u_A = -\frac{k}{\mu_A}(\nabla p_A - \rho_A g \nabla h)$$

are used. Note that the relative permeability is not required in this model. The single phase compressible flow model forms the basis of much of the theory of well and flow-based formation testing.

Incompressible Single Phase Flow: In an incompressible flow without sources, and with constant permeability the pressure is a solution of Laplace's equation.

Formulations: The equations of flow through porous media are generally too complicated to have exact solutions in analytical form and numerical methods must be used. There are special cases, however, of great theoretical interest that are valuable for benchmarking numerical methods and validating computer implementations. As part of the process of deriving a numerical method it is sometimes useful to reformulate the differential equations to clarify the mathematical structure. It is also useful to relate the theory to diffusion, convection or wave propagation, for which there are models with canonical interpretations. To illustrate this idea, and to derive results needed in a later section, the reformulation of *incompressible*, two-phase flow will be developed. For simplicity, the case without sources is studied.

For reference, the equations as they arise naturally from the balance law and constitutive equations, are called the *natural formulation.* The derived formulation will be called the *pressure-saturation formulation.*

In terms of the average pressure, p, define

$$p_A = p - \frac{p_c}{2}$$

and

$$p_B = p + \frac{p_c}{2}.$$

Then the phase fluxes are given by

$$u_A = -\lambda_A k(\nabla p - \frac{1}{2}\nabla p_c - \rho_A g \nabla h)$$

and

$$u_B = -\lambda_B k(\nabla p + \frac{1}{2}\nabla p_c - \rho_B g \nabla h),$$

where the notation $\lambda_A = \frac{k_A}{\mu_A}$ and $\lambda_B = \frac{k_B}{\mu_B}$ has been introduced. λ_A and λ_B are known as *mobilities*.

Multiplying the expression for u_A by λ_B and the expression for u_B by λ_A, subtracting one equation from the other and substituting $u_B = u - u_A$ where the total flux, u, is defined by $u = u_A + u_B$, it follows that

$$u_A = f_A u + f_A \lambda_B k \nabla p_c + f_A \lambda_B k(\rho_A - \rho_B) g \nabla h,$$

where the *fractional flow* of phase-A,

$$f_A = \frac{\lambda_A}{\lambda_A + \lambda_B},$$

has been introduced.

By adding the mass balance equations, and substituting Darcy's law, one finds that

$$\nabla \cdot u = 0$$

and

$$u = -k(\lambda_A + \lambda_B)\nabla p + k(\lambda_A - \lambda_B)\frac{1}{2}\nabla p_c + k(\lambda_A \rho_A + \lambda_B \rho_B) g \nabla h.$$

Substitution of the last equation for the total flux into the zero divergence condition provides an elliptic equation for the total pressure, p.

Substitution of the equation for the flux of phase-A into the phase-A mass balance gives the equation

$$\phi \frac{\partial S}{\partial t} + \nabla \cdot (f_A u) + \nabla \cdot (f_A \lambda_B k(\rho_A - \rho_B) g \nabla h) = -\nabla \cdot (f_A \lambda_B k \nabla p_c).$$

This last equation, for a given flux u, is a parabolic equation for the saturation, S. In the limit of high flow rates, the flux term on the left dominates the capillary diffusion term on the right, and so the diffusion term is often neglected. There are, however, subtleties involving boundaries and discontinuities that merit further investigation regarding the neglect of capillary effects at high flow rates.

In the yet further specialisation of neglecting gravity effects one arrives at the celebrated Buckley-Leverett equation, [28],

$$\phi \frac{\partial S}{\partial t} + \nabla \cdot (f_A u) = 0.$$

There is not enough space to describe the properties of this equation. Suffice it to say that the Buckley-Leverett equation has shock-like solutions, where the saturation *front* is a wave propagating through the reservoir. This combination of an elliptic equation for the total pressure and a parabolic, but nearly hyperbolic equation for the saturation, gives rise to great mathematical interest in two-phase flow though porous media.

For further discussion of the pressure-saturation formulation see [128].

Notes on Dual Porosity and Dual Permeability Models: Systems with two or more distinct types of flow path such as (i) fractures and matrix blocks or (ii) high porosity, low permeability inclusions in a moderate porosity and permeability background, can be modelled as a single spatially rapidly varying, heterogeneous continuum or as two interpenetrating slowly varying continua. In *dual porosity* models, all flow takes place in one of the continua and the other acts as a spatially distributed source or sink. In a *dual permeability* model, flow can take place in both continua with mass transfer between them. Such models are a practical way for predicting behaviour in systems that have small length scale heterogeneities possessing different characteristic *time* scales from their surroundings. The idea was introduced in the classical paper [14]. Dual porosity and dual permeability models are usually used for fractured reservoirs, but they could have application in a more general way [23] to reservoirs displaying two or more local time scales.

Numerical Methods for Fluid Flow Simulation: The main classes of traditional numerical methods are: finite difference methods, finite element methods, finite volume methods, spectral methods and pseudo-spectral methods. When applied to the saturation equation in the pressure-saturation formulation, the important device of a moving coordinate system has sometimes been tried.

Most numerical methods work by introducing a grid of points or cells in the 3D space occupied by the reservoir model, and a discrete series of time steps, t^n.

It is assumed that the reader has at least a basic knowledge of numerical methods. The next paragraphs give an overview of the properties of the different methods.

The finite difference technique replaces the differential operators in the partial differential equation formulations with difference operators. For porous media flow studies, this is almost never done, except possibly in situations where the properties are constant. As there exists no variational principle for the full two-phase flow equations, the finite element method must be used in the Galerkin approach. In its simpler forms, this uses a piecewise linear trial function and orthogonalises the residuals against localised basis functions. Finite element methods can be made to work, but there are suspicions regarding the robustness of a pure finite element method when applied to a hyperbolic equation. Spurious wiggles appear near shock fronts and produce non-physical values which, in turn, cause problems when passed as arguments to relative permeability or capillary pressure functions.

The finite volume method, which returns to the balance equation form of the equations, where one level of spatial derivatives are removed is the method of choice; always for the pressure equation and nearly always for the saturation equation. Commercial reservoir simulators are, with the exception of streamline simulators, entirely based on the finite volume method. See [11] for some background on the finite volume method, and [26] for an introduction to the streamline method. The robustness of the finite volume method, as used in oil reservoir simulation, is partly due to the diffusive nature of the numerical error, known as *numerical diffusion*, that arises from upwind difference methods. An interesting research problem would be to analyse the *essential* role that numerical diffusion might play in the actual physical modelling process; particularly in situations with unstable flow. In the natural formulation, where the character of the problem is not clear, and special methods applicable to hyperbolic, or near hyperbolic problems are not applicable, the finite volume method, in the opinion of the author, is the most trustworthy approach.

Finally there are spectral and pseudo-spectral methods [75] which use a global basis related to Fourier or Chebyshev expansions. The order of accuracy is $O(N)$ in the number of nodes. In simple geometries the method is of great power, at least for the hyperbolic equation. It has not to our knowledge been tried on the pressure equation of reservoir modelling. One might expect difficulties relating to the discontinuous nature of permeability, but these might be surmountable. The pseudo-spectral method mixes working in Fourier space and real space. In this way it is possible to apply spectral-like methods to problems with non-polynomial nonlinearity. The pseudo-spectral method, which uses spectral interpolation on a grid, is slightly less accurate than a spectral method although it is of the same order of accuracy. However, the pseudo-spectral method is more efficient and also simpler than the spectral method [24].

Classification of Simulation Methods by Time Stepping Scheme: Commercial flow simulators generally discretise time derivatives using a first order finite difference formula (Euler's method). The time derivative thus involves the difference of functions at the end and at the start of each time step. All other terms in the equations are discretised to involve functions evaluated at the start and the end of each time step. The pressure *always* appears at the end of the time step and one says that the pressure is *implicit*. Saturations appear at the end of the time step in the *fully implicit* approach. The saturation *only* appears at the start of the time step in the *implicit in pressure, explicit in saturation* method (IMPES). When some regions of the flow domain are IMPES and some are fully implicit, and the regions are modified during the simulation, then the time stepping is said to be the *adaptive implicit* method (AIM).

Generally speaking, the more implicit is the simulation method, the more stable is the scheme. The more stable schemes can take longer time steps without spurious, growing, transients. However, stability is usually accompanied by loss of accuracy.

There are many variants on these ideas. For example, one can form a scheme in the pressure-saturation formulation where the pressure is implicit, as usual, but in the pressure equation the saturation is evaluated at the start of the time step, and in the saturation equation the scheme is fully implicit in all variables. This gives improved stability compared to the IMPES scheme, but it is not as stable as the fully implicit method.

Finite Volume Methods: The finite volume method, when the permeability tensor is diagonal in the selected coordinate system, approximates the pressure and saturation functions as piecewise constant in each grid block. The flux components are assumed constant in their related half-cells. Thus when two cells are joined by a face, the related component of flux is assumed to be the same each side of the face. The balance laws are invoked separately on each grid block, and are discretised in time either by an explicit or fully implicit first order Euler scheme or other variant as discussed in the previous subsection.

The final expression for the numerical flux uses the central difference approximation to $\nabla p_A - \rho_A g \nabla h$. When the permeability tensor is diagonal, for the x-component say, Darcy's law can be arranged to read

$$\frac{u_A^x}{k^x} = -\lambda_A\left(\frac{\partial p_A}{\partial x} - \rho_A g \frac{\partial h}{\partial x}\right),$$

where a superscript notation is used to denote the x-components of flux and the tensor. In the central difference approximation λ_A is assumed constant over the two half-cells adjoining a face with a value equal to that in the cell which is *upstream* of the face. The upstream direction is determined from the sign of $\frac{\partial p_A}{\partial x} - \rho_A g \frac{\partial h}{\partial x}$. The device of using the upstream mobility is necessary for numerical stability. However, it does not take into account any vector attributes of the information flow direction. For this reason the numerical scheme can exhibit grid orientation effects, where results are sensitive to the grid used, and may not converge under grid refinement. This is particularly so when the flow is physically unstable. Some further detail about the finite volume scheme is given in [58], and for a full description, [11] is recommended.

Streamline Methods: For some years, originating with [69] there has been interest in applying the method of characteristics to the solution of the hyperbolic equation. Such methods do not possess a local mass conservation property, as is the case with the finite volume method, but, nevertheless, when applicable they are spectacularly more accurate than finite volume techniques. This is particularly the case for miscible displacement, which is analogous to using *straight line* relative permeabilities. Here the characteristics are the streamlines of the velocity field, and so points distributed in the volume of interest move in the direction of the local total velocity, and carry a constant, or perhaps slowly changing, saturation value. For a detailed review of these moving point methods see [54].

The paper [130] describes a method of characteristics accounting for shock waves. In one dimension the two-phase flow problem can be solved using characteristics even with capillary pressure effects. In two dimensions the situation is more difficult as a result of diffraction of characteristics at permeability discontinuities. Other than [130] there was no progress in applying methods using characteristics on the full two-phase flow case, with general relative permeability curves. However in the paper of [26], the idea of simply solving the Buckley-Leverett equation along pre-calculated streamlines was introduced. This has led to a whole new class of numerical methods, the *streamline methods.* There has not been much rigorous analysis of the convergence of these schemes, and some people are of the view that they are not *proper* numerical methods. However, their simpler and older cousins the moving point methods can be brought into the fold of rigorous approaches (even when diffusive effects are included) and so it may not be long before the streamline method is also rigorously analysed. (For a proof that variants of the moving point method do converge see [53].)

Streamline methods are very fast compared to fully implicit and IMPES methods. This, however, has little to do with their streamline character but more to do with their close similarity to implicit-in-pressure, implicit-in-saturation methods, with the additional feature of very large time steps between updates of the pressure, and many small time steps that update the saturation between the pressure updates.

Instability and Viscous Fingering: The main method of secondary oil recovery injects water into oil reservoirs to push oil towards the producing wells. When the viscosity of the water is less than that of the oil, any small perturbation in the shape of the oil-water front is unstable. Indeed, the shorter wavelength disturbances can grow faster than longer wavelengths, until damped by nonlinear effects. This phenomenon, known as *viscous fingering*, makes the prediction of the fluid flow even more difficult. The viscous fingering instability is closely related to the famous *Hele-Shaw* problem upon which there is a large mathematical literature [86].

The Effects of Heterogeneity: Natural rocks are generally very heterogeneous in the spatial distribution of their properties. There are usually at least three length scales in the variations of permeability. At the lamination scale of centimeters, at the layer scale of meters, and on areal scales of 10's of meters.

It is impossible to resolve all of these scales in detail, and so it is necessary to choose a *cut-off* length scale in any particular modelling exercise, and all phenomena on a scale below the cut-off must be modelled by the appropriate assignment of the properties that appear in the equations.

Different values of the cut-off imply, in general, different values of the properties in the equations. Sometimes, even the form of the equations must be modified for different values of the cut-off length scale.

Sometimes there might be actual or interpreted data that provides a description on some small scale. Then it becomes essential to find a model that only predicts the larger, tractable, scales by a process of upscaling.

The Structure of a Reservoir Simulator: A reservoir simulator is software for solving the porous medium flow equations with detailed models of the spatial distribution of rock properties, detailed models of the thermodynamics of phase behaviour, of the wells and how the wells connect with each other through surface networks. Further, a reservoir simulator will have built-in support for optimisation software in that derivatives of specified flow diagnostics, such as well rates, or masses of chemical components in specified volumes, with respect to a variety of parameters can be computed.

Thus a simulator is a complicated item of software. The simulation of fluid flow through rocks is just a small part.

The Assembly Stage: The simulator starts from the discrete nonlinear equations relating the state of the reservoir at one time step to the values at a previous step. These equations are linearised using Newton's method for nonlinear equations. Experience shows that simulators are more robust when linearisation derivatives are found analytically, rather than numerically. The derivatives have to be chained through the thermodynamics in the reservoir, up the wells and into the separators.

The result of linearisation is a large, mainly sparse, system of linear equations which are generally non-symmetric. Any significant departure from sparsity arises from coupling of geometrically distant parts of the reservoir along wells or large, explicitly modelled, fractures [98].

Linear Solvers: Although general purpose solvers exist, major improvements in efficiency can be gained by exploiting structural features of the linear equations that are particular to reservoir simulation. The key parts of any linear solver are (i) a method of pre-conditioning, essentially an approximation to the system of equations that can be solved directly, but without storage or speed problems (ii) an iteration scheme.

For many years the standard preconditioner has been the method of nested factorisation due to [8]. The iteration scheme of choice, based on well known approaches to nonlinear optimisation, has been the orthomin method of [153]. Only recently have there been challenges to nested factorisation from multigrid methods and domain decomposition methods. Deeper analysis may lead to improvements on basic orthomin. (See [134] for a complete overview and [150] for a review of multigrid.)

Multiscale Simulation Techniques: Recent developments in geological modelling have led to a need for simulations on models so large that linear solvers cannot operate without virtual memory. One approach to this problem is upscaling, another is the use of multiscale simulation. Upscaling is essentially an averaging method that can be performed separately from simulation. Thus upscaling is usually done in another application, the model generator itself or a special purpose model translator. Multiscale simulation is more drastic, at least in its most promising forms. Here instead of assuming in each grid cell that the pressures and fluxes are constant, or possibly linear plus a constant, the variables are complicated functions of position, calculated from small-scale simulations. This technique, introduced by [91] is the current focus of world-wide research efforts [94].

Monte Carlo Simulation: Under most circumstances, models that are implied by data are non-unique. Many different models integrate the same data. Hence a probabilistic description is appropriate. This is reviewed in detail later. However, it requires the results from multiple simulations to assess the implied statistics of the diagnostic functions.

As of now, the only practical approach when dealing with nonlinear systems with stochastic input, is to sample the input and then run the simulator to gather a sample of the diagnostics. In practice people use rather small samples, but numerical experiments show that many samples are needed if the mean and the standard deviations of the output are to be estimated with any accuracy.

One answer is to use smaller, less detailed models, so that many samples can be obtained. More research is needed to establish optimal workflows regarding this problem.

The Bayesian Analysis of the Output from Large, Complex, Computer Codes: In practice one is faced with having to *guess* the answers to difficult technical problems. This is formalised in the Bayesian approach to statistics. Some background knowledge is needed to enable results to be guessed to some reasonable level of accuracy. For example, this knowledge might be based on large numbers of coarse grid simulations. Then one can form a rational view of the statistics of the output. A few detailed simulations can then be performed, that through the application of Bayes' rule can be used to update our prior probability assessment. This is a new, exciting approach, to the analysis of simulation output and has been explored in [38, 99, 100]. Note that this application of Bayesian statistics is quite distinct from the application to the evaluation of uncertainty arising from measurement error and under-determination.

3 Grid Generation

A grid is a decomposition of some volume, Ω, into a finite set of subvolumes, Ω_b such that (i) the union of the subvolumes is approximately equal to the parent volume and (ii) most of the intersection of the surfaces of any pair of subvolumes is included in the intersection of their respective volumes. The subvolumes are called cells or blocks. Thus one is allowing some finite volume of overlap between subvolumes and also some holes provided that the offending volumes are small. Ideally the volume of overlap and of holes will be zero, but in complicated geometries and with the possible constraint to hexahedral cells this is not always practical.

For example, *the uniformly tessellated cube* or *logical grid* is the tessellation

$$\Omega = \bigcup_{i,j,k} \Omega_{i,j,k},$$

where

$$\Omega_{i,j,k} = \{\mathbf{x} \mid ih_x \leq x \leq (i+1)h_x, jh_y \leq y \leq (j+1)h_y, kh_z \leq z \leq (k+1)h_z\},$$

and where h_x, h_y, h_z are the grid sizes in the x,y,z directions and i,j,k are integers.

3.1 Structured Grid Generation

A structured grid is a cell-by-cell mapping of a given volume into a part of, or the whole of, the logical grid. When mapped into only part of the logical grid, it is said that the unmapped cells are *inactive*. An example is shown in Figure 1 where the inactive cells are shown with dotted lines.

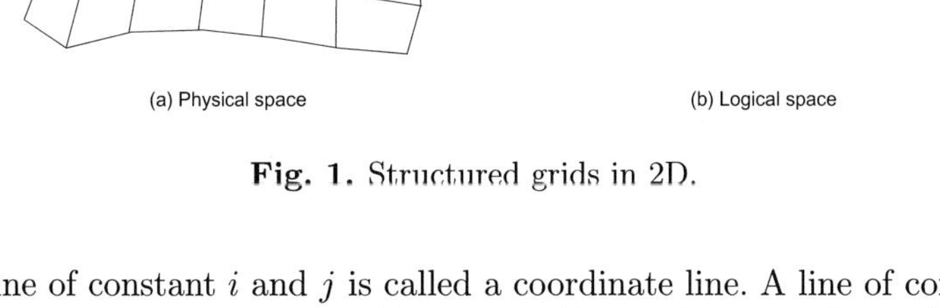

Fig. 1. Structured grids in 2D.

A line of constant i and j is called a coordinate line. A line of constant i and k or j and k is called a grid line. Roughly speaking coordinate lines are *vertical* lines and grid lines are *horizontal* lines (in physical space, not on the page).

Strategies: Most literature on structured grid generation concerns boundary conforming grids. In many cases, such as flow in a pipe, the boundary is external, and in other cases such as flow around a vehicle, an internal boundary. In both cases the boundary is a closed surface.

Grid generation requirements in the geosciences are more demanding. Not only must the grid conform to an external boundary, but it must also honour the layers and the faults. Layers tend to divide the volume or area defined by the external boundary into well-defined segments, but the faults are not generally introduced in such a convenient form. In a geological system there may be many of these *internal* boundaries due to faults and layer boundaries. Many of the internal boundaries will not divide the system into pieces; for example they mark a region where rocks were fractured and then displaced that is quite local.

Standard structured grid generation methods position the grid nodes according to some criterion of smoothness. This can be expressed in the form of an energy functional to be minimised. Sometimes the method is presented as that of solving a partial differential equation, but this is normally equivalent to minimising an energy or smoothness function or functional. In traditional engineering grid generation the constraint on the minimisation is that the grid conforms to the external boundary, and this formulation naturally generalises to one in which the individual internal grid nodes associated with internal boundaries are constrained to particular values. Thus if one can identify which grid nodes are to be related to which geological surfaces one has solved most of the structured grid generation problem.

There are three main strategies for building structured grids which are as follows:

Zigzagging: The first approach is to build a structured grid of coordinate lines without reference to the internal boundaries. The coordinate lines, in some sense nearest to the internal boundaries, are then located and the internal boundary is displaced to the lines. That is, the geometric data is itself modified. This process is called *zigzagging.* This is the most robust technique and can generate a grid for any object if the initial grid has sufficient resolution.

Snapping: The second approach is to build a structured grid that conforms to the external boundary. An algorithm then identifies the coordinate lines nearest to the internal surfaces and displaces them to the surfaces. This process is called *snapping.* If the initial snapping only involves a small deformation, the subsequent grid may be of sufficient quality. However, the snapping might cause badly deformed cells or even cell-folding, where a grid cell is self-intersecting. Thus the snapped position of the nodes might be used as constraints and the grid generation algorithm is then executed again, but this time applying the internal constraints. There is no theory behind this strategy that guarantees a good quality grid. Nevertheless, if the grid smoothness algorithm involves curvature terms, then the results can be good.

In practice each geological object is introduced in a sequence in which snapping is followed by smoothing. In other words, one does not introduce all of the internal boundaries in a single step. Success in grid generation may be sensitive to the order in which the internal boundaries are applied as constraints.

Block Splitting: The third approach is to divide the system into large pieces so that the internal boundaries are included in the boundaries of the large pieces. This requires that the internal surfaces are extended to the boundary in some way. This is quite difficult to do, but with some help from the user, algorithms for this can perform quite well. This particular strategy is reducing the problem of grid generation with internal components of the boundary to a set of problems equivalent to pure boundary grid generation problems. A good implementation would, however, only use the topological information arising from the splitting process, and in the grid generation it would minimise the smoothness functional subject to the constraint that the grid nodes associated with the internal boundaries are constrained to lie on the internal boundary surfaces.

The second and third approaches are usually combined with the first approach, so that the most distorted internal boundaries are zigzagged and the smoother internal boundaries, that align with parts of the external boundaries and amongst themselves, are modelled by splitting or snapping.

Some methods require the internal boundaries to be classified as lining up with the i- or the j-direction. This means that the external boundary has four main parts, corresponding to the four sides of the logical cube. Once the external boundary is deformed there are two families of internal coordinate lines that can be mapped onto the internal boundaries. Software needs help to make this classification. Users can help further, by cutting the more distorted internal boundaries into pieces, and mapping the pieces onto different families of coordinate lines.

Partial Differential Equation Methods: The classical, beautiful, paper of [158] is of particular significance to structured grid generation. The key idea, given by Winslow in an appendix to his paper, is to solve *heat conduction* problems - *in the physical space* - without sources or sinks and with boundary conditions chosen so that each isotherm divides the region into two pieces. Since the maximum principle applies, that the maximum and minimum temperatures must occur on the boundary, the isotherms cannot cross. In two-dimensions, by solving two problems with two sets of isotherms that intersect one another broadly at right angles, one can construct the required grid cells. Of course, to solve these problems one needs a grid. To escape from the obvious dilemma, Winslow transformed the equations into the logical space and solved for the coordinates of the cell corners, rather than the temperatures. This procedure

paper an unstructured triangulation was constructed using a *slicing approach* followed by building vector fields and then building the structured grid.

An initial triangulation is made by dividing the squares in a structured rectangular grid. Then the boundary polygons, for both the external and internal boundaries are processed, line-segment by line-segment. That is, each line segment is used to divide any triangles which they cut into three triangles (two in degenerate cases). At the ends of the internal boundaries, a line is drawn at right angles to its end-point. This extra line is first absorbed into the triangulation so that the end-point of the internal boundaries are included amongst the nodes of the triangulation in subsequent steps.

The lines from the boundaries are then used as boundary conditions for two vector fields, defined on the vertices of the triangulation so that they minimise a measure of smoothness of the fields. The smoothness condition enables an interpolation of the boundary vectors to be constructed.

Once the vector fields are constructed, they can be used to extrapolate the internal boundaries to the external boundary. This divides the volume inside the external boundary into large pieces whose boundaries can be used as the external boundary of a standard structured grid generation method such as the area-orthogonality method. See [57] for the details and some illustrations.

3.2 Unstructured Grid Generation

Conventional Unstructured Grid Methods: In general any grid that is not structured is an unstructured grid. Of particular importance are Voronoi tessellations and their dual the Delaunay tessellation. In three dimensions Voronoi cells are convex polyhedra and Delaunay cells are tetrahedra. In two dimensions Voronoi cells are convex polygons and Delaunay cells are triangles.

The Voronoi grid consists of points, called centres, that are scattered over the domain. The Voronoi cell, Ω_b, with centre x_b consists of the points

$$\Omega_b = \{x \mid \|x - x_b\| \leq \|x - x_{b'}\|, \forall b' \neq b\},$$

where the index b' ranges over all the other centres. In words, the cells correspond to those points that are closest to their respective centre. The faces of a Voronoi cell, that do not lie on the surface of the entire volume, Ω, are flat. The points on the interface between two grid cells are equidistant from the centres of each cell. The lines joining the centres of cells that share a common face, define the dual, Delaunay grid. The tetrahedra of the Delaunay grid possess the remarkable property that their circumsphere does not contain any of the centres other than the centre at the centre of the circumsphere. See [123] for a comprehensive review of Voronoi and Delaunay tessellation theory. See [145] for applications to unstructured grid generation.

A problem with unstructured grid techniques, particularly in 3D, is to build grids honouring the surface of a given volume, or internal surfaces such as faults and horizons. This can be achieved but not without considerable difficulty in programming and in computation. In 2D the method is robust, but in 3D it can fail and require *user intervention* to help the algorithms achieve good results.

Unstructured grids have found widespread application in many areas of computational fluid dynamics and elasticity theory but have not been used very much in the geosciences. See [66] for a selective example in the geosciences. Unstructured grids have, however, found application in the context of 2.5D unstructured grid generation as explained next.

Corner Point Unstructured Grids: The idea of the corner-point unstructured grid is to first *verticalise* all of the faults in the model. The fault traces of the verticalised faults are then projected onto a plane, and a conforming Voronoi or Delaunay grid constructed. A coordinate line is then introduced at each vertex in the planar unstructured grid. These coordinate lines extend vertically until they cut an upper and a lower plane, both parallel to the projection plane. The coordinate lines that lie on the corners of each of the unstructured cells in the plane define a coordinate tube analogous to the tubes defined for the structured grid case. This is essentially an unstructured logical space. The task of building the final grid is now to deform these logical tubes to conform to the original (non-verticalised) fault surfaces and to impose the condition that, at least in the region of space occupied by the reservoir, none of the tubes are self-intersecting or inside-out.

The coordinate lines that lie on faults are then sloped until they lie optimally on the fault surface. The intersections of these coordinate lines are found with the upper and lower planes. The intersection points are then used as boundary conditions in two smoothing problems on the upper and lower planes. The coordinates of the other, non-fault, coordinate lines are determined by a condition that relates their coordinates to the average of their neighbours. This is essentially an unstructured version of the Laplace equation method in logical space. On the upper and lower planes the resulting unstructured grid can be very complicated, with many self-intersecting cells. Miraculously, in most cases anyway, the grid when examined in the region of the reservoir is perfectly satisfactory.

The layers and faults are finally sampled onto the coordinate lines in the same way as for the same task for the structured grid case. A detailed explanation can be found in [77].

Notes on the literature. In the oil reservoir simulation literature the papers of [82, 83, 141] are of note.

For reviews of the grid generation literature see [59, 102, 146, 148] for information concerning structured grids or [145] for reviews of both unstructured and structured approaches.

4 Spatial Statistics: Interpolating Scattered Data

Spatial statistics, often called *geostatistics* and sometimes *property modelling* is concerned with problems of *interpolation under conditions of uncertainty.*

Consider, for example, interpolating a scalar valued function $\varphi = \varphi(x)$, in some region of D-dimensional space, $\mathcal{R}^D$, where the values of φ, $\{\varphi_i\}$, at the points $\{x_i\}$ have been measured with only small errors. Further data is abstracted from some *prototype* or analogue that could be said to *look like* the property that φ is to model. To be specific; given detailed information about a rock outcrop or a densely sampled reservoir of the same depositional environment as the target reservoir, construct an interpolant of the actual measurements that is qualitatively the same as the prototype. Where there are manifest differences between the prototype and the system to be modelled it is necessary to devise methods of transforming the data relating to the prototype in response to expert judgement. The prototype is used to assign realistic estimates of statistical measures such as correlation functions (equivalent, essentially, to variograms). It is a mistake to only use the data available from the target reservoir to ascertain the correlation structure, unless the data are sampled on a scale smaller than the correlation length.

In general, other types of data will be available for many different properties and on several scales. However, this level of generality will not be considered until later sections.

There are obviously many possible interpolants of the data that look like the prototype. Uncertainty quantification is the characterisation of the variation between these different, but data consistent, interpolants. Sometimes only one of these interpolants is selected. For example, the one that is, in some sense, the *smoothest* or the *most probable.* Some methods, such as kriging, allow an estimate of uncertainty to be assigned to these single estimates. See Section 4.6 for an explanation of the dangers in using a single representative from the set of all possible interpolants.

There are several approaches to this interpolation problem; approaches that are not equivalent. It is, however, generally agreed that some probabilistic element is required. Having said that, it is also the case that deterministic interpolation procedures are in widespread use. Thus, before reviewing statistical and stochastic methods, a survey of deterministic methods is given. Later sections show that these methods are closely related to kriging. This is not a new result [89] but does not seem to be widely known.

Note that the problem of scattered data set interpolation is discussed in many other contexts: for example in weather forecasting [39] and in oceanography [160].

4.1 Deterministic Methods

Delaunay Triangulation: Assuming a mappable, continuous surface project the data points onto a datum plane. Applying Delaunay triangulation build an unstructured grid in the plane with the data points located at the nodes. Then perform linear interpolation. If the data points are located at the points $\{\mathbf{x}_i \mid i \in [0, M-1]\}$ then introducing piecewise linear basis functions $\{b_i(x) \mid b_i(x_j) = \delta_{ij}\}$, where $\delta_{ij} = 1$ if $i = j$ or $\delta_{ij} = 0$ if $i \neq j$, the interpolant is of the form $z(x) = \sum_i z_i b_i(x)$. This method works well if the data points are uniformly distributed, uniformly or fully at random, in the region of interest. When the data points are clustered, and large triangles occur, then the method is insufficient. It is difficult to inject any interpretation or style into the method other than by manually inserting control points. This is only a local means of control and not based on any principle.

The method does not easily generalise to problems with overturned surfaces or discontinuities.

(Note that in Delaunay triangulation the data points form the vertices of the triangles in such a way that the circumcircles of the triangles do not contain any data points. Data points can only lie on the circumcircles of the triangles.)

Search Radius Methods: The search radius technique assumes a single valued surface, but accounts for discontinuities. A circle is specified by its radius and, centred on the sampling point at which the height (or other property) is required, all scattered data points inside the circle are located. A polynomial, of some specified degree, is then determined by a least-squares method. Many variants are possible, such as weighting data points by their distance from the sampling point. Discontinuous surfaces are accommodated by ignoring data points which are not on the same side of the discontinuity as the sampling point. A more sophisticated version might use a restored surface estimate from the other side of the discontinuity.

Search circle techniques, with user controls allowing increasing or decreasing circle radii and changes in the order of the polynomials in response to the number of scattered data within the circle, have been used for many years in commercial mapping packages. However, these methods do not have much support from theory, can be slow, and without extensive use of surface patches and help from the user, do not work for reverse faults. See [95] for an extended discussion of search circle techniques.

The Method of Minimum Curvature: Briggs, [27], introduced this method which has found widespread application in commercial mapping software (for example, see [84]). The method is known to be closely related to spline interpolation [90, Chapter 9]. As shown later, minimum curvature methods are a special case of the radial basis function approach and also of the kriging method.

The technique is based on the requirement that interpolating surfaces render the functional,

$$\int_{\Omega} (\nabla^2 z)^2 d^2 x$$

stationary, subject to the constraints that $z(x_i) = z_i^*$, where the $\{x_i \mid i \in [0, M-1]\}$ are data points at which the height, or other, data is z_i^*.

Briggs studied the case where the surface was single valued and without discontinuities. Applying the calculus of variations one obtains the problem

$$\nabla^2(\nabla^2 z) = 0, \qquad z(x_i) = z_i^*, \qquad i = 1, \ldots, N-1,$$

with the natural boundary conditions

$$\nabla^2 z = 0, \qquad \nabla^2\left(\frac{\partial z}{\partial n}\right) = 0, \qquad x \in \partial\Omega.$$

On a rectangular grid, with scattered data points located on the grid vertices it is straightforward to solve this problem using a finite difference method. A slight modification, providing the user with some qualitative control on the character of the surface, is to use the functional

$$\alpha \int_\Omega (\nabla^2 z)^2 d^2x + (1-\alpha) \int_\Omega (\nabla z)^2 d^2x$$

with $\alpha \in [0, 1]$, which implies the partial differential equation

$$\alpha\nabla^2(\nabla^2 z) - (1-\alpha)\nabla^2 z = 0,$$

but with slightly more complicated boundary conditions. It is unnecessary to use the natural boundary conditions for this composite problem, and using the natural conditions for the pure biharmonic problem might be preferable, as long as α is greater than zero. This functional contains a single parameter that can be used to inject some interpretation or style. As α decreases to zero from unity, the peaks and troughs that occur at the data points become sharp, and the surface comes to resemble an elastic material suspended on a set of vertical poles. When α is near unity the peaks and troughs are rounded and it is often difficult to see where the data points were located.

The author has found that in applications of the minimum curvature method to surface modelling that a linear solver (a preconditioned conjugate gradient method) greatly increases the speed of solution, compared to a simple relaxation method such as the Gauss-Seidel iteration, particularly with a non-zero value of α. A similar method, but ignoring cross-derivatives, has been discussed at length by [10]. These authors use a fast alternating-direction-implicit solver. An interesting and, from a practical point of view, a very important generalisation of the Briggs method of minimum curvature is to perform a transformation of coordinates before the surface modelling. When the transformation is implemented via a set of, unstructured, or structured, coordinate lines all faults appear *vertical* in the new coordinate system, but without error. It must be emphasised that this particular *verticalisation* is very different from the *active* verticalisation that can be made (by default or as an option) in some modelling systems. Once the transform is applied then a Briggs type method can be implemented. Care must be taken, however, now that a curvilinear coordinate system is in use. This idea is used, with great effect, in the application described in [64].

Discrete Smooth Interpolation: DSI is as much a philosophy of mathematical modelling as it is an interpolation method [113, 114]. A topological structure, in the form of a set of nodes with a pre-assigned neighbourhood structure, is assumed. A *local roughness function* is then defined for each node. An essential feature of the method is the assumption that each roughness function is a quadratic form in the values to be interpolated. Note that the interpolated function can be vector valued.

A *global roughness function* is then defined as the sum of the local roughness functions. The global function, too, is then a quadratic function of the unknowns. A simple node-by-node minimisation technique can be quite effective if the number of cells in the grid is not too large. Otherwise simultaneously setting all of the derivatives of the global function with respect to the unknowns, to zero defines a linear system. This system can then be solved efficiently by one of the many methods for solving large, sparse, symmetric linear systems.

[113] and [114] presents DSI as an essentially discrete approach although sometimes the choice of local roughness function is motivated by discretising a derivative. DSI is a variational method that restricts the roughness function to be quadratic in the unknowns.

There is a long history of using variational methods in structured grid generation - see Section 3 - in which quadratic and non-quadratic functionals are used. During the last few years (although interest has now diminished) there were many papers on direct discrete formulations of the grid generation problems [33]. In the case of structured grid generation, where the problem is to map a given shape onto a rectangular grid (in 3D and 2D) the quadratic method, introduced by [6] is known to perform badly unless the shape is close to rectangular. This is a very simple application of a method, equivalent to DSI, that is known to be of limited usefulness. The technique of [158], which solves Laplace type problems on the physical region, showed that *non-quadratic* functionals are required for robust interpolation to be possible. Thus it would seem preferable to focus on problem formulations using a continuum approach with subsequent discretisation in the implementation.

Another disadvantage of direct discrete methods is their requirement for the assignment of many weights. When the discrete variational form arises from a continuum variational principle most of the weights (except those controlling the global balance of the major terms of the functional such as total area or curvature) are deduced via the discretisation.

Radial Basis Functions: In these methods a continuously differentiable, *radial basis function*, ϕ, is introduced and interpolants of the form

$$z(x) = a + b \cdot x + \sum_{j=0}^{M-1} \lambda_j \phi(x - x_j)$$

are sought, where the $M+1+D$ unknowns, the λ_j's, a and b, are coefficients to be determined using the conditions

$$z_i^* = a + b \cdot x_i + \sum_{j=0}^{M-1} \lambda_j \phi(x_i - x_j), \qquad i = 0, \ldots, M-1,$$

$$\sum_{j=0}^{M-1} \lambda_j = 0,$$

$$\sum_{j=0}^{M-1} \lambda_j x_j = 0,$$

noting that b and x are D-vectors. These conditions give $M+1+D$ conditions for the $M+1+D$ unknowns. Generalisation to higher dimensions (in the x-variable) and to generalisations which include higher order polynomials other than the affine terms $a + b \cdot x$ are sometimes used. Specialisation to one or two dimensions is straightforward, and often the affine term, a, is omitted. The function φ can be chosen from a large list [31], where r, c and α are real parameters, r satisfies $r > 0$, and $\|x\| = \sqrt{(\sum x_i{}^2)}$,

$\phi = r = \|x\|$	linear
$\phi = r^2 \log r$	thin-plate spline
$\phi = e^{-\alpha r^2}$	Gaussian
$\phi = (r^2 + c^2)^{\frac{1}{2}}$	Multiquadric.

See [31] and [117] for specific applications and interesting further details.

The classical *multiquadric* method of [79] has often been used in the geosciences. See [63] for a review of *classical* methods of scattered data set interpolation.

Radial basis function methods are elegant, do not need a grid, and possess excellent convergence and robustness properties. However when the number of scattered data points is large it requires sophisticated iteration methods to determine the coefficients which, though in itself not an obstacle, when coupled with a requirement to evaluate the interpolant - a sum of a large number of products - on a large number of points, such as on a grid, reduces the attractiveness of the method.

There is very little literature that accounts for discontinuities in radial basis function interpolation. However, a search radius method, in which a radial basis interpolant is used, rather than a polynomial, would be practical and would generalise to discontinuous cases. The paper [78] considers the detection of vertical faults using a radial basis function technique.

Overturned surfaces are a greater problem, although application of a search radius method in combination with sloping coordinate lines could lead to a practical solution.

It is of interest that $r = \|x\|$ is the Green's function for the biharmonic operator in 3D and $r^2 \log r$ the Green's function for the biharmonic operator in 2D. With these particular choices of ϕ, the radial basis function method is an exact solution of the pure Briggs minimum curvature method, or, depending on the implementation, a close approximation.

Thus for large sets of scattered data and with a need to evaluate the interpolant at a large number of points on a grid one might be better served, when D is small, by the numerical solution of a partial differential equation as in the Briggs technique. However, the recent research into radial basis functions with compact support [61] and application of the fast multipole method [29] do provide efficient methods.

Recent reviews of radial basis function methods can be found in [29] and [92].

4.2 Statistical Approaches

Two classes of probabilistic approach are possible. One class is the direct probability density functional (pdf) approach, often generalising the multivariate Gaussian (normal) distribution. The other class consists of assumptions and rules that define a stochastic process. In this second class of method it is not usually possible to state an explicit pdf for the interpolants; the process must be studied via its sample realisations and their properties. The derivation of standard geostatistical results often appears to be rule based but, as shown in the next few pages, can be derived from an explicit pdf. More research using explicit pdfs could lead to new results and insights into the methods of spatial statistics.

4.3 Random Functions

There is insufficient space here to give a full treatment of the theory of random functions. Thus the following specialises the treatment and proceeds formally. There are many texts on basic multivariate probability theory and statistics, and so a basic knowledge of these subjects is assumed on the part of the reader. Texts on the theory of turbulence, such as [65], present the theory in sufficient detail for applied geostatistics. Background in statistical field theory as reviewed by [18, 19, 73] is very relevant.

Review of Some Basic Theory: This subsection reviews some fundamental properties of *Gaussian random fields* in D-dimensions. The case $D = 2$ is needed for surface modelling, and the $D = 3$ case for property modelling.

The following needs the notion of the *functional derivative*,

$$\frac{\delta F}{\delta \varphi(x)}$$

of a functional $F[\varphi]$.

To define this, first introduce the *first functional differential*

$$DF[\varphi : \delta\varphi] = \frac{d}{d\epsilon} F[\varphi + \epsilon\delta\varphi]|_{\epsilon=0}$$

for abitrary functions $\delta\varphi$. If the expression for the differential can be written as

$$DF[\varphi : \delta\varphi] = \int \xi(x)\delta\varphi(x)d^D x,$$

then the function valued functional, $\xi(x)$, is called the *functional derivative* of F and the notation

$$\xi(x) = \frac{\delta F}{\delta \varphi(x)}$$

is used. Higher order functional derivatives are then defined by applying functional differentiation to the lower order functional derivatives, as all functional derivatives are themselves functionals. For more information concerning the functional differential calculus see [18, 19].

A later theorem needs the well known result that

$$\int_{-\infty}^{\infty} e^{-y^2/2} dy = \sqrt{2\pi} \tag{1}$$

and the further expression, obtained by completing the square that

$$\int_{-\infty}^{\infty} e^{-\frac{\lambda}{2}\gamma^2 + j\gamma} d\gamma = \sqrt{\frac{2\pi}{\lambda}} e^{\frac{j^2}{2\lambda}} \tag{2}$$

for real λ, γ and j. The functional probability density of a general Gaussian random field, $\gamma(x)$ with zero mean is of the form

$$\pi(\gamma) = C \exp\left(-H[\gamma]\right), \tag{3}$$

where

$$H[\gamma] = \frac{1}{2} \int \gamma(x) a(x-y) \gamma(y) d^D x d^D y \tag{4}$$

and the integral is over Ω, the volume, or area, of interest. C is a normalisation constant such that

$$\int_S \pi(\gamma) D[\gamma] = 1, \tag{5}$$

where $D[\gamma]$ denotes integration over some suitable space of functions, S. A general Gaussian random field with non-zero mean is written as $\varphi(x) = h(x) + \gamma(x)$, where $h(x)$ is the expectation value, or mean of φ and γ has an average of zero.

One way to make sense of functional integrals such as (5) is to discretise on a finite grid of N cells, with γ_i a uniform value in the i-th cell. Then, using the same symbol for the approximate γ function,

$$\pi(\gamma) = C_N \exp\left(-\frac{1}{2}\sum \gamma_i a_{i,j} \gamma_j\right), \tag{6}$$

and the summation is over i and j and $a_{i,j} = \int_{x\in\Omega_i, y\in\Omega_j} a(x-y)d^D x d^D y$ is an integral over the cells, Ω_i and Ω_j. (6) is just the usual expression for the multivariate Gaussian distribution. The coefficient C_N is chosen so that the integral of the distribution over all N variables is unity.

Introducing Green's function, $g(x-y)$, defined as the solution of the integral equation

$$\int a(x-y)g(y-z)d^D y = \delta(x-z), \tag{7}$$

where $\delta(x-z)$ is the usual Dirac δ-function, the following result holds:

$$\langle \gamma(x)\gamma(y)\rangle = g(x-y). \tag{8}$$

That is, the Green's function is the *correlation function*, where the angular brackets denote the average obtained by integrating over all functions in the space, S, with the probability measure, $\pi(\gamma)$.

To prove this result, first define the *moment generating functional*

$$Z[J] = \int_S \exp\left(-H[\gamma] + \int \gamma(x)J(x)d^D x\right) D[\gamma].$$

Before giving meaning to this last formal expression note that the correlation functions can be derived via functional derivatives of Z with respect to J evaluated at $J = 0$. Thus

$$\langle \gamma(x)\gamma(y)\rangle = \frac{1}{Z[0]}\frac{\delta^2 Z[J]}{\delta J(x)\delta J(y)}.$$

To define the functional integral and to prove the result (8), expand all functions as infinite superpositions of eigenfunctions $\psi_n(x)$ defined by the equations

$$\int a(x-y)\psi_n(y)d^D y = \lambda_n \psi_n(x).$$

Then set

$$\gamma(x) = \sum \gamma_n \psi_n(x), \qquad J(x) = \sum J_n \psi_n(x),$$

assuming the eigenfunctions are normalised so that $\int \psi_n(y)\psi_m(y)d^D y = \delta_{nm}$.

By substitution into the integral equation (7), it follows that

$$g(x-y) = \sum_n \frac{\psi_n(x)\psi_n(y)}{\lambda_n} \tag{9}$$

is a representation of Green's function. Let us also note the standard result that

$$\delta(x-y) = \sum_n \psi_n(x)\psi_n(y),$$

(for arbitrary $f(x)$, $f(x) = \sum f_n \psi_n$, $\int f(x) \sum_n \psi_n(x)\psi_n(y) d^D x = \sum f_n \psi_n(y)$).

Substitution into the generating functional gives

$$Z[J] = \int_{-\infty}^{\infty} \prod_n d\gamma_n e^{-\frac{1}{2}\lambda_n \gamma_n^2 + J_n \gamma_n}.$$

Exchanging the order of the product and the integral leads to

$$Z[J] = \prod_n \int_{-\infty}^{\infty} d\gamma_n e^{-\frac{1}{2}\lambda_n \gamma_n^2 + J_n \gamma_n},$$

and using (1) and (2)

$$Z[J] = \prod_n \sqrt{\frac{2\pi}{\lambda_n}} e^{-\frac{J_n^2}{2\lambda_n}}.$$

Finally using the expression (9), gives

$$Z[J] = Z[0] \exp\left(\frac{1}{2} \int J(x) g(x-y) J(y) d^D x d^D y\right),$$

where

$$Z[0] = \prod_n \sqrt{\frac{2\pi}{\lambda_n}}.$$

It thus follows that

$$\frac{1}{Z[0]} \left(\frac{\delta^2 Z[J]}{\delta J(x) \delta J(y)}\right)_{J=0} = g(x-y).$$

4.4 Kriging

In this section the theory of kriging is reviewed along the lines of its usual presentation. See, for example, [85, 96].

In some cases the variance of a property is infinite, so geostatisticians use the *variogram*,

$$\nu(x-y) = \frac{1}{2}\langle[\varphi(x) - \varphi(y)]^2\rangle,$$

which is an affine transformation of the correlation function. As the treatment in this chapter is formal the correlation will be used throughout.

Simple Kriging: Consider the problem of interpolating a *random* function. That is, given M sample values $\varphi_i, i = 1, \ldots, M$, at each of M spatial points x_i, estimate the value of the function, φ at the point x.

In simple kriging it is *assumed* that the interpolant is a linear superposition of the data,

$$\varphi = v_0 + \sum_j v_j \varphi_j,$$

where the scalar parameters, v_i are to be found.

Conditions for the determination of the v_i's are motivated by the arguments (i) the interpolant is correct on average and (ii) if the true value, φ_0 were known then for each realisation the interpolant should minimise the variance of the error. Note the v_i's are *not* random, and have the same values in all realisations; they are, however, functions of x.

Thus it is asserted that

$$\bar{\varphi} = v_0 + \sum_j v_j \bar{\varphi}_j,$$

and that the *error*, $\langle(\varphi - \varphi_0)^2\rangle$ is a minimum, considered as a function of the v_i's. Where the overbar notation and the angular bracket notation both indicate the expectation of the expression within their scope.

If it is also assumed that the mean of the process is a known constant, m, say, then it follows that

$$v_0 = m - \sum_j v_j m.$$

Minimising the expression for the error gives the linear equations

$$\sum_j g_{ij} v_j = g_{0i}, \tag{10}$$

where $g_{ij} = \langle(\varphi_i - m)(\varphi_j - m)\rangle$ and $g_{0i} = \langle(\varphi_0 - m)(\varphi_i - m)\rangle$ are the assumed known covariances of the random function φ. In applications the covariances are determined from sample averages of an analogue system, possibly inferred from the data, or just guessed. The values of the covariances can also be adjusted as part of the process of integrating further data into the model.

By solving the system of equations, (10), to give $v_i = \sum_j [g_{ij}^{-1}] g_{0j}$ the interpolant

$$\varphi_0 = m + \sum_{ij} [g_{ij}^{-1}] g_{0j} (\varphi_i - m)$$

is obtained, where $[g_{ij}^{-1}]$ denotes the elements of the inverse of the $[g_{ij}]$ matrix.

Ordinary Kriging: In ordinary kriging the mean is an *unknown* constant. As before, an interpolant of the form $\varphi = v_0 + \sum_j v_j \varphi_j$ is sought. The conditions for the v_i's are that (i) the mean error is zero and (ii) the mean square error is a minimum. Thus

$$m = v_0 + \sum_i v_i m$$

for *arbitrary* m. This can only hold if $v_0 = 0$ and if $\sum_i v_i = 1$. The minimum error variance is then sought, subject to the constraint that $\sum_i v_i = 1$. By introducing a Lagrange multiplier, τ say, and looking for the stationary point of $\langle(\varphi_0 - \varphi)^2\rangle - 2\tau(1 - \sum_i v_i)$, the set of equations

$$\sum_j g_{ij} v_j + \tau = g_{0i},$$

$$\sum_i \lambda_i = 1,$$

is deduced. These can be solved for the v_i's so that $\varphi = \sum_i v_i \varphi_i$. It is interesting that the mean, m, is not required in evaluating the interpolant.

Universal Kriging: In universal kriging the mean is unknown but a function of x in the form of a linear superposition of a small number of basis functions, $\psi^k(x)$ which are here assumed to be orthonormal. Thus $\int \psi^k(x)\psi^i(x) d^D x = \delta_{ki}$, and $h(x) = \sum_k b_k \psi^k(x)$, where the coefficients b_k are to be determined by the interpolation procedure. In this method, once again the interpolant is *assumed* to be a linear superposition, $\varphi = v_0 + \sum_j v_j \varphi_j$, of the values at the data points. The usual conditions (i) that the mean error should be zero and (ii) the mean square error should be a minimum are imposed.

The zero mean error condition implies that

$$\sum_k b_k \psi^k(x) = \sum_{k,j} v_j b_k \psi^k(x_j).$$

This can only be satisfied for arbitrary b_k, provided that

$$\psi^{k0} = \sum_i v_i \psi^{ki},$$

for $k = 1, \ldots, K$, where $\psi^{ki} = \psi^k(x_i)$ and $\psi^{k0} = \psi^k(x_0)$. These conditions are known as the *universal kriging* constraints and to compute the v_i the expression $\langle(\varphi_0 - \varphi)^2\rangle$ is minimised subject to the universal kriging constraints.

Introducing Lagrange multipliers τ_k, and setting the derivatives of

$$\langle(\varphi_0 - \varphi)^2\rangle + 2\sum_k \tau_k \left(\sum_i \psi^{ki} - \psi^{k0}\right)$$

to zero, the set of equations

$$\sum_j g_{ij} v_j + \sum_k \tau_k \psi^{ki} = g_{0i}, \tag{11}$$

$$\sum_j v_j \psi^{kj} = \psi^{k0}, \tag{12}$$

$$\varphi_0 = \sum_i v_i \varphi_i, \tag{13}$$

is deduced. These can be solved for the v_i and the interpolant evaluated. (11)-(13) are the *universal kriging equations* introduced by Matheron in 1969. See [34] for a sympathetic exposition.

Indicator Kriging: This technique transforms the observations into integer valued quantities with the transformation

$$\begin{aligned} I_\nu(\varphi) &= 1, \quad \varphi < \nu, \\ &= 0, \quad \varphi \geq \nu, \end{aligned}$$

for some cut-off value ν. One then performs simple kriging [42] on these transformed values for a sequence of ν values. The interpolants, f_ν, say, are not integers, but real numbers that are interpreted as the probability that the value is less than ν. This seems plausible, but unfortunately there is no guarantee [85] that this honours the necessarily true requirements that $f_\nu \geq f'_\nu$ if $\nu > \nu'$. Apparently the response [85] on failure to satisfy the conditions is to impose them as constraints. [42] provide a detailed discussion of corrections to indicator kriging so that the conditions are satisfied. The main application of indicator kriging is to the numerical simulation of stochastic processes. The ad hoc nature of this method implies the need for a better technique.

Disjunctive Kriging: The methods just described make an implicit assumption that the underlying random process is Gaussian (this is shown in detail in the next section). Often the data are manifestly non-Gaussian. The approach of disjunctive kriging or nonlinear geostatistics is a response to this in which (i) a *normal scores* transformation is performed on the original data before interpolation (that is, the single point probability density is transformed to that of the normal distribution) (ii) the correlation structure and mean values are assigned or estimated from the transformed data (iii) the interpolation is performed (iv) the results are back transformed using the inverse of the normal scores transform. There exists some elegant theory using Hermite polynomials to construct the forward and inverse normal scores transform [85]. There is more, however, to a Gaussian random function than just its single point probability distribution. The other key property is that the process is completely specified by the mean and the correlation function.

Failure of this property is not addressed in disjunctive kriging. For reasons of space, non-Gaussian spatial statistics are not discussed in this chapter. However see [56] or [97] for examples of non-Gaussian techniques.

4.5 Maximum Probability Interpolants

If one assumes that a given property is characterised by a general Gaussian pdf of the form (3) then a natural interpolation technique is obtained by posing the question: Given data $\{\varphi_i^*\}$ on a discrete set of points $\{x_i\}$, which function $\varphi(x)$, with the values $\{\varphi_i^*\}$ at the data points, maximises the probability functional (3)?

In general the mean will be unknown, and following the lead of universal kriging, it is assumed that the mean is given by $h(x) = \sum_k b_k \psi^k(x)$, where the basis functions ψ^k are orthonormal with $\int \psi^k \psi^l d^D x = \delta_{kl}$.

This formulation implies the task of minimising the functional argument

$$\sigma = \frac{1}{2} \int a(x-y)(\varphi(x) - h(x))(\varphi(y) - h(y)) d^D x d^D y$$

subject to the constraints $\varphi(x_i) = \varphi_i^*$. Introducing Lagrange multipliers, λ_i one can write

$$\tilde{\sigma} = \sigma - \sum_i \lambda_i (\varphi(x_i) - \varphi_i^*)$$

or

$$\tilde{\sigma} = \sigma - \sum_i \lambda_i \int (\varphi(x) - \varphi_i^*) \delta(x - x_i) d^D x.$$

Taking variations with respect to φ the equations

$$\int a(x-y) \left(\varphi(y) - \sum_k b_k \psi^k(y) \right) d^D y = \sum_i \lambda_i \delta(x - x_i) \tag{14}$$

are deduced.

Differentiation with respect to λ_i gives $\varphi(x_i) = \varphi_i^*$ and differentiation with respect to b_k gives

$$\int a(x-y) \varphi(y) \psi^k(x) d^D x d^D y = \int a(x-y) \sum_j \psi^k(x) b_j \psi^j(y) d^D x d^D y. \tag{15}$$

Multiplying (14) by $\psi^k(x)$, integrating over x and using (15) together with the usual properties of the δ function, leads to the expression

$$\sum_i \lambda_i \psi^k(x_i) = 0,$$

which is the condition used in the general radial basis function method.

Introducing a Green's function, $g(x-y)$, satisfying

$$\int a(x-y)g(y-z)d^D y = \delta(x-z)$$

reduces the calculation to solving

$$\sum_k b_k \psi^{kj} + \sum_i \lambda_i g_{ij} = \varphi_j^*, \tag{16}$$

$$\varphi(x_j) = \varphi_j^*, \tag{17}$$

$$\sum_i \lambda_i \psi^k(x_i) = 0, \tag{18}$$

where $\psi^{kj} = \psi^k(x_j)$. These equations determine the λ_i and the b_k and also provide the interpolant

$$\varphi(x) = \sum_k b_k \psi^k(x) + \sum_i \lambda_i g(x - x_i). \tag{19}$$

The following theorem can now be proved:

The radial basis function system (19) *is the adjoint of the universal kriging system* (11) *and* (12).

Before the proof is given the adjoint method will be explained. Here one has a linear system $A_{ij}x_j^u = r_i^u$ (using the usual convention regarding repeated indices) with a symmetric matrix, A_{ij}, and multiple right hand sides r_i^u thus generating multiple solutions, where x_i^u is the i-th component of the u-th solution. The aim is to compute scalar products of the form $s^u = \sum_j c_j x_j^u$. That is, the s^u rather than the solution are of primary interest. It is easy to prove that this problem is equivalent to solving the *adjoint problem* $A_{ij}^T \alpha_j = c_i$ and evaluating $s^u = \sum_j \alpha_j r_j^u$ This replaces many direct linear systems with one indirect adjoint system. Adjoint methods are thus vastly more efficient than the original formulation.

To prove the equivalence of radial basis functions and the universal kriging method, the universal kriging system will be now transformed into the radial basis function equations. Multiply (11) by λ_i and sum over i to give

$$\sum_{ij} \lambda_i g_{ij} v_j + \sum_{ik} \tau_k \psi^{ki} \lambda_i = \sum_i \lambda_i g_{0i}.$$

Then using (17) and substituting (16) obtain

$$\sum_j \varphi_j^* v_j - \sum_{kj} b_k \psi^{kj} v_j = \sum_i \lambda_i g_{0i}.$$

Finally substituting from (13) and (12), (18) is derived, which proves the equivalence.

A corollary of this result is that both the radial basis function method and the kriging method are maximum probability interpolants.

Further results can be obtained using the maximum probability formalism. For example, by discretising the functional and exploiting well-known properties of the multivariate Gaussian distribution [115] one can derive closed form expressions for the single point probability density.

4.6 Stochastic Sampling Techniques

The Reason for Stochastic Sampling ('Simulation' or 'Monte Carlo'): Attention now turns to using random function theory in geological modelling and reservoir simulation. For example, a fluid flow simulator will compute, say, the production of oil from a producing well over some time interval. This can be written formally as $q = Q[\varphi]$, where Q represents the simulator and any post-processing, and q is the numerical value obtained from the simulation results. If φ is a random function (such that porosity and permeability functions are specified functions of φ) then q is a random variable - or in general a random function. Functionals, such as Q, are called *diagnostic functionals*, or just *diagnostics*. The central task in applications of spatial statistics is to determine pdfs of diagnostics, or at least to provide a summary of the properties of such pdfs. The main quantity of interest might be just the average $\langle q \rangle$ or $\langle q^2 \rangle$. Thus it is necessary to evaluate integrals of the form

$$f = \int_S F[\varphi]\pi(\varphi)D[\varphi],$$

where F denotes a general diagnostic functional. F is often a very expensive functional to calculate - it might take several hours of computer time to make a single evaluation. So how is one to calculate a function space integral, essentially an infinite dimensional integral?

One case where the problem is simple to solve, is when the probability density functional is sharply peaked about its maximum value, φ^*. This is of course the value obtained from the maximum probability method. When π is sharply peaked it follows that $f = F[\varphi^*]$ is a good approximation to the exact value. Thus the equivalent methods of kriging, radial basis functions or minimum curvature are all valid approaches to providing input for simulation when the functional probability density of the geological properties is vary sharply peaked about its maximum value.

Now, what is to be done when $\pi(\varphi)$ is *not* sharply peaked, as might be expected to happen in an actual reservoir simulation study? In such cases kriging or its equivalents should not be used, as the errors can be very large.

Monte Carlo Methods and the Metropolis Algorithm: If the diagnostic functional is discretised on a spatial grid the dimension of the integral is then of the order of N, the number of grid blocks. (It could be more as there may be several random fields involved in a model.) N can be of the order of 100's of thousands or even millions. This is, in terms of integration, a large dimension. It would clearly be very difficult to perform a conventional numerical integration over so many variables, bearing in mind that the diagnostic is so difficult to evaluate. In such circumstances it is widely accepted that methods using *random numbers* - Monte Carlo methods - are a suitable approach.

The basic principle can be explained by discretising the values of φ into, say, G levels. There are then G^N possible arrangements of values on the grid. Most of these arrangements approximate very rough functions, with wildly varying values, a few will be smooth. Most of the arrangements have a very low probability as given by the value of π. Suppose now that M sample functions, $\varphi^r, r = 1, \ldots, M$, have been generated by simply choosing a value independently at random in each cell uniformly from the possible discrete values.

It can be proved that as M increases, the estimate

$$\langle f \rangle \approx \frac{1}{M} \sum_r F[\varphi^r] \pi(\varphi^r) \tag{20}$$

will converge to the average of the diagnostic. Although the above method will work *in principle*, it is very inefficient because most of the realisations have negligible probability density of occurring. Thus methods which take into account the structure of the pdf, π, and weight the selection of the realisations according to π will be a vast improvement.

At least three subject areas make use of stochastic sampling: statistical physics, statistics and geostatistics. Each area has its own favourite methods, partly a result of the different application areas, but, one suspects, largely due to cultural reasons. There are many texts in these areas, so the following only gives an outline and some selected references. A relevant review may be found in [55].

The statistical physicists were the first to use Monte Carlo methods in the celebrated paper of [118]. The Metropolis algorithm has the following ingredients. First there is a starting state, perhaps a random state with each cell value chosen at random. Then there is an update method which is stochastic, in that (i) the probability of choosing φ' from φ is the same as choosing φ from φ' (ii) all states, φ', are accessible from any other state φ. The algorithm, where $\pi' = \pi(\varphi')$, then visits each cell of the discrete grid and performs the operations:

- Generate a new field φ' from the current state φ;
- Evaluate the *energy* difference $\Delta E = \ln \pi' - \ln \pi$;
- Calculate $P = \min(1, e^{-\Delta E})$;
- Generate a uniform random number r on $[0, 1]$;
- Accept the update, φ', if $r < P$.

Note that if $P = 1$, i.e., $\Delta E < 0$, evaluation of r is not required as the condition, $r < P$, is automatically satisfied.

To use this algorithm it is necessary to calculate the energy difference, and the new energy very efficiently. In physics applications the energy is a *local* function. That is, the energy is a sum over cells, where the energy associated with any cell only depends on the state of the cell and its immediate neighbours. Thus only a few arithmetic operations are needed to update each cell. Nevertheless, many Monte Carlo iterations are needed. Each cell must be visited thousands of times before equilibrium is achieved. After that there must be many Monte Carlo iterations between samples if the selected realisations are to be statistically independent. Careful monitoring of convergence is required. See [121] for a detailed evaluation of the method. There is scope for simultaneous updating of multiple sites and this too, is discussed in [121]. Parallelism is easy to achieve if the model is local. Once equilibrium is achieved and multiple independent realisations are obtained the diagnostic functional is calculated from

$$\langle f \rangle \approx \frac{1}{M} \sum_r F[\varphi^r],$$

noting that the factor $\pi(\varphi^r)$ is no longer needed as in (20). An important property of the Metropolis algorithm is that it is unnecessary to compute the normalisation factor in the pdf.

The Gibbs Sampler: This technique was introduced to the image processing literature by [71] and is a stochastic simulation method of key interest in mathematical statistics, particularly Bayesian Statistics. [67] and [122] review the statistics literature with sections on the *Gibbs sampler*, the next algorithm to be described.

The Gibbs sampler, after r iterations, updates to the $(r+1)$-th iteration by visiting each cell and setting

$$\begin{aligned}
&\varphi_1^{r+1} \text{ by sampling from } \pi(\varphi_1|\varphi_2^r, \varphi_3^r, \ldots, \varphi_N^r)\\
&\varphi_2^{r+1} \text{ by sampling from } \pi(\varphi_2|\varphi_1^{r+1}, \varphi_3^r, \ldots, \varphi_N^r)\\
&\ldots\\
&\varphi_i^{r+1} \text{ by sampling from } \pi(\varphi_i|\varphi_1^{r+1}, \varphi_2^{r+1}, \ldots, \varphi_{i-1}^{r+1}, \varphi_{i+1}^r, \ldots, \varphi_N^r)\\
&\ldots\\
&\varphi_N^{r+1} \text{ by sampling from } \pi(\varphi_N|\varphi_1^{r+1}, \varphi_2^{r+1}, \ldots\ldots, \varphi_{N-1}^{r+1}),
\end{aligned}$$

where $\pi(\varphi_i|\varphi_1^{r+1}, \varphi_2^{r+1}, \ldots, \varphi_{i-1}^{r+1}, \varphi_{i+1}^r, \ldots, \varphi_N^r)$ is the conditional probability density of observing φ_i given $\varphi_1^{r+1}, \varphi_2^{r+1}, \ldots, \varphi_{i-1}^{r+1}, \varphi_{i+1}^r, \ldots, \varphi_N^r$. It is defined by the expression

$$\pi(\varphi_i|\varphi_1^{r+1}, \varphi_2^{r+1}, \ldots, \varphi_{i-1}^{r+1}, \varphi_{i+1}^r, \ldots, \varphi_N^r) = \frac{\pi(\varphi)}{\int \pi(\varphi) d\varphi_i}.$$

As with the Metropolis method, it is necessary to make many iterations before the realisations are suitable for use in (20). [122] shows that the Gibbs sampler does produce realisations with π as their pdf.

Using pdf functionals with local energy functions enables very efficient algorithms to be constructed. The single site conditional pdfs used in the Gibbs sampler can be derived by integrating out just the i-th variable. This is straightforward for the Gaussian distributions, but also feasible for more general, non-Gaussian probability density functionals. As with the Metropolis method it is not necessary for us to compute the normalisation factor in the functional pdf.

Sequential Simulation Methods: In applications of geostatistics the method of *sequential simulation* has achieved great popularity. Geostatisticians do not usually exploit locality. The pdfs are non-local as they are correlation function based models rather than models using local coupling constants.

The sequential method is based on the following exact result that can be proved by induction [34].

Suppose that the site labels, $1, \ldots, N$ are ordered so that the observations are labelled by $1, \ldots, M$. Then the joint conditional probability of the unknown values, conditioned on the M observed values is a product of single point conditional probabilities as follows.

$$\pi(\varphi_{M+1}, \varphi_{M+2}, \ldots, \varphi_N | \varphi_1, \ldots, \varphi_M) = \prod_{i=M+1}^{N} \pi(\varphi_i | \varphi_1, \ldots, \varphi_{i-1}). \tag{21}$$

This is proved by starting with the equation

$$\pi(\varphi_{M+1}, \varphi_{M+2}, \ldots, \varphi_N | \varphi_1, \ldots, \varphi_M) =$$
$$\pi(\varphi_N | \varphi_1, \ldots, \varphi_{N-1}) \pi(\varphi_{M+1}, \varphi_{M+2}, \ldots, \varphi_{N-1} | \varphi_1, \ldots, \varphi_M),$$

and repeatedly using the definition of conditional probability. The final expression is then written in reverse order to give (21).

A sequential simulation proceeds by performing the following operations:

Randomly order the cells after the first M cells with conditioning data.
Visit the first cell and draw a random value from $\pi(\varphi_{M+1} | \varphi_1, \ldots, \varphi_M)$.
Visit the second cell and draw a random value from $\pi(\varphi_{M+2} | \varphi_1, \ldots, \varphi_{M+1})$
...
At the i-th cell, draw a random number from $\pi(\varphi_{M+i} | \varphi_1, \ldots, \varphi_{M+i-1})$
...
Finally draw a random number from $\pi(\varphi_N | \varphi_1, \ldots, \varphi_{N-1})$.

In principle this will work, and will only require a single pass through the unconditioned cells. In practice the difficulty is in calculating the conditional probability densities. Modifications introducing approximate locality, so that conditional pdfs are only dependent on geometrically close points, are needed.

As soon as this is done the elegance of the exact statement just given is lost. The resulting methods are fast, the images that result from visualising the results are convincing - provided a random path is taken through the unconditioned cells - but the statistics of the results are unknown. The statistics of the results have to be determined by analysis of the numerical results. It is likely that further analysis will show that *sequential* simulation is a first pass of an approximate Gibbs sampler. Further passes are needed for the statistics to converge to the correct values.

There is thus a basic problem in geostatistics: find a method that is as fast as a sequential method but which has a sound theoretical basis as in the Metropolis or Gibbs sampling methods.

Note on the literature. For further information about functional methods see [19, 73, 164]. Reviews of geostatistics are to be found in [34] or [42].

5 Forward and Inverse Modelling

5.1 Introduction

Physical systems are modelled by postulating a relationship between three objects - the properties, φ, the state ψ and the auxiliary data, ψ_a. The auxiliary data and the properties together are referred to as *input.* The state is sometimes called the *output.* The properties characterise the unchanging aspects of the system; the state characterises the aspects of the system that respond to different selections of the auxiliary data. The auxiliary data corresponds to those aspects of the system that are under human, or other, control. An example is that of single phase fluid flow in a porous medium; the properties are the permeability, the state is the pressure and the flux. The auxiliary data are the boundary conditions imposed on the flow system. In a time dependent problem, the auxiliary data will also include the initial conditions.

In a *discrete system,* φ will be a vector of values, one for each cell in the system. The values themselves might be vectors, so that φ is a vector of vectors. In a continuum system φ will be a function of x and sometimes also of t. Sometimes φ will be a vector or tensor valued function and can be discontinuous. In computer simulation the continuum system is approximated by a discrete system. For the purposes of *deterministic* mathematical modelling, the properties are supposed given as specific vectors or functions. In *stochastic* modelling, properties are specified by probability density functions, or perhaps implicitly by some other stochastic model.

Similarly, the state and auxiliary data are also described by vectors or functions with the same possibilities of space or time dependence. The auxiliary data which models control parameters such as flow rates in wells, boundary and initial conditions, can be deterministic or stochastic. The state is deterministic if both the properties and the auxiliary data are deterministic.

If one or both of the properties and auxiliary data are stochastic, then the state is stochastic.

A *mathematical model* then takes the form of a postulated relationship between φ, ψ_a and ψ which is written formally as

$$\mathcal{N}(\varphi, \psi_a, \psi) = 0.$$

In many models of practical use, it is assumed, and can sometimes be proved, that for any particular choice of φ and ψ_a, the state exists and is a unique solution of the equation, $\mathcal{N} = 0$.

In most cases, the state is not itself the main item of interest. Various *diagnostic functionals* are needed. In the deterministic setting these will be quantities such as the oil in place in some volume, the total production of oil over some period of time, the rate of water production, and so on. In the stochastic setting the expectation values of such quantities and their associated variances will be required.

The problem of determining the output given the input is known as *forward modelling.* The task of determining an unknown input from the values (or expectation values) of one or more diagnostic functionals is called *inverse modelling.* This inverse problem has to be solved when the values of the diagnostics contain errors, thus rendering the values mutually inconsistent. In addition, there may be far too little data for the problem, as stated, to be able to determine the input even if the input is error-free.

Deterministic and stochastic approaches to such inverse problems are discussed in the following two subsections.

References of particular value, that go further than the following brief review, are [131], [142] and [151].

5.2 Deterministic Inverse Problems

Consider the problem of finding a function (or vector) φ given the values of the vector valued diagnostic,

$$f^* = F[\varphi, \psi, \psi_a],$$

such that

$$\mathcal{N}(\varphi, \psi_a, \psi) = 0. \tag{22}$$

Three possible situations are of interest; (i) there are no functions, φ (ii) there is exactly one function, φ (iii) there are many functions, φ, possibly an infinite number, that satisfy the equation $\mathcal{N} = 0$ and are approximately consistent with the values of the diagnostics.

Interest is only in functions that have a continuity property. That is, small changes in f^* should imply only small changes in φ. A problem is said to be *well-posed* if it has a unique solution that is continuously dependent upon the faces defining the boundary between two layers are called *horizons*. Next in complexity is the case where layers are (i) discontinuous or (ii) *pinchout* where there was a change in the mode of deposition. A further increase in complexity is when a geological event leads to erosion of part of a layer cake before recommencement of deposition. With just non-uniform deposition and erosion, complex patterns of rock can evolve. An example of this is the Grand Canyon [110]. When the horizons are relatively smooth and continuous, it is easy to model them with parameterised *maps*. These can consist of an array of depth values on a regular rectangular grid. At points away from the grid points, values can be interpolated. The various methods described in the section on spatial statistics are employed in the making of computer maps.

Structural Geology: Structural geology studies rock deformation during and after deposition. Faulting greatly complicates matters. When faults are vertical, horizons are torn along the fault lines and this leads to displacement of the maps. When faults are *normal*, meaning the fault surface slopes such that any vertical line intersects a particular horizon once or not at all, description is fairly straightforward. In more general situations, such as intersecting faults, or reverse faults (where a vertical line may intersect the same horizon twice) it is not possible to describe the geometry in a simple way.

Another complication occurs when horizons are folded, as in the case of salt domes, so that again vertical lines exist that intersect horizons more than once. These are also difficult to model.

Operational Geology: The standard approaches to geological modelling attempt to build the model as directly as possible. That is, the sequence of operations involved in the model building does not attempt to mimic actual geological processes (although horizons and unconformities may be ordered in a time sequence, so as to specify the erosion rules). In operational geology a sequence of the basic geological processes (sedimentation, deformation, jointing, diagenesis and erosion) that is a feasible sequence is postulated. One might even conjecture a sequence of operations as the actual history of a particular formation.

Reconstructive Geology: In reconstructive geology, the attempt is made to provide a model of the actual sequence of events that caused a particular rock pattern, in as realistic a manner as is useful. Reconstructive geology is a special case of operational geology, in that operational geology may not suggest that the sequence of events is the actual sequence. An introduction can be found in [76].

Basin Modelling: In basin modelling, an operational approach is applied to a very large volume of space, in which there may be many oil reservoirs. By looking at this larger scale it is possible to postulate the source of the sediment that has been deposited, and to model the actual formation of the oil - an example of a diagenetic process, [5].

Notes on the literature. Books that introduce geology, aimed at people involved in petroleum studies, are [43, 108, 110, 138].

6.2 Seismic Acquisition, Processing and Interpretation

Seismic acquisition involves sending acoustic energy into the subsurface and analysing the echoes. This is an inverse problem on the wave equation. Data are gathered from many independent experiments, as a sound source is moved over the land or sea surface. The results are preprocessed using a wide range of techniques designed to correct for topographical features and noise. A complete review of these techniques can be found in [139]. Clear explanations of the older techniques can be found in [3].

After preprocessing, the main inversion step is known as *migration.* The amount of data, and the size of the system under investigation are both very large. This precludes the use of simple numerical techniques for solving the forward model, as described for the fluid flow forward problem. Instead a range of semi-analytical approximations are used. The approximations vary in accuracy, complexity and applicability. The simplest methods are known as *time migration,* using Fourier processing on wave equations, linearised about a simple background spatial distribution of properties.

The more sophisticated and more accurate methods known as *depth migration* are based on high-frequency ray-tracing methods. Occasionally, in very complicated situations, full finite difference or pseudo-spectral approximations are used, but the resolution requirements are very demanding. A major decision, relating to the amount of spatial variation in the acoustic properties, is to migrate pre- or post-stack. This refers to one of the major preprocessing steps where an averaging technique is applied to the raw measurements that combines the observations from different sources and receivers. This indicates the complexity of the seismic processing activities. The mathematics is explained in detail in [21].

The results are usually presented in the *time domain* which, to a first approximation, removes the effects of any layer cake background model. Users of the inverted results are able to apply their own *time-to-depth* conversions as more data becomes available, or as the seismic data is integrated with other data. The time-to-depth conversion is performed using alternative layer cake models of the sound speed and rock density. Note that depth processed seismic results using depth migration are often displayed in the time domain by applying a depth-to-time conversion.

After migration, and display in the time domain, the observations are equivalent to a numerical experiment that measures the response that would be observed in an ideal experiment where the sound waves are assumed to propagate in vertical, straight lines, and that any reflectors that are encountered are locally horizontal. The resulting calculations are then displayed as seismograms of the virtual experiment.

maximising the information recovered, minimising the size and weight of the apparatus and the energy requirements [7].

Parameter inference. The forward model or theory relates the input properties and geometry to the instrument response. In the simplest case the theory will predict the response as a simple function of the input, such that the inverse formula, stating the input parameter as a function of the response can be obtained analytically. An example is calculating the resistance of a conductor as the ratio of potential gradient to electric current, or permeability as the ratio of fluid flux to pressure gradient. In a more complicated case there may be several parameters such as the radii of the different zones of mud invasion. When there are just a few parameters it may be possible to infer the parameters from one or more instrument responses using a least squares procedure. This is an application of the usual methods of the theory of measurement as reviewed in Section 5.

In general there are naturally occurring heterogeneities that may require hundreds or thousands of parameters for a realistic model. Such inversions are computationally demanding, but using the adjoint method, which is reviewed in the later section on history matching, it might be possible to make further progress in improving the realism of inverse modelling.

Transforming the results. The end result of a well logging measurement will be values of parameters such as electrical resistivity. This is not the information that is actually required. Instead, porosity, permeability, elastic constants and so on are needed. Some of these can be inferred by combining the results of several different logs and using a law, such as Archie's law (see [45, 138] for an explanation) to infer the property of direct interest.

Some Notes on the Different Types of Log: Temperature log. Many other measurements are sensitive to temperature, so this is always recorded in a suite of logging measurements.

Spontaneous potential (SP). The low resistance water or mud used in drilling evokes a spontaneous potential difference between a point on the well surface and a reference point at the well head. This potential varies along the well bore. High values correspond to high hydrocarbon saturations or high shale density. The SP log was the first type of well log (see [112] for the history and historical references). The method can only be used in uncased holes.

Resistivity logs. The introduction of a source of electrical energy leads to reliable and controllable measurements. There are many types of electrical resistivity log. They differ by their depth of penetration and their vertical resolution. The electrical resistivity of the different invasion zones can be inferred from these measurements. For a review of inversion procedures see the review of [7]. Resistivity logs measure the effective resistivity on scales of meters down to centimeters, depending on the characteristics of the instruments.

Dielectric logs. The dielectric *constant* of a material is a measure of its ability to store electric charge in an imposed electric field. The dielectric properties are fairly insensitive to the water salinity, unlike the resistivity. The dielectric constant for the vacuum is unity and that for gases only marginally greater. However, water has a dielectric constant of around 50 when fresh and 80 when saline. Oil has a value of about 2.2 and sedimentary rocks range from 4 to 10 [138]. Thus the dielectric measurement can be used to determine the porosity and the water saturation. The log does not distinguish between connate water, mud filtrate or the water content of shales. (Connate water is the original formation water, and a shale is a rock made of mud particles smaller than 0.06mm [22].) By altering the frequency of electromagnetic waves, different depths of investigation are possible.

Passive Gamma-ray logs. Radioactive elements are naturally present in the rocks. The most commonly occurring radioactive elements are potassium, uranium and thorium [43]. Potassium is prevalent in clays, and present to some extent in some other minerals, and so measurement of the gamma-ray emissions can be used to identify the chemical or mineralogical composition, estimate the concentration of shale and, like other logs, has an important role in well picking. More detailed measurement uses the energy wavelength spectrum and can identify minerals with more precision.

Active gamma-ray logs. By emitting gamma radiation from a source in the sonde, and measuring the backscattered gamma radiation reflected from electrons in the formation, it is possible to determine the total or bulk density of the medium. This includes the rock and fluid in the pore space. The device, often called the gamma-gamma tool [138] can be used to estimate the bulk rock density and provide an upper bound on the flow related porosity. The gamma-gamma log works close to the well, in the invaded zone.

Neutron logs. A radioactive source is transported within the sonde, and the neutron bombardment causes emission of gamma radiation in proportion to the hydrogen content of the formation. As nearly all the hydrogen is present in the fluids, but not in the minerals, the neutron log determines the porosity of the rock. Since not all the pore space is open to flow (some pores contain fluid sealed in by diagenesis) the neutron derived porosity is an upper bound on the flow-related porosity. Neutron logs can be used in cased holes.

NMR logs. Nuclear magnetic resonance logs are another way of measuring the hydrogen content of the formation. They work by first applying a steady and strong magnetic field, about 1000 times stronger than the Earth's magnetic field. The nuclei of the hydrogen atoms carry a small magnetic moment and so there is an induced field of slightly larger magnitude than the strong static field. A second magnetic field, in the form of short high bursts of high frequency magnetic fields disturbs the induced magnetisation. By observing the decay of these responses it is possible to determine the hydrogen content. Remarkably there is sufficient information to enable estimates of the environments experienced by the hydrogen atoms. Thus the fluid saturations of different fluids can be determined as can the size distribution of the pores of the rock. The histogram of pore sizes enables estimates of permeability to be made. It would appear that the scale of these measurements is of the order of 10 cms. Only in recent years has the NMR method become available in well logging. It would seem to be a measurement of particular power and usefulness in reservoir characterisation.

[112] gives a quite detailed review of NMR logging and [138] provides a short introduction, though longer than that just given. The book of [87] contains background about NMR in general.

Sonic logs. The sonic log provides in situ measurement of the speed of sound in the formation over the scale of the sonde. It can only be used in open, uncased holes. The measurement is useful in seismic processing, it can be used to infer porosity but not that accurately and can be used to identify the rock type since the sound speed is a definite characteristic of each type of rock. Further detail can be found in [45, 138].

Dipmeter log. The dipmeter, introduced in the 1940's, uses small electrodes to measure the resistivity along three or more parallel tracks along the well bore surface. Analysis of correlations of the resistivity between the tracks and fitting of a plane through the correlated points determines the tangent plane to the layers passing through the well. The geologists characterise this plane by the angle made to the horizontal by the line of steepest descent on the plane (the dip angle) and the angle made to true north of the projection of this steepest descent line when projected onto the horizontal plane (the azimuth). Assuming that layers of resistivity correspond to lithological layers, this is thus a very useful measurement. [45] and [112], amongst others, are enthusiastic about applying dip meter measurements. Both these authors give examples of interpretations of dipmeter data best done, as usual, in concert with other data. According to [112] the modern dipmeter, with a vertical sampling rate of 0.25 cm, and an electrode size of 0.5 cm has a resolution of 1cm. That is, the measurements can be used to distinguish features separated by a distance of 1cm. It would appear that careful geological consideration needs to be given as to the scale of the layers whose dip is measured. That is other data is needed along with interpretation (i.e., choosing one of many models that fit the data) to decide if the layers are cross bedding, or small scale laminations.

Borehole Imaging. The dipmeter has evolved into the borehole imager. Here there are 100-200 electrodes measuring the resistivity around the well bore. These are processed into convincing images which look like photographs of rocks. In principle such data can replace the dipmeter but, at present, according to [112] the dipmeter is still widely used. The borehole image can be used to determine dip, much as with the dipmeter log. However, via a borehole image it is also possible to determine the presence of fractures, faults, unconformities and sedimentary structures [138].

Rock-type analysis. Rock samples, from cores or drilling cuttings, of known depositional environment can be correlated with a suite of well logs. Then in other wells, the logs can be used without reference to rock samples to infer the depositional environments along the well. The processed logs (that is those with ascertained depositional environment) are used to make well picks. Then the geologist can build maps or 3D models of the spatial variation of rock type. The rock type variation can then be used in more complicated interpolations of properties such as permeability.

Notes on the literature. An extensive and modern review, that concentrates on reservoir model building is [112]. This also has useful references to the more general well logging literature. There is also an interesting history of well logging.

A general background, at a very accessible level may be found in [108]. A similar, but slightly more detailed review can be found in [43]. The modern review of petroleum geology by [138] is a mine of interesting and useful information. The book [72] gives case studies on how well logs are used in building geological models.

Articles, frequently on logging topics, appearing in the *Oilfield Review* are of general interest.

6.5 Well Testing, Production Logs and History Data

Observation of processes that directly involve fluid flow are of particular importance in building a model aimed at simulating fluid flow. Well tests provide data on pressure and flow in individual wells when such wells are disturbed from their normal state. For example, in a shut-in test, a well is closed to flow, and the transients in pressure and flow are used to infer permeability in the vicinity of the well. Tests in which the well is shut and then a very high pressure is induced in the well and allowed to decay, serve a similar purpose. The integration, into geological and simulation models, of production and other flow data is discussed in more detail in Section 9.

7 Geological Modelling

The aim in geological modelling is often to integrate all available data into a single representation for subsequent decision making. Increasing use is being made of multiple representations, in the process of quantifying uncertainty. The historical method for model building was to use maps; maps of the surfaces and maps of the properties in the regions close to the surfaces. Of course, the subsurface is a volume, and not a surface and so, in tandem with developments in computer technology, particularly in 3D graphics technology, geologists have turned to the use of the available 3D modelling packages. Until recently 3D models, too, were built using maps as input. This is still the most common method for modelling, and so it will be reviewed in some detail.

All interpolation schemes specify that the value of some property at a point is a weighted, generally nonlinear, average of neighbouring or nearby values. The weighting depends on direction and distance from the point to its neighbours. Properties generally depend more strongly upon neighbours within the same layer than in a layer above or below.

Geology shows this is not an arbitrary interpolation problem. Geological systems show marked patterns of layering, deformation, faulting and diagenesis. The aim of *structural modelling* is to provide a geometric framework incorporating the major discontinuities. Such discontinuities are boundaries or *horizons* between layers or fault discontinuities within layers. This geometric, structural, framework guides and constrains the development of a geocellular grid that, in turn, defines the neighbourhood structure used in property interpolation.

7.1 Structural Modelling

All software packages for geological modelling currently take the step of building a large scale structural framework.

Historically the first step in structural modelling was to focus on the horizons (the boundaries between adjacent layers). The faults were modelled via some indication of the fault-horizon intersections. Usually the assumption of *mappability* - that the horizon surfaces are single valued height fields above a flat datum - was made and the fault-horizon intersections were defined by the projections onto the datum plane. The projections of the fault-horizon intersection lines are known as *fault traces.*

More recently it has become the custom to model fault *surfaces* (not just their fault traces) before any attempt to model horizons. This approach (see [84]) helps ensure consistency between all horizons in a particular structural model.

Types of Structure to Model. In this section, illustrations with explanatory text are used to specify the main types of geological structure. The illustrations are all drawn as cross-sections. In three dimensions the geometry is much more complicated than shown, but the pictures that are shown are sufficient to indicate the difficulty of geometric modelling and subsequent or associated grid generation problems.

Simple Layered Systems: The simplest case is a system of layers, not necessarily flat, but mappable and continuous. This particular system is often called a *layer cake* and is illustrated in Figure 2.

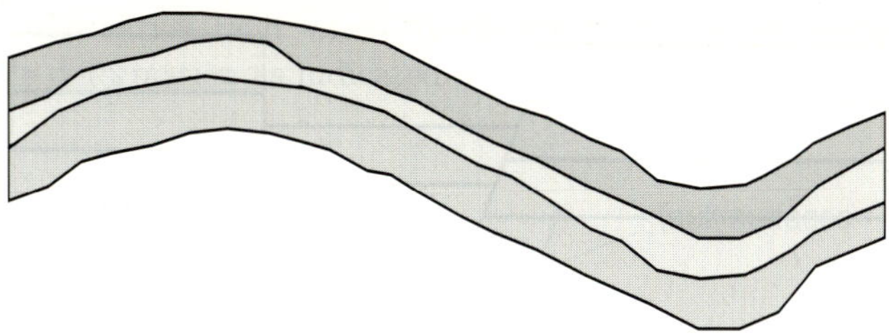

Fig. 2. Layer cake geological model.

Layered Systems with Overturned Surfaces: Overturned surfaces, or folds, which cannot be represented as single valued height fields relative to any flat reference surface, may be caused by, for example, (i) compressive deformation of sufficiently plastic layers from the boundaries or (ii) deformation from below by the upward movement of subterranean salt bodies under the influence of buoyancy forces. See Figure 3.

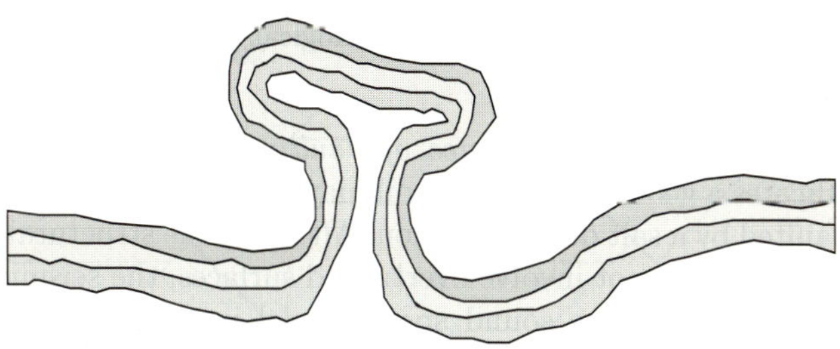

Fig. 3. Geological model with overturned surfaces.

Normally Faulted Layered Systems: Normal faulting, in which a gap opens up in the layers, may be caused by extensional forces. Each horizon, starting as a single mappable height field, remains as such but with holes appearing where the fault surface intervenes. Figure 4 also contains a vertical fault which may be considered a limiting case of a normal fault.

Reverse Faulted Layered Systems: A compressional force can cause a layer to fracture and then ride over itself. The resulting surface is not mappable, in any simple sense, as the height field is overturned. See Figure 5.

three-dimensional seismic data set, it is impossible to extract all of the surfaces of discontinuity in a purely manual fashion. *Autotracking* or *autopicking* techniques [81], implemented as part of seismic interpretation packages extract large sets of three dimensional points conjectured to lie on a particular horizon. Autotracking works by starting from a manually digitised line of points thought to lie on a common surface. The algorithm then seeks points on neighbouring seismic traces that have a similar pattern of seismic amplitudes. Points that lie on fault surfaces can also be identified by autotracking.

Identifying seed points for the autotracking algorithm, and assigning values for the autotracking parameters are ways of injecting geological expertise into model construction.

Well Data: Examination of cores or borehole image logs makes the position of some horizons obvious in a particular well. In the absence of cores or images petrophysicists and geologists must rely on a set of logs. Software exists to help segment the logs into layers [112]. The expert can identify the depositional environment from the patterns in selected well logs. When combined with core data or drill cuttings, the pattern of layers or *zonation* can be assigned with high confidence.

When there are several, or many wells the geologists and petrophysicists can correlate the picks in different wells by grouping them to belong to a particular surface. Combining this data with data on a seismic picked surface increases confidence in the well-to-well correlations and uses the seismic data to help interpolate the horizon surfaces between the wells.

Decisions made in correlating the layers in wells calls for considerable geological expertise. It would seem that the people called upon to make such judgements must imagine what the reservoir and surrounding geology must look like, in advance.

At this stage in the model building process, the model is a list of sets of scattered 3D points. Each set of points represents a geological horizon or fault surface. Since horizon surfaces are conventionally regarded as continuous, away from faults, (although horizons may touch other horizons at *pinchouts*) a natural requirement is to construct a surface interpolating these points. Techniques for doing this were reviewed in Section 4.

The Classical Map Based Methods: One structural model building approach is to first make contour maps or mesh maps (height fields on a grid). Recent reviews by [76] and [143] discuss such approaches in detail.

The simplest technique is to guess the projections of the fault-horizon intersections and to treat these as discontinuities. The surface data can thus be interpolated treating the fault traces as internal barriers. Although simple, such a technique can lead to a strange fault surface when traces are used to define a fault surface. A better method, but requiring more work, is to first build a surface model for each of the fault surfaces. When working by hand

([76, 143]) an iterative approach is adopted whereby fault traces are explicitly calculated by intersecting map contours with the fault surface contours.

Fault Framework Modelling: It is now generally agreed that the best approach for modelling a faulted geological structure is to first focus upon the fault surfaces. Explicit interpolations of the fault surface are needed. This approach is now used in the most commercially successful geological modelling software packages ([64, 84]). A particular difficulty is presented by fault-fault intersections - should they be an input or an output?

Manual Approach for Fault Framework Modelling: One approach, a manual technique, is a possibility [64] where the intersections and the fault surfaces are built together, interactively in a 3D visualisation environment, with visual cues provided by seismic data, interpolated horizons or both. In such an approach lines are digitized onto the observed intersections. Further lines are digitized that manually construct a ruled surface. These surfaces can be interpolated using bilinear patches or splines. Thus in this approach the fault-fault intersections are input to the modelling, provided by the user. A schematic is provided in Figure 9.

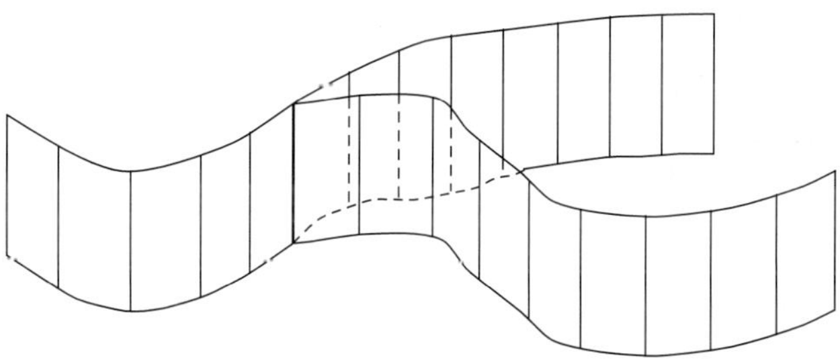

Fig. 9. Fault-fault intersection framework.

Automatic Approach for Fault Framework Modelling: The second approach is to interpolate the fault data for each fault separately and then to compute the fault-fault intersections. This may require some manual assistance to the algorithms to help extend surfaces so that the surfaces do intersect if there is insufficient data. In principle, surface-surface intersections are easy to compute. In practice, however, the intersections may have a very complicated topology largely arising from artefacts of the interpolation. The intersection calculations can be very delicate. At best a great deal of computer time is needed, and in the end a lot of user interaction may be required to *clean up* the results. Ironically it appears that in practice such an *automatic* approach is only suited to models with a small number of faults. The most practical approach, that inserts the intersections directly - not needing any intersection calculations - is the easiest and fastest method. It is difficult to provide references to support these observations as the science of structural modelling is dominated by commercial companies who do not publish descriptions of their algorithms in the open literature.

Fault Block Splitting (FBS): One approach to the structural modelling problem imbeds the fault surfaces into an extended set of surfaces that divide the volume of interest (VOI) into a set of closed compartments. When there are only a few faults this is a simple extension of the fault framework modelling activity. Some extra surfaces need to be added, and parts of the fault surfaces on which there is a throw, and the parts on which there is not must be specified. There is however an extra burden on the user of the software; the order of the faults has to be chosen. The reason is that a convenient approach (for the software implementation) to fault block splitting is to build a *binary FBS tree.* Thus the volume of interest is defined, the first fault is extended so that it divides the VOI into two pieces. This operation is then repeated with each fault in turn. The extended faults must cut one or more fault blocks - the compartments in the binary FBS tree - into two pieces. In principle this is simple but in practice very difficult.

In two dimensions, with vertical faults, it is possible to automatically extend the faults and to carry out automatic subdivision [135]. This requires delicate programming but has been commercially available for some time. When the faults are sloping this automatic method can still be used. The approach is able to deal with very large numbers of faults.

When the faults have shallow dips, that is they are nearly horizontal surfaces, as in a thrust fault, then much user editing is required. When faulting is complex, the process of model building is essentially manual. As a general rule, software must allow user interaction to enable interpretation and assumptions of geological style to be imposed by a geologist. However, at the time of writing all software requires extensive user intervention. There is clearly much room for mathematical innovation in providing better algorithmic tools to assist the geological model builder.

The Corner Point Grid Method.

Geometry, Topology and Construction of Corner Point Grids: The main components of a corner point grid are:

- The bounding box.
- The boundary.
- The internal control surfaces (usually faults).
- The coordinate tubes.
- The tube dividing surfaces (usually horizons).
- The grid blocks.

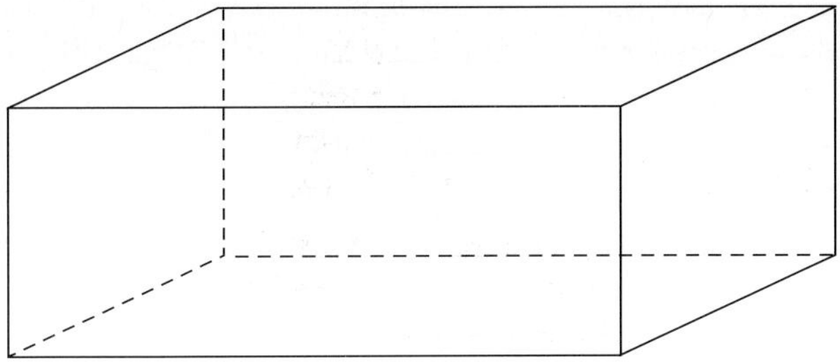

Fig. 10. The bounding box.

Imagine a finite, rectangular bounding box containing all relevant point, line and surface data, Figure 10.

Introduce the boundary; a surface that is broadly vertical or aligned with the sides of the bounding box. In practice the boundary is chosen to guide the slope of the final grid and plays an important role in guiding the general shape of the grid.

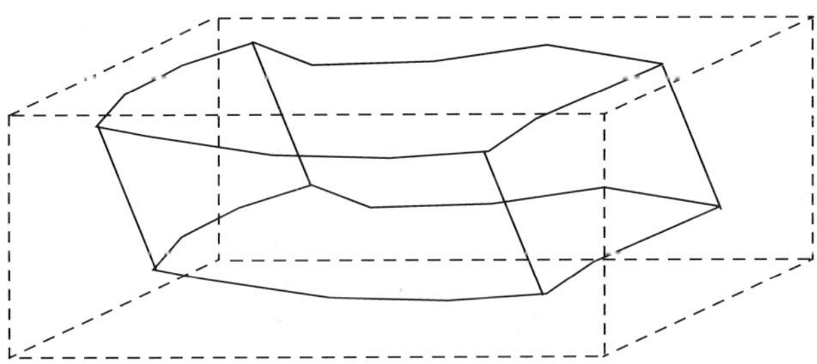

Fig. 11. The boundary inside the bounding box.

The internal control surfaces are surfaces that are again broadly vertical. These surfaces are divided into two sets: the *honoured surfaces* and the *zigzag surfaces*. Most of the time the control surfaces are fault surfaces. In some situations however, the control surfaces might be the flanks of a salt dome or other intrusion.

The structured coordinate tubes are the cells of an N_x by N_y by 1 grid that conforms to the boundary and the honoured surfaces. The edges of the coordinate tubes are called the coordinate lines. On a structured grid each tube has four coordinate lines, each one located at a corner of the tube.

In technical terms the tubes define a regular Cartesian N_x by N_y by 1 grid in a *logical cube* that deforms to fit the boundary and honoured surfaces. Often the grid will also be subjected, through the grid generation algorithms,

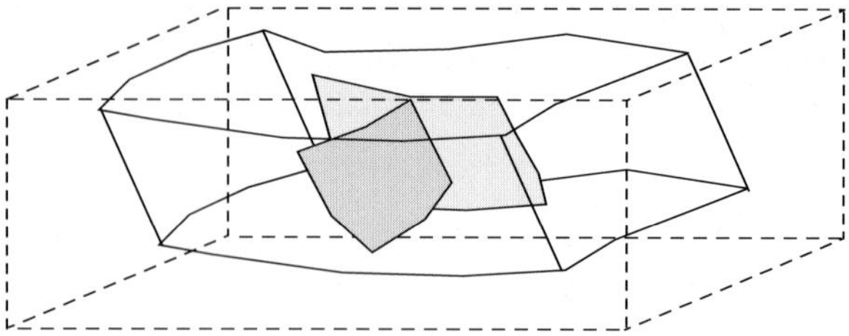

Fig. 12. Some honoured surfaces inside the boundary.

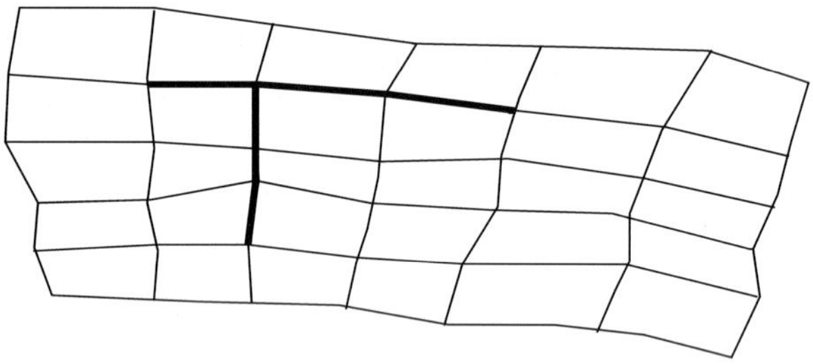

Fig. 13. Areal view of a structured corner point grid with honoured surfaces in bold.

to general trends in the zigzag surfaces. The construction difficulties concern the choice of coordinate lines to go with a particular honoured surface. Also the honoured surfaces need to be extended, automatically by an algorithm, so that the imbedding of the honoured surface in the grid of tubes is smooth and natural looking to the eye. Algorithms exist for the automatic assignment of coordinate lines to particular honoured surfaces [57].

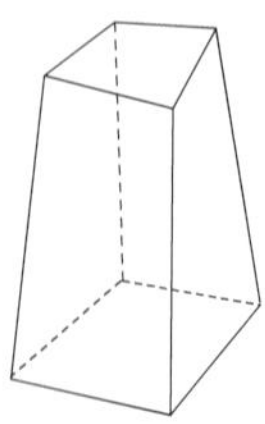

Fig. 14. A typical coordinate tube.

The sides of the coordinate tubes need not be flat. The only essential requirement is that the tube is a singly connected volume. In practice the sides are often bilinear surface patches and the edges are straight lines. In such a case the tubes are hexahedra. The most advanced packages allow the edges to be piecewise linear or *segmented* lines [64]. The tubes are then stacks of hexahedra.

Surfaces that are not honoured by the tubes are the zigzag surfaces. In the case of zigzag surfaces the surfaces are moved to the nearest sequence of tube sides. This is analogous to the process of rasterisation that occurs on a digital display, but is in a 3D space. The geometric accuracy of the grid is proportional to the size of the tubes.

In the case of unstructured coordinate tubes an initial unstructured grid with vertical coordinate lines is built. This unstructured grid might be a Voronoi grid, a Delaunay grid or a grid made from aggregated cells of one of these types of grid. The control surfaces for the initial grid are obtained by essentially verticalising the original data. These vertical lines are then sloped to honour the actual control surfaces. Coordinate lines that do not lie on the control surfaces are positioned using an interpolation procedure.

The next stage in building a corner point grid introduces the *tube dividing surfaces*. In the simplest approach the tubes are collected into groups and a set of tube dividing surfaces is assigned to each group of tubes. It is assumed that each surface is continuous within the group and forms part of a consistent stack of surfaces. This is illustrated in Figure 15. Pinchouts are allowed but surfaces are not allowed to cross.

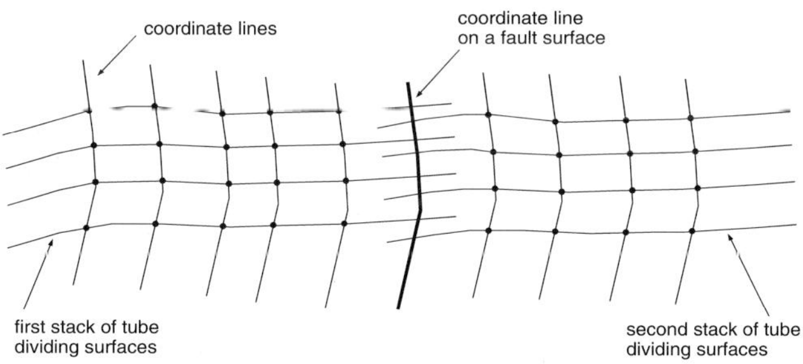

Fig. 15. Cross section of some tubes with two sets of tube dividing surfaces.

The simplest tube dividing method just samples the surfaces onto the coordinate lines at the intersection points of the coordinate lines with the surfaces. When coordinate lines are on the boundary of a tube dividing group there is a multiple intersection; one for each group. This leads to *split-nodes* on the coordinate lines. This is illustrated in Figure 16.

The corner point grid blocks are then defined by interpolating the cell corners with straight lines. On structured grids the blocks are generally hexahedra and on unstructured grids the blocks are general polyhedra. A cross section of a corner point grid is shown in Figure 17.

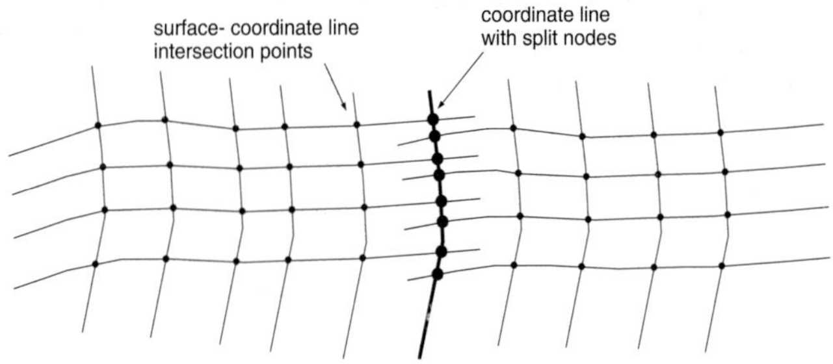

Fig. 16. Grid block corners at tube-dividing-surface/coordinate-line intersections.

The Slicing Method: An alternative approach for building corner point grids has been suggested by [107]. In this method the surfaces of the horizons and faults are assumed known and the coordinate tubes are simple vertical sided tubes with an orthogonal areal pattern. The simulation or geological grid cells are then defined as the intersection sets of the tubes with the layers. The cells can thus be constructed by slicing the tubes using the fault and horizon surfaces. In [107] the implementation is based on the assumption that the surfaces are height fields with respect to a flat reference plane. The method is promising and could be generalised to use overturned surfaces and any pattern of coordinate tubes. A disadvantage of this method is that the surfaces need to be constructed in an external application.

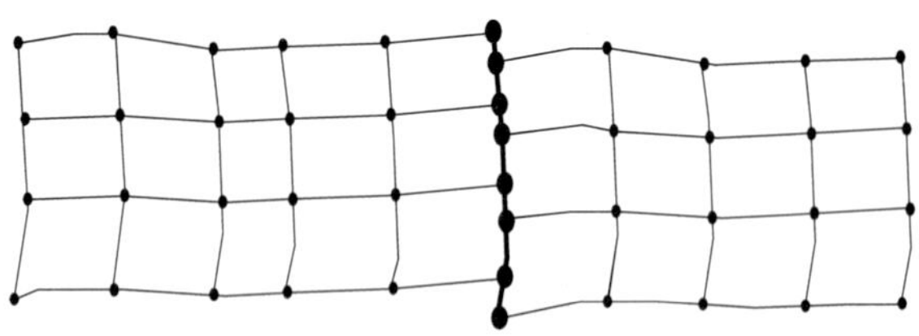

Fig. 17. Final stage of grid block construction.

Input Data for Corner Point Grids: In the earliest corner point grid building applications, the input data were in the form of *horizons* and *fault traces.* The fault traces were used to drive the construction of the tubes. The sides of a particular sequence of coordinate tubes then defined the fault surfaces. Horizons were used as the tube dividing surfaces. This workflow is still widely used, although it can involve a lot of work in editing fault traces so that they lie on their associated fault surface. An early case study is [74]. Often the horizon surfaces are inconsistent as they are constructed separately from one another, and the fault traces are assigned separately to each surface. If the traces arrive from a mapping package that knows about fault surfaces then these difficulties can be avoided.

In the most recent corner point gridding applications there is a move toward building a framework of fault surfaces before any other model building operation. This is exemplified by the application described in [64], where it is even possible to perform scattered data interpolation of the tube dividing surfaces (most often horizons) directly onto the coordinate lines. One way to understand this step further is to regard a conventional mapping package as moving surface nodes up and down on vertical *coordinate* lines that are arranged in a regular Cartesian grid. In a mapping package the *tubes* have vertical, flat, sides. All faults are essentially zigzagged, although some packages perform local unstructured grid refinement, by slicing the tubes with the fault traces. Detail from a mapping package can be high, and use is convenient, provided all faults are normal and there are no intrusions such as salt domes.

When building surfaces directly inside the corner point grid application there are no problems with reverse faults, as the coordinate lines conform to the fault surfaces, and in the local coordinate system defined by the coordinate tubes there are no *reverse* faults. This is very convenient and powerful. The approach has only been partially exploited so far, and one looks forward to improvements in efficiency and quality. Much more remains to be done with regard to the proper evaluation of statistical uncertainty in structural modelling. Some discussion of this can be found in [152].

Problems with Corner Point Grid Modelling: For many situations the corner point grid approach is natural, accurate and fast. Problems occur when there are geometric or topologically complicated systems to model. One particular advantage of the corner point grid approach is that in simpler problems there are no performance limitations regarding the number of faults that can be included in the grid. Clearly, eventually memory becomes an issue, but this is much more of a difficulty in other approaches. A corner point grid can have 1000's of faults and yet be constructed on a relatively modest computer.

Geometric complexity often involves overturned surfaces such as found in an intrusion. If this is not too complicated then editing of the scattered data that lies on the surface, splitting it into groups that control the base surface, the sides, and the top can lead to a model. It is already possible for skilled users to build such a grid using the most sophisticated corner point grid applications. A schematic cross section of a salt dome grid is shown in Figure 18.

Another topological complication occurs when two or more faults intersect. An example of the geological nature of the problem is indicated in Figure 8.

These *y-faults* or *λ-faults* cause severe difficulties for current state-of-the-art geological modelling and grid generation systems. One promising approach, successfully implemented in [136], is able to build a grid as long as

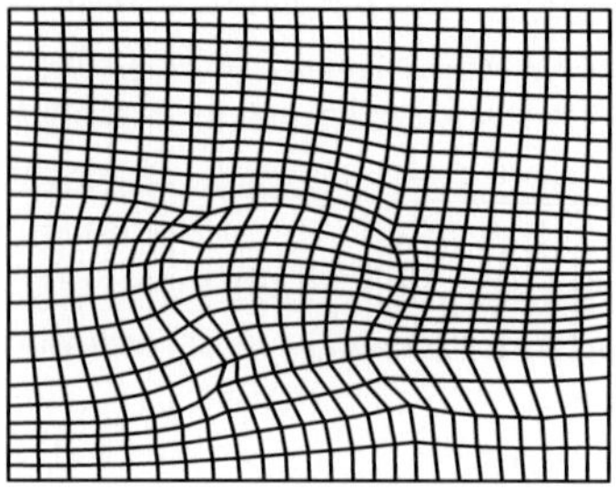

Fig. 18. Cross section of a grid that honours a salt dome.

the surfaces are provided as in the right hand side part of Figure 8. That is, the surfaces must be trimmed and related to one another so that all reasonable topological and geometric queries can be answered. [136] constructs an *ijk* or *vertically zigzagged* grid (see also Section 3.1). This means, essentially, that the geological model is digitised onto a grid, rather in the way that a 2D picture is rasterised onto a 2D raster of rectangular pixels. Here, a 3D *raster* is built, made of a 3D adaptive grid with general curvilinear hexahedral cells. The sophistication lies in the fact that the raster is adapted to the local surfaces. This increases the accuracy whilst reducing the number of grid blocks. A more natural and accurate solution can be constructed by exactly slicing the coordinate tubes with the tube dividing surfaces, a technique that has been commercialized by [107]. Another approach might use a vector field technique to build a system of coordinate lines that can then be used to sample the surfaces.

The final difficulty to be highlighted, regarding corner point grid generation methods, is the listric or thrust fault problem. A schematic of this geometry is shown in Figure 7. Again sophisticated use of curved coordinate lines, breaking up of scattered data into tube dividers and honoured surfaces could build adequate models, but this requires extensive work by a user of any software. The state-of-the-art for such structural modelling problems is to ask the users of software to construct the grid in a sequence of manual operations. Clearly, further research is needed for these difficult modelling problems.

7.2 Property Modelling

Spatial Statistics in Curved Spaces. Topics in spatial statistics have been reviewed in Section 4. There the coordinate system was assumed to be rectangular Cartesian. This assumption is often made in the geostatistical literature and in geostatistical software. See, for example, [42].

Geological systems are manifestly non-Euclidean. Layers are twisted, pinched-out, or faulted. At best there is a local Cartesian coordinate system.

The situation is reminiscent of continuum mechanics. There the requirement to describe the deformation of material objects has led to the development and application of differential geometric and tensor analysis techniques, that make it possible to describe systems without any particular assumption of a coordinate system until one has to actually perform a calculation. Indeed, if the calculation is performed numerically, by for example a finite volume method, global coordinate systems are unnecessary. One simply requires co-ordinates in the vicinity of various objects so that quantities such as areas and volumes can be computed.

This section outlines common approaches to statistical interpolation in curved layers. The last section indicates a solution along the lines of that used in continuum mechanics.

The usual approach assumes the geostatistical algorithms are satisfactory as they are, and that the problem is to find new coordinates (u, v, w), given by a transformation, f,

$$(x, y, z) = f(u, v, w)$$

with the property that for any two points, (x_1, y_1, z_1) and (x_2, y_2, z_2) in the same layer with,

$$w_1 = w_2$$

then the distance

$$((u_2 - u_1)^2 + (v_2 - v_1)^2)^{\frac{1}{2}}$$

is the *intrinsic* distance between the points in the layers. (The intrinsic distance is defined as the length of the shortest curve joining the two points where all the points on the curve lie on the surface.) These coordinates are a generalisation of isometric coordinates to a faulted surface. They are very difficult to construct and do not generally exist. However, one can treat them as an approximate goal; one tries to find a coordinate system as close to the isometric ideal as possible [114].

Stratigraphic Coordinate Interpretation.

Mapping Method: To introduce this method - the *straw man* of property modelling - let us remind ourselves about the geometric structure of a map of a faulted surface. Suppose that the faults are normal faults. In such a case each (x, y) point on a horizontal reference plane is mapped into the horizon, or into a fault surface. The boundaries of the fault surfaces are the fault traces which are polygons in the reference plane. The horizon is not itself a continuous surface. However, the union with the fault surfaces *repairs* the horizon surface to a continuous surface. That this can be done is the main reason for the historical success of mapping software and its main restriction - one cannot describe reverse faults without extensive extra mathematical (and thus software) objects.

However, having repaired a faulted horizon to a continuous surface it is possible to apply any 2D interpolation algorithm for any property such as the well values of, say, porosity. This completely ignores the effect of the faults. When the post faulting diagenesis has not markedly altered the continuity of the rock properties, this must be incorrect. The properties on either side of the fault, where corresponding points in the horizon were originally joined, should have the same value, but in this mapping method this is not generally the case. Although the approach is clearly defective, and is not used in any of the packages which have a stratigraphic 3D grid, it provides a useful example that contrasts in an interesting way with other, more sophisticated methods.

(i, j, k)-*Method:* Many geological modelling applications use a structured grid. Thus there is a natural mapping from (x, y, z)-space to a logical, rectangular (i, j, k) space. This mapping is a natural by-product of the grid generation process. The simplest technique for applying an algorithm that requires rectangular coordinates is to apply the algorithm in (i, j, k)-space. To do this, it is necessary to transform the scattered data into the (i, j, k)-space. Any correlation functions have to be defined in the (i, j, k)-space.

Advantages of this method are (i) it is fast (ii) it is qualitatively correct across faults when diagenetic effects, related to the fault are negligible and the pre-deformation properties of the rock are appropriate. The method does not properly account for the true distance between points, but does honour the connectivity across a fault surface.

(x, y, k)-*Method:* In some structural modelling approaches a 3D corner point grid is used to carry the properties. As part of model building, an intermediate 2D grid is built, assuming that the faults are verticalised. (This is done for purposes of analysing the topology.) This means that each grid cell, after the coordinate lines have been sloped or even bent into segmented lines, inherits an (x, y) coordinate from the intermediate areal grid. If the scattered data are transformed into this coordinate system and the properties interpolated in this space, then the true distance between the cell centres is more properly accounted for than in the (i, j, k)-method. In other words, because the faults are *verticalised* at this intermediate stage it is valid to use the mapping approach. Further, because the verticalisation is all through the model, it is possible to map the curvilinear grid into a 3D model space throughout the model. Three-dimensional property interpolation procedures can then be applied and posted back into the true geometric configuration [137].

The method is one of the best in current use but must be regarded as an interim solution, while research continues.

Mapping Methods. In the only book to have been published on 3D geological modelling [114], there is a section on building an isometric coordinate system such that given the coordinates of any two points they provide the distance using the Euclidean metric. It is possible to construct a best possible approximating set of isometric coordinates. This is the best one can do.

Coordinate Free Methods. The need to construct an approximately isometric coordinate system is an avoidable problem. All one needs to do is to use a functional probability density function that has a local form. That is of the form

$$\pi(\varphi) = Ce^{-H[\varphi]},$$

where $H[\varphi]$ is a functional that is an integral of a local function of φ and its derivatives of low order. (See Section 4.3 for an explanation of the symbols in the previous equation.) Such functionals are widely studied in the statistical physics literature. By discretising such a functional one can generate realisations or construct maximum probability interpolants that do not require any non-local coordinate system. Further research is urgently needed to fully explore such an approach.

8 Upscaling and Upgridding: Controlling Scale Dependence

8.1 Effective Medium Theories

Deterministic Version of Effective Medium Theory: Using the notation introduced in Section 5 consider the problem of: Find $\tilde{\varphi}$ belonging to the subspace $\tilde{S}$ of S such that selected diagnostic functionals, $\tilde{f} = \tilde{F}[\tilde{\varphi}, \psi_a, \tilde{\psi}]$ of the problem

$$\tilde{\mathcal{N}}(\tilde{\varphi}, \psi_a, \tilde{\psi}) = 0$$

approximate as closely as possible the diagnostic functionals $f = F[\varphi, \psi_a, \psi]$ of the problem

$$\mathcal{N}(\varphi, \psi_a, \psi) = 0,$$

where φ belongs to the space S. In many situations it is assumed that the *effective medium* equations $\tilde{N} = 0$ and diagnostics $\tilde{F}$ are *of the same form* as the original equations $N = 0$ and diagnostics F.

However, it is usually the case that the equations change form under the effective medium transformation. For example, in a layered system fine scale permeability scalars can become tensors, two-phase flow models become dual-porosity models, and so on. Further references and discussion may be found in the book [127] and the review [58].

Stochastic Version of Effective Medium Theory: Following the pattern of inverse problem theory, in the study of effective media there is a stochastic version where the problem is to find the best approximating functional probability density. Thus the problem is: find $\tilde{\pi}(\tilde{\varphi})$ over the subspace $\tilde{S}$ of S such that the expectation values of selected diagnostic functionals, $\langle \tilde{f} \rangle = \langle \tilde{F}[\tilde{\varphi}, \psi_a, \tilde{\psi}] \rangle$ of the problem

$$\tilde{\mathcal{N}}(\tilde{\varphi}, \psi_a, \tilde{\psi}) = 0$$

approximate as closely as possible the expectation values with respect to $\pi(\varphi)$ of the diagnostic functionals $\langle f \rangle = \langle F[\varphi, \psi_a, \psi] \rangle$ of the problem

$$\mathcal{N}(\varphi, \psi_a, \psi) = 0.$$

This formulation is close to that studied in the theory of the *renormalisation group* [19, 157, 164].

8.2 The Upscaling Approach

Full Upscaling Methods: In the deterministic effective medium problem it is assumed that a property model φ is available, in the stochastic problem the upscaling approach works on realisations of the stochastic process, π.

Upscaling methods have two aspects. First some *fine scale* experiments are performed. These can be *local*, in that they work on a very small part of the whole domain or *global* when the experiments are performed on a large part, or even all of the domain. Then the coarse model is defined by a coarse parameterisation - usually by introducing a coarse grid and assuming the properties to be piecewise constant on the coarse grid. Given the fine scale experiments and the coarse grid, a *coarse grid calibration* is performed. This too, can be local or global. Finding the coarse scale properties is an inverse problem with computationally generated data. The inverse problem requires solution of a subsidiary forward problem; when this is local, it is said to be a *local* calibration, and when the calibration inverse problem solves a forward problem on a large portion of the domain it is said to be a *global calibration.* A more detailed discussion can be found in [58]. The most common approach is the local-local method, where both experiment and calibration involve a single coarse grid block, perhaps including some influence from the nearest neighbours.

Generally speaking the more global is the calibration and experiment the more accurate will be the resulting upscaled model, but at increasing cost.

Upscaling is expected to work best when (i) the fluid process is stable (that is the displacing fluid has a lower mobility than the fluid that is displaced) (ii) the length scales of the heterogeneity are small compared to the averaging scale. In general averaging is performed by seeking the best approximation in the smoother space $\tilde{S}$. This idea is justified rigorously in the limit of infinite scale separation in the theory of homogenisation.

Preconditioning by Adaptive Coarse Grid Generation: It seems necessary for the fine grid properties within each coarse grid cell to be either constant or possess a very small length scale compared to a characteristic length scale of the averaging process. Given a model with fine scale detail it might be possible, through a process of adaptive gridding, to build coarse grids with the desired properties.

[159] was the first to suggest a method. [106] continued to investigate the problem along the same lines as [159]. Methods in which a flow equation is solved on the fine grid and the equipotentials and streamlines are used to construct *optimal* coarse grids have been developed further by [2]. The heuristic approach, based on solving flow problems and then grouping layers according to the flux through the layers, introduced in [47], is of interest. The alternative of minimising the variance of some property in a coarse cell has received some attention in [13, 59, 60, 68, 133].

A moving finite element approach to variance minimisation was proposed by [13]. In [59] and [60], combinatorial aspects of the problem were tackled using a global optimisation algorithm. [133] made this method faster by restricting coarse grid cells to consist of strict subsets of fine grid cells.

There are many opportunities for further work in this area. By using unstructured grids, success might be easier. In two-dimensions progress in *image segmentation*, a very similar problem, could inspire a new attack on this *upgridding* problem. Deep results are available, which look as though they generalise to three-dimensions (see [120] for references).

Multiscale Simulation: Standard numerical methods, such as the finite volume method, seek approximate solutions in the form of piecewise constant or piecewise linear expansions. The basis functions are simple, and as the number of grid blocks is increased - known as *h-refinement* - so the error is reduced. Alternative methods, such as the spectral methods, seek expansions as finite superpositions of basis functions; for example a Fourier expansion in trigonometric functions. Spectral methods do not require a grid, although they are restricted to simpler geometries as a result. A compromise method is the finite element *p-refinement* method, where higher order polynomial expansion is used on a discretisation. In the interesting paper of [91] a numerical technique for constructing finite element basis functions that respond to local heterogeneity was described. This method is closer to the local-local upscaling method than one might at first think. Indeed a finite volume version of Hou's method was shown to be a form of local-local upscaling in [58].

The multiscale finite element and finite volume method is difficult to generalise to multiphase flow (however see [104] for a paper which discusses an approximate technique for modelling rate dependency effects). The novel, and very stimulating idea of using (i) an upscaling method for the average pressure, and (ii) a reconstruction stage to find mass-conserving approximate fine-scale fluxes was described in [94]. The reconstructed fluxes are used in an explicit method for the saturations. The method is very much faster than one solving the fine grid pressure equation with, apparently, little reduction in accuracy.

It will be interesting to observe research on multiscale simulation over the next few years. Will local upscaling methods be improved so that the two-scale method of [94] looses some of its apparent advantage?

8.3 The Homogenisation Approach

Homogenisation is a multiscale perturbation method in which the ratio of length scales is used as a small parameter. Consider a two-scale problem, with (i) a large-scale parameter, L, characterising, for example, the well spacing and (ii) a heterogeneity scale l, characterising the size of high permeability zones. In this case the expansion parameter is chosen to be $\epsilon = l/L$. Then the property model may be considered a function of the *slow* scale x and the fast scale $y = x/\epsilon$, that is $\varphi = \varphi(x, y)$. The gradient operator becomes

$$\nabla f = \nabla_x f + \frac{1}{\epsilon}\nabla_y f$$

for arbitrary functions, f.

The method then seeks expansions of the form $s = \sum_{i=0}^{\infty} \epsilon^i s_i(x, y)$ or even more general expansions with non-integer powers in ϵ or possibly general functions of ϵ. It seems that considerable skill is required to pre-scale the equations with ϵ's inserted at various strategic places. For examples see, [44] and [127].

The procedure makes the expansion, gathers up terms of equal orders of ϵ and then a variety of arguments based on averaging are used to derive an equation for the order zero term. In the limit as ϵ goes to zero this lowest order term can, by more rigorous analysis (see [88]), be shown to be an exact model for the volume-averaged behaviour.

The method can be made rigorous, but as far as one can tell the degree of intuition required to find an appropriate expansion is similar to that required to set up an upscaling method. In many cases the homogenised equations are identical to those obtained by simpler and more intuitive upscaling arguments.

There are clearly opportunities for further research in this area, particularly in relating the homogenisation methods to the upscaling and multiscale methods. The results of [23] that show the need for dual permeability models as effective media for 2-phase flow are particularly interesting. Dual permeability behaviour has been observed in the field, [35], in apparently unfractured reservoirs.

8.4 The Method of Stochastic Equations

A method of considerable historical interest is the method of direct solution of stochastic equations. Thus for a stochastic problem of the form

$$\mathcal{N}(\varphi, \psi_a, \psi) = 0,$$

where the input, φ is a random function is studied using analytical perturbation techniques. Analytical methods for stochastic porous medium problems have been reviewed recently by [162]. Such methods appear to be limited to situations where a Green's function at zeroth-order can be found in closed form. Similar techniques have been reported in the turbulence literature for many years, where they have reached a higher level of sophistication (see [109]).

9 History Matching: Integrating the Production Data

9.1 The Problem of Integrating Production Data

Production data are defined to be any flow rates or pressures measured in the well bore other than measurements made on a scale of centimeters or less. Thus well testing and measurements made during field operation are included.

The problem of integrating production data involves either (i) constructing a deterministic property model that leads to flow diagnostics that broadly agree with the production data or (ii) modifying some prior probability model so that the posterior distribution is consistent with Bayes' rule and the data. The problem of integrating production data is often called *history matching.*

The general problems of forward and inverse modelling were discussed in Section 5 and the problem of integrating production data is, in principle, of the same type. However, in practice the large computer times needed to perform flow simulations and the strongly time dependent nature of the data to be processed makes the problem very difficult, compared to the inversion of small scale measurements.

Problems in other parts of the geosciences are very similar. In particular the problem of assimilating observations into weather forecasting [39] and ocean circulation models [160] are conceptually close to the oil reservoir forecasting problem. Indeed, there is probably much to be learned by cross-fertilisation between these different fields of geoscience.

Comparison studies of several different methods have been performed as reported in [15, 62, 111, 163].

Two key references in the petroleum simulation area are [124] and [116].

9.2 The Bayesian Formulation

The Bayesian formulation is not universally used in history matching, but there is a general trend toward thinking in this way. One advantage that the Bayesian formulation provides is that it places all methods in a common framework. Within this framework many methods are viewed as approximations to other more *ideal* approaches or as summary methods of the posterior distribution.

As the Bayesian formulation was described in Section 5 it is sufficient to recall the main uses of the formulation in the *maximum a posteriori* (MAP) mode or in the *stochastic sampling* mode. The *maximum likelihood estimation* method is obtained by setting the prior to unity in the MAP method. The MLE method is essentially the least squares method. Without a suitable choice of prior it may be necessary to introduce further ad hoc regularisation in the case of MLE. A carefully chosen prior should regularise the problem in a satisfactory way.

9.3 Deterministic Algorithms

Deterministic algorithms are methods for solving the MLE or, preferably, the MAP equations. That is a unique, global minimum of the MLE or MAP functionals is to be computed.

Parameterisation Methods: If the objective functional is not suitably chosen, or the number of unknowns is considered too large, then it is necessary to *reparameterise* the unknowns.

The most common way of doing this is via *zonation.* That is, the domain of interest is divided into a relatively small number of subdomains, and either the values treated as piecewise constant by subdomain or a scalar multiplier is applied to the values in each subdomain.

Techniques for choosing the division into subdomains are assignment by (i) subjective decision (ii) regions of maximum sensitivity [20] and (iii) streamlines of the total flux as introduced by [155]. A clear example of the streamline method is [1]. Several papers [30, 40, 80, 103, 119, 161] have appeared recently using variants of streamline methods for history matching.

A generalisation of the subdomain method is to use a general basis, and represent the unknown parameters as a superposition of the basis functions. This method was used in an early and impressive paper by [70] in which all the ingredients for MAP are in place.

A special case of the basis function method works by choosing a subset of grid blocks - the *pilot points*, and setting the values in all other blocks using kriging. The pilot points might be chosen subjectively or using sensitivities. It was observed in [116] that the pilot point method is a version of a basis function method. The results in Section 4 are relevant to this observation.

A very interesting variant of the basis function method, introduced in [132], uses two or more realisations from the prior as basis functions.

Minimisation Algorithms: Production data integration must, in some way, involve forward simulation of the fluid flow model. In MLE or MAP, even with reparameterisation, an optimisation method must be used. There are two choices; to use a derivative-free method, or to use derivatives.

Derivative-Free Methods: Derivative-free methods can simply call the simulator and use the results. A simple technique is simulated annealing which was investigated in [126] and [41]. Using a fast simulator such as a streamline method (fast by virtue of the IMPES approximation and the one-dimensional approximation along the streamlines) or a coarse grid simulator, this might be practical.

The recent developments in streamline zonation use a simple iterative, derivative free, update of the permeability field along the streamlines.

Gradient Methods: If the aim is to find the MLE or MAP estimates then the method of choice must be a derivative method. Once the effort is made to modify the simulator so that derivatives can be calculated exactly (that is without using numerical differentiation) then the higher order convergence of a gradient method can easily outperform a derivative free method.

Gradient methods fall into two classes, depending on the number of derivatives to be calculated [124]. If the number is small, then direct calculation of the objective function gradient is best. The reason for this is that with knowledge of the gradient vector, the Gauss-Newton method (where the Hessian is approximated as the direct self product of the gradient) can be applied. This tends to be a second order convergent technique. Simulators with a fully implicit formulation calculate many derivatives as part of the Newton step at each time step. The extra derivatives needed for the gradient (when the number of unknowns is small) can be obtained with a small overhead.

When the number of unknowns is large, as for example in the case that all grid block property values are regarded as unknown parameters, then it is impractical to find the derivatives of the objective function. By treating the fluid flow model as a constraint, the adjoint method can be used. This is, conceptually, the same adjoint method as described in Section 4.5. The method exploits the fact that the dot product of the gradient with a small number of fixed vectors can be obtained at relatively low cost. However, this means that a lower order optimisation method, such as the conjugate gradient technique must be used. This is why for small numbers of unknowns the direct technique is best.

9.4 Stochastic Methods

As explained in Section 4, unless the prior pdf is very sharply peaked about the maximum, the MLE or MAP estimates are not actually all that useful. The real problem is to compute diagnostic functional integrals of functionals, using the posterior distribution as the probability measure.

With the ever increasing power of computers, this is now just feasible. Monte Carlo integration is likely to become increasingly important and, in the opinion of the author, should be regarded as the method of choice and the method in which to invest research and development. That is to say, current approaches to Monte Carlo need to be improved, but Monte Carlo is a fruitful line of investigation.

ally by manual editing, but more recently with assistance from optimisation methods [93] implemented in software.

10.2 The 3D Workflow

The 3D workflow is very similar to the 2D workflow. The differences are quantitative, rather than qualitative. More detail is captured, and there are growing efforts to quantify the uncertainty or non-uniqueness in the models that are consistent with available measurements. The attempt is made, subject to computer constraints, to map all of the *important* layers that can be seen at the wells. This, if successful, will capture the low and high permeability structures which influence the flow. The consequence is that the size of the model can be of the order of 20 million active cells. In the study of [93], there were 25 million cells, of which 19 million were active. The cells were of a size, 75m by 75m by 0.5m. These are actually very large objects. Nevertheless simulation is not practical on such a grid, and so upscaling is performed. The resulting simulation cells in [93] were 90m by 100m with thicknesses ranging from 2-4m in a 104-layer model to tens of meters in a 19-layer simulation grid. The areal upscaling factor is quite small in this instance, with the majority of the averaging taking place in the vertical direction.

In [93] it is stated that an advantage of this workflow is that multiple simulation models can be built from a single *shared earth* geological model. It was found that by including the extra detail in the model the initial model was a good starting point for history matching. In [93] five realisations were generated, and the structural framework was deterministic.

In practice it would seem that *fine models* are actually quite coarse, and that attempts to quantify uncertainty are still limited by inability to handle large numbers of realisations.

In [17] it is stated that oil companies often underestimate risk. Of the many possible causes of such underestimates (from the point of view of this chapter) clearly (i) insufficient numbers of realisations in Monte Carlo studies (ii) suppression of fluctuations caused by upscaling methods are contributors.

The question is: can one do better? One answer is to wait for improved hardware - and this will help in some ways. The next section speculates on possible ways in which different workflows and new algorithms could help.

10.3 Workflow Possibilities

There are six main classes of flow simulation model relevant to the question of scaling. The first distinction is between stable and unstable flow. Instability causes multiple length scale features to evolve in saturations and other state variables. Knowledge about modelling such unstable behaviour is rather scant. Stable flow is much better understood, although there are many unanswered questions.

An important characteristic is ϵ, the ratio of the smallest heterogeneity length scale to the flow modelling scale. The flow modelling scale is in practice set by the grid block size and for accuracy should be a few times the grid block size. The separation of length scales is then measured by the value of ϵ. Length scales can (i) be well separated (the $\epsilon \ll 1$ case) (ii) have large fluctuations on the averaging length scale (the $\epsilon \sim 1$ case) or (iii) have no small scale fluctuations on the averaging scale so that properties are essentially constant on the averaging scale (the $\epsilon \gg 1$ case).

Thus there are six cases in all and three cases to be discussed, as the unstable case is not considered in this chapter.

Stable and $\epsilon \ll 1$ Case: For stable flow and $\epsilon \ll 1$ there are no insuperable difficulties. The aim of upscaling is to transform a stable and $\epsilon \ll 1$ problem into a stable and $\epsilon \gg 1$ problem. If this works then it was not strictly necessary to build a finely gridded model for the whole reservoir in the first place. By building a detailed model, in the vicinity of the wells, for example, the large scale simulation model can be built directly. In fact, in this way one could build models that effectively contain billions of cells. One could go further and say that the attempt to build fine models, but with far from sufficient computer memory available, is a cause of error and unnecessary computation. Indeed in some cases *fine grids* are being used as an alternative to tensor modelling. A tensor requires three vectors, in addition to, if it is symmetric and second order, six numbers. The vectors are defining the bedding planes and they do not require explicit models of their geometry.

One factor preventing the increased use of tensors is the complete absence of geostatistical theory for interpolating vector and tensor properties. (The view expressed in [46] that "there are few areas where new algorithms need to be developed" is perhaps too optimistic. There are in fact many outstanding problems; for example, defining non-Gaussian geostatistics without explicit grids, interpolating vectors and tensors and constructing realisations that are reliable samples of the underlying distributions.)

Stable and $\epsilon \sim 1$ Case: When the problem is stable and $\epsilon \sim 1$ there are severe difficulties with upscaling. One possible improvement follows from the observation that the heterogeneity is *almost* resolved by the grid. Rather than upscale the realisations, an alternative is to resample the original pdf, but on the simulation grid. The change in workflow, although slight in implementation, is rather considerable in concept. That is, one realises that the geological model is the pdf, and not realisations of that pdf. It is in this case that shortcomings in geostatistical interpolation techniques are driving the design of the workflow. Interpolation methods that *rely* on a simple grid structure *must* be used in combination with an upscaling method. A grid-free geostatistical method could be used, on demand, *directly* on a corner point grid, for example.

Stable and $\epsilon \gg 1$ *Case:* In this case one need not upscale. Simple sampling is adequate. That is the geological value at the simulation cell centre is used for the whole simulation cell. Of course, in such a model there are always accuracy improvements to be obtained from grid refinement, as the pressure distribution and flow paths may be complicated as a result of the well pattern. One must not forget that the grid has to resolve the state variables as well as the input variables.

Unstable Flow: Unstable flow requires a review of its own. There are few results of substance, and many opportunities for useful and interesting research.

11 Concluding Discussion

11.1 Outstanding Problems in Mathematical Geoscience

As the various component disciplines have been discussed, with a focus on the mathematical structure, several problems have been highlighted. (The author must stress that the following remarks are expressions of personal opinion, and must be viewed as tentative and speculative. It is, however, useful to summarise such conclusions at the end of a review of mathematical geoscience.)

Flow through porous media and reservoir simulation: The method of streamline simulation is growing in importance. However, it has not received much attention from numerical analysts. Convergence analysis is needed, and insights gained could be valuable to streamline simulation and to more conventional finite volume methods.

Multiscale methods have started to appear, and show considerable promise. However, the link with upscaling approaches has not been clarified, and it may be the case that, as yet undiscovered upscaling workflows, are competitive. Multiscale methods are an area of growing academic interest, with substantial academic projects starting at several universities around the world.

Improved large scale models of the average behaviour of unstable flow are needed. The existing models have not been derived from fundamental theories at the small scale even in special cases. The phenomenological models do not themselves possess proven stability properties on the smaller length scales.

Grid Generation: Even when a model of the structural geometry of a reservoir is available, in three dimensions there are no universally applicable and robust grid generation methods - even in cases where the geometric input is perfect. (Perfection here means that all intersections are well-defined and there are no overlapping regions.) One might think that unstructured grid generation holds the key, with elegant Voronoi or Delaunay grids. However, even for problems with just an external boundary this is a challenge. Geological problems, however, possess complicated internal boundaries and this is essentially uncharted territory. There is thus a real need for innovation in this area. But, as mentioned later, the problem is even harder than this: the geometry has to be built first.

Spatial Statistics: Interpolating Scattered Data: Many geostatisticians regard their subject as mature and the main task as being one of public education. From the point of view of general applications this is far from being the case. Spatial statistics, when stochastic realisations are required are overly dependent upon structured, and even regular grids with cells all the same size and shape. Ad hoc devices are needed to circumvent this problem. Grid-free methods may be possible and are a fruitful area for research.

The link between kriging, radial basis functions and maximum probability interpolants could be investigated in a much deeper way than in Section 4. The huge effort to analyse and develop radial basis function methods would be made more valuable if the participants in the growing radial basis function literature were more aware of the need for *statistical* considerations in the scattered data problem. Deterministic approaches to problems with sparse data are not applicable in most of the problems encountered in the geosciences.

The sequential sampling methods - popular with users because of their high speed - are suspect when compared with methods such as the Gibbs sampler which possess rigorous proofs of statistical convergence.

Methods that provide stochastic models of vector or tensor quantities are essentially absent.

Forward and Inverse Modelling: Many inverse problems in the geosciences involve sparse data used to constrain functions representing three-dimensional heterogeneous property fields, such as porosity or permeability. It is not that clear which parameters are to be determined in the inversion. Is it properties as functions of position, or is it the parameters in the correlation functions summarising the heterogeneity? It may be some mixture of the two, with the parameters in the correlation functions displaying the most sensitivity to the measurements. There is room for much research here: model problems should be studied so that our intuitions can be further developed.

The growing influence of the Bayesian viewpoint, is something to be welcomed, but with caution. The relationship between traditional *deterministic* solutions of inverse problems, maximum probability solutions and Monte Carlo approximation of output statistics needs further clarification and analysis. Studies of model problems, once again, would help us all to understand the issues better.

There seems to be no review of methods for inverse problems that compares and contrasts the various theoretical approaches in the way that was outlined in earlier sections of this chapter. A substantial review, going into far more detail than provided here, would be of enormous value.

The role of length scales in the prior, as part of the inversion process, could be much clearer. This is related to the problems of upscaling, and clarification might require new insights into the general methodology of mathematical modelling.

The different inverse problems in the geosciences, particularly seismic inversion, resistivity inversion, and the history matching of production data are usually treated in quite different ways: is this an accident of history, or should there be changes in all areas, so that a unified approach is used?

Geological Modelling: The characterisation of uncertainty of geometric properties has not received the same level of attention as uncertainty in properties such as porosity. This needs to be rectified.

In general the subject of geological modelling is fraught with difficulty. Only relatively simple systems can be parameterised in a satisfactory way - using the corner point grid approach. Finding a way of building the geometric aspects of geological models, so they can be easily modified, by users and by software performing automated inversion algorithms is an open problem. A breakthrough is needed in this area if the dream of a *shared earth model* is ever to be achieved in a practical way. In situations involving complicated geometric features, such as intersecting faults, the problems are exceptionally difficult and interesting.

Upscaling and Upgridding: Controlling Scale Dependence: The upgridding problem is unsolved in the general case. Sometimes the problem could be avoided, but users are demanding better methods, as the workflow that builds fine grids and then upscales them, demands a solution to this problem.

The ideas involved in dual-porosity and dual-permeability models could be more widely applicable to upscaling than is generally realised. By using models that are designed to characterise behaviour involving two time scales, the need for fine scale models can be reduced.

History Matching: Integrating the Production Data: A solution of the history matching problem requires (i) a method of sampling that reduces the numbers of realisations needed for accurate Monte Carlo calculations (ii) better ways of generating realisations that are consistent with both the prior and the measured data.

Workflow Analysis: Deeper analysis of workflow possibilities is needed. Current methods are applied in an uncritical fashion, and are not examined with a view to establishing an optimum workflow for a particular engineering objective.

11.2 Concluding Remarks

The main mathematical techniques used in building geological models for input to fluid flow simulation have been reviewed. The subject matter concerns the entire geological and reservoir simulation modelling workflow in the subsurface. Seismic acquisition, processing and interpretation, well logging and geology have only been reviewed in outline. However, the topics of grid generation, geometric modelling and spatial statistics have been covered in considerable detail. A few new results in the area of geostatistics were proved. In particular the equivalence of radial basis functions, general forms of kriging and minimum curvature methods was shown. A Bayesian formulation of uncertainty assessment has been outlined. The discussion of upscaling was brief, consisting of a summary, as a recent, detailed review in this area, [58], is available.

The classical approach, using maps, is giving way to an approach that builds 3D geocellular models. Some of the methods for this are still map based - but some now work directly on the basic data - seismic picks, well logs and cores.

Accompanying this move to 3D is a strong tendency to build as detailed a model as computer memory allows. However computer memory is far too limited for properly detailed models of a whole reservoir to be built, and so geological detail is grossly under-resolved. Intuitive analysis of upscaling leads to the conclusion, when the fluid flow is stable and the length scales are well separated, that upscaling can be very accurate. By applying knowledge of the scales at which the length scales *are* well separated, one can design multiscale models that *do* upscale. When this is possible one can build fine scale - but now truly fine scale - models near the wells. Then a second application of geostatistics on the larger scale, conditioned on the upscaled near-well data, can suffice.

Acknowledgements

I would like to thank the Royal Society for the award of an Industry Fellowship at the University of Oxford. I would also like to thank Zhijie Cai (Fudang) and John Ockendon (Oxford) for stimulating discussions during the writing of this review. Discussion with many Schlumberger colleagues has been very helpful in understanding the various parts of the workflow. Particular thanks are due to Martyn Beardsell, Alberto Malinverno and Jonathan Morris. I am grateful to Michael Thambynayagam for support in initiating and completing the project of writing this review. Thanks are also due to Daniel Busby and the referees for their useful suggestions. Any errors or omissions are, of course, my full responsibility.

Notes on the References

Many papers listed in the bibliography are from the Society of Petroleum Engineers. Most of these, where they are not explicitly from an SPE journal, are available in electronic form from the SPE at `http://www.spe.org/`. *FloGrid* is a trademark of Schlumberger.

References

1. B. Agarwal and M. J. Blunt (2003) Streamline-based method with full-physics forward simulation for history-matching performance data of a North Sea field. SPE Journal **8**:, 171–180.
2. R. Agut, M.G. Edwards, S. Verma, and K. Aziz (1998) Flexible streamline-potential grids with discretization on highly distorted cells. Proceedings of the 6th European Conference on the Mathematics of Oil Recovery (ECMOR VI). Peebles, Edinburgh, Scotland, 8-11 September, 1998.
3. H.N. Al-Sadi (1982) *Seismic Exploration.* Birkhäuser, Basel.
4. J.R.L. Allen (1982) *Sedimentary Structures: Their Character and Physical Basis, v1–2.* Elsevier, Amsterdam.
5. P.A. Allen and J.R. Allen (1990) *Basin Analysis - Principles and Applications.* Blackwell, Oxford.
6. A.A. Amsden and C.W. Hirt (1973) A simple scheme for generating general curvilinear grids. J. Comput. Phys. **11**, 348–359.
7. B. Anderson, V. Druskin, et al. (1997) New dimensions in modeling resistivity. Oilfield Review **9**, 40–56.
8. J.R. Appleyard and I.M. Cheshire (1983) Nested factorization. SPE 12264. Proceedings of the 7th SPE Symposium on Reservoir Simulation, San Francisco, 1983.
9. J.S. Archer and C.G. Wall (1986) *Petroleum Engineering: Principles and Practice.* Graham and Trotman, London.
10. E. Arge and A. Kunoth (1998) An efficient ADI-solver for scattered data problems with global smoothing. J. Comput. Phys. **139**, 343–358.
11. K. Aziz and A. Settari (1979) *Petroleum Reservoir Simulation.* Applied Science Publishers, London.
12. M. Bacon, R. Simm, and T. Redshaw (2003) *3-D Seismic Interpretation.* Cambridge University Press, Cambridge.
13. M.J. Baines (1994) *Moving Finite Elements.* Oxford University Press.
14. G.I. Barenblatt, I.P. Zheltov, and I.N. Kochina (1960) Basic concepts in the theory of seepage of homogeneous liquids in fissured rocks (strata). J. Appl. Math. Mech. **24**, 1286–1303.
15. J.W. Barker, M. Cuypers, and L. Holden (2001) Quantifying uncertainty in production forecasts: Another look at the PUNQ-S3 problem. SPE Journal **6**, 433–441.
16. J. Bear (1988) *Dynamics of Fluids in Porous Media.* Dover Publications Incorporated, New York.
17. S. Begg, R. Bratvold, and J. Campbell (2002) The value of flexibility in managing uncertainty in oil and gas investments. SPE 77586. Proceedings of the SPE Annual Technical Conference and Exhibition, San Antonio, Texas, 2002.

18. M.J. Beran (1968) *Statistical Continuum Theories.* Interscience Publishers, New York.
19. J.J. Binney, N.J. Dowrick, A.J. Fisher, and M.E.J. Newman (1992) *The Theory of Critical Phenomena.* Oxford University Press, Oxford.
20. R. Bissell, Y. Sharma, and J.E. Killough (1994) History matching using the method of gradients: Two case studies. SPE 28590. Proceedings of the 69th SPE Annual Technical Conference and Exhibition, 1994, 275–289.
21. N. Bleistein, J.K. Cohen, and J.W. Stockwell (2001) *Mathematics of Multidimensional Seismic Imaging, Migration, and Inversion.* Springer, New York.
22. S. Boggs (1995) *Principles of Sedimentology and Stratigraphy.* Prentice Hall, Englewood Cliffs, New Jersey.
23. A. Bourgeat and M. Panfilov (1998) Effective two-phase flow through highly heterogeneous porous media: capillary nonequilibrium effects. Computational Geosciences **2**, 191–215.
24. J.P. Boyd (2001) *Chebyshev and Fourier Spectral Methods.* 2nd edition, Dover Publications Inc., Mineola, New York.
25. J.U. Brackbill and J.S. Saltzman (1982) Adaptive zoning for singular problems in two dimensions. J. Comput. Phys. **46**, 342–368.
26. F. Bratvedt, K. Bratvedt, C. Buchholz, H. Holden, L. Holden, and N.H. Risebro (1989) A new front-tracking method for reservoir simulation. SPE 19805. Proceedings of the 64th SPE Annual Technical Conference and Exhibition, San Antonio, 1989.
27. I.C. Briggs (1974) Machine contouring using minimum curvature. Geophysics **39**, 39–48.
28. S.E. Buckley and M.C. Leverett (1942) Mechanism of fluid displacement in sands. Trans. of the A.I.M.E. **146**, 107.
29. M.D. Buhmann (2000) Radial basis functions. Acta Numerica **9**, 1–38.
30. J. Caers, S. Krishnan, Y. Wang, and A.R. Kovscek (2002) A geostatistical approach to streamline-based history matching. SPE Journal **7**, 250–266.
31. J. Carr, W.R. Fright, and R.K. Beatson (1997) Surface interpolation with radial basis functions for medical imaging. IEEE Transactions on Medical Imaging **16**, 96–107.
32. J.E. Castillo (1991) Discrete variational grid generation. *Mathematical Aspects of Numerical Grid Generation*, J.E. Castillo (ed.), SIAM, Philadelphia, 33–58.
33. J.E. Castillo, ed. (1991) *Mathematical Aspects of Numerical Grid Generation.* SIAM, Philadelphia.
34. J.-P. Chilés and P. Delfiner (1999) *Geostatistics: Modeling Spatial Uncertainty.* John Wiley, New York.
35. B. Choudhuri and S.K. Khataniar (1997) Characterisation of a complex high permeability reservoir through transient analysis: A case study of the lower Eocene reservoirs of Upper Assam basin, India. Paper no. 145. Proceedings of the *National Seminar on Reservoir Management and Exhibition*, Baroda, India, Oct. 6–7, 1997.
36. J.A. Coffeen (1990) *Seismic on Screen.* Pennwell Books, Tulsa.
37. R.E. Collins (1961) *Flow of Fluids through Porous Materials.* Reinhold Publishing Corporation, New York.

38. P.S. Craig, M. Goldstein, A.H. Seheult, and J.A. Smith (1997) Pressure matching for hydrocarbon reservoirs: a case study in the use of Bayes linear strategies for large computer experiments. Case Studies in Bayesian Statistics III, Springer, New York.
39. R. Daley (1991) *Atmospheric Data Analysis.* Cambridge University Press, Cambridge.
40. A. Datta-Gupta, K.N. Kulkarni, S. Yoon, and D.W. Vasco (2001) Streamlines, ray tracing and production tomography: generalization to compressible flow. Petroleum Geoscience **7**, S75–S86.
41. A. Datta-Gupta, L.W. Lake, and G.A. Pope (1995) Characterizing heterogeneous permeable media with spatial statistics and tracer data using sequential simulated annealing. Mathematical Geology **27**, 763–787.
42. C.V. Deutsch and A.G. Journel (1992) *GSLIB: Geostatistical Software Library and User's Guide.* Oxford University Press, New York.
43. P.A. Dickey (1986) *Petroleum Development Geology.* PennWell Publishing Company, Tulsa.
44. J. Douglas, Jr., M. Kischinhevsky, P.J. Paes-Leme, and A. Spagnuolo (1998) A multiple-porosity model for a single-phase flow through naturally-fractured porous media. Computational and Applied Mathematics **17**, 19–48.
45. J.H. Doveton (1986) *Log Analysis of Subsurface Geology: Concepts and Computer Methods.* John Wiley Sons, New York.
46. O. Dubrule and E. Damsleth (2001) Achievements and challenges in petroleum geostatistics. Petroleum Geoscience **7**, S1–S7.
47. L.J. Durlofsky, R.C. Jones, and W.J. Milliken (1994) A new method for the scale up of displacement processes in heterogeneous reservoirs. Proceedings of the 4th European Conference on the Mathematics of Oil Recovery, Røros, Norway, 1994.
48. Encyclopædia Britannica (2003): `http://www.britannica.com/`.
49. H.W. Engl, M. Hanke, and A. Neubauer (1996) *Regularization of Inverse Problems.* Kluwer, Dordrecht, 1996.
50. R. Ewing, ed. (1983) *The Mathematics of Reservoir Simulation.* SIAM, Philadelphia.
51. S.W. Fagin, ed. (1991) *Seismic Modeling of Geologic Structures:Applications to Exploration Problems.* Society of Exploration Geophysicists, Tulsa.
52. J.R. Fanchi (2002) *Shared Earth Modeling.* Butterworth-Heinemann, Amsterdam.
53. C.L. Farmer (1985) A moving point method for arbitrary Peclet number multi-dimensional convection-diffusion equations. IMA Journal of Numerical Analysis **5**, 465–480.
54. C.L. Farmer (1987) Moving point methods. *Advances in Transport Phenomena in Porous Media*, J. Bear and M.Y. Corapcioglu (eds.), NATO ASI Series E: Applied Sciences no. 128, Martinus Nijhoff, Dordrecht, 953–1004.
55. C.L. Farmer (1988) The generation of stochastic fields of reservoir parameters with specified geostatistical distributions. Proceedings of the 1987 IMA Conference on *Mathematics in Oil Production*, Sir Sam Edwards and P.R. King (eds.), Oxford University Press, Oxford, 235–252.
56. C.L. Farmer (1992) Numerical rocks. *Mathematics of Oil Recovery*, P.R. King (ed.), Oxford University Press, Oxford, 437–447.

57. C.L. Farmer (1998) An application of triangulations to the building of structured grids. Proceedings of the 6th European Conference on the Mathematics of Oil Recovery (ECMOR VI). Peebles, Edinburgh, Scotland, 8-11 September, 1998.
58. C.L. Farmer (2002) Upscaling: a review. International Journal for Numerical Methods in Fluids **40**, 63–78.
59. C.L. Farmer and D.E. Heath (1990) Curvilinear grid generation techniques. Proceedings of the 2nd European Conference on the Mathematics of Oil Recovery (ECMOR II). Arles, France, 11-14 September, 1990.
60. C.L. Farmer, D.E. Heath, and R.O. Moody (1991) A global optimisation approach to grid generation. SPE 21236. Proceedings of the 11th SPE Symposium on Reservoir Simulation, Anaheim, USA, 1991.
61. M.S. Floater and A. Iske (1996) Multistep scattered data interpolation using compactly supported radial basis functions. Journal of Computational and Applied Mathematics **73**, 65–78.
62. F.J.T. Floris, M.D. Bush, M. Cuypers, F. Roggero, and A.-R. Syversveen (2001) Methods for quantifying the uncertainty of production forecasts: a comparative study. Petroleum Geoscience **7**, S87–S96.
63. R. Franke (1982) Scattered data interpolation: tests of some methods. Math. Comp. **38**, 181–199.
64. N.P. Fremming (2002) 3D geological model construction using a 3D grid. Paper E01. Proceedings of the 8th European Conference on the Mathematics of Oil Recovery (ECMOR8), Freiberg, Germany, 3-6 September, 2002.
65. U. Frisch (1995) *Turbulence.* Cambridge University Press, Cambridge
66. C.W. Gable, H.E. Trease, and T.A. Cherry (1996) Geological applications of automatic grid generation tools for finite elements applied to porous flow modeling. *Numerical Grid Generation in Computational Fluid Dynamics and Related Fields*, B.K. Soni, J.F. Thompson, H. Hauser, and P.R. Eiseman (eds.), Engineering Research Centre, Mississippi State University Press.
67. D. Gamerman (1997) *Markov Chain Monte Carlo: Stochastic Simulation for Bayesian Inference.* Chapman and Hall, London.
68. M.H. Garcia, A.G. Journel, and K. Aziz (1992) Automatic grid generation for modeling reservoir heterogeneities. SPE Reservoir Engineering, May 1992, 278–284.
69. A.O. Garder, D.W. Peaceman, and A.L. Pozzi (1964) Numerical calculation of multidimensional miscible displacement by the method of characteristics. Society of Petroleum Engineers Journal **4**, 26–36.
70. G.R. Gavalas, P.C. Shah, and J.H. Seinfeld (1976) Reservoir history matching by Bayesian estimation. Society of Petroleum Engineers Journal **16**, 337–350.
71. S. Geman and D. Geman (1984) Stochastic relaxation, Gibbs distributions and the Bayesian restoration of images. IEEE Transactions on Pattern Analysis and Machine Intelligence **6**, 721–741.
72. J. Gluyas and R. Swarbrick (2004) *Petroleum Geoscience.* Blackwell Publishing, Oxford.
73. N. Goldenfeld (1992) *Lectures on Phase Transitions and the Renormalization Group.* Perseus Books, Reading, Massachusetts.
74. W.H. Goldthorpe and Y.S. Chow (1985) Unconventional modeling of faulted reservoirs: A case study. SPE 13526. Proceedings of the SPE Reservoir Simulation Symposium, Dallas, 1985.

75. D. Gottlieb and S.A. Orszag (1977) *Numerical Analysis of Spectral Methods.* SIAM, Philadelphia.
76. R.H. Groshong Jr. (1999) *3D Structural Geology: A Practical Guide to Surface and Subsurface Map Interpretation.* Springer-Verlag , Berlin.
77. D. Gunasekera, J. Herring, and J. Cox (1998) Segmented coordinate line based unstructured gridding. Proceedings of the 6th European Conference on the Mathematics of Oil Recovery (ECMOR VI). Peebles, Edinburgh, Scotland, 8-11 September, 1998.
78. T. Gutzmer and A. Iske (1997) Detection of discontinuities in scattered data approximation. Numerical Algorithms **16**, 155–170.
79. R.L. Hardy (1971) Multiquadric equations of topography and other irregular surfaces. J. Geophys. Res. **76**, 1905–1915.
80. Z. He, S. Yoon, and A. Datta-Gupta (2002) Streamline-based production data integration with gravity and changing field conditions. SPE Journal **7**, 423–436.
81. D. A. Herron (2000) Horizon autopicking. J. Geophys. Res. **19**, 491–492.
82. R.V. Higgins and A.J. Leighton (1962) A computer method to calculate two-phase flow in any irregularly bounded porous medium. J. Pet. Technology, June 1962, 679–683.
83. G.J. Hirasaki and P.M. O'Dell (1970) Representation of reservoir geometry for numerical simulation. Soc. Pet. Eng. J., December 1970, 393–404.
84. K.S. Hoffman and J.W. Neave (1996) Horizon modeling using a three-dimensional fault restoration technique. SPE 56445. Proceedings of the SPE Annual Technical Conference and Exhibition, Houston, Texas, 1996.
85. M.E. Hohn (1988) *Geostatistics and Petroleum Geology.* Van Nostrand Reinhold, New York.
86. G.M. Homsy (1987) Viscous fingering in porous media. Annual Review of Fluid Mechanics **19**, 271–311.
87. P.J. Hore (1995) *Nuclear Magnetic Resonance.* Oxford University Press, Oxford.
88. U. Hornung, ed. (1997) *Homogenization and Porous Media.* Springer, Berlin.
89. F.G. Horowitz, P. Hornby, D. Bone, and M. Craig (1996) Fast multidimensional interpolations. *26th Proceedings of the Application of Computers and Operations Research in the Mineral Industry* (APCOM26), R.V. Ramani (ed.), Soc. Mining, Metall., and Explor. (SME), Littelton, Colorado, USA, 53–56.
90. J. Hoschek and D. Lasser (1993) *Fundamentals of Computer Aided Geometric Design.* A.K. Peters, Wellesley, Massachusetts.
91. T. Hou and X.H. Wu (1997) A multiscale finite element method for elliptic problems in composite materials and porous media. J. Comput. Phys. **134**, 169–189.
92. A. Iske (2002) Scattered data modelling using radial basis functions. *Tutorials on Multiresolution in Geometric Modelling*, A. Iske, E. Quak, and M.S. Floater (eds.), Springer, Heidelberg, 205–242.
93. T. Jacobsen, H. Agustsson, J. Alvestad, P. Digranes, I. Kaas, and S.T. Opdal (2000) Modelling and identification of remaining reserves in the Gullfaks field. SPE 65412. Proceedings of the SPE European Petroleum Conference, Paris, 2000.

94. P. Jenny, S.H. Lee, and H.A. Tchelepi (2003) Multi-scale finite-volume method for elliptic problems in subsurface flow simulation. J. Comput. Phys. **187**, 47–67.
95. T.A. Jones, D.E. Hamilton, and C.R. Johnson (1986) *Contouring Geologic Surfaces with the Computer.* Van Nostrand Reinhold, New York.
96. A.G. Journel (1989) *Fundamentals of Geostatistics in Five Lessons.* American Geophysical Union, Washington, USA.
97. A.G. Journel and F.G. Alabert (1989) Non-Gaussian data expansion in the Earth sciences. Terra Nova. **1**, 123–134.
98. P. Kaufman and P. Fjerstad (2002) Explicit simulation of conductive faults. Paper E64. Proceedings of the 8th European Conference on the Mathematics of Oil Recovery (ECMOR8), Freiberg, Germany, 3-6 September, 2002.
99. M.C. Kennedy and A. O'Hagan (2000) Predicting the output from a complex computer code when fast approximations are available. Biometrika **87**, 1–13.
100. M.C. Kennedy, A. O'Hagan, and N. Higgins (2002) Bayesian analysis of computer code outputs. *Quantitative Methods in Current Environmental Issues*, C.W. Anderson, V. Barnet, P.C. Chatwin, and A.H. El-Shaarawi (eds.), Springer, New York.
101. J.W. King, P.F. Naccache, O. Nichols, R.K. Pollard, D.K. Ponting, and J. Rae (1982) The use of non-neighbour connections in reservoir simulation. EUR 281. Proceedings of the European Petroleum Conference, London, October, 1982.
102. P. Knupp and S. Steinberg (1993) *Fundamentals of Grid Generation.* CRC Press, Boca Raton.
103. K.N. Kulkarni, A. Datta-Gupta, and D.W. Vasco (2001) A streamline approach for integrating transient pressure data into high-resolution reservoir models. SPE Journal **6**, 273–282.
104. A. Kumar, C.L. Farmer, G.R. Jerauld, and D. Li (1997) Efficient upscaling from cores to simulation models. SPE 38744. Proceedings of the Annual Technical Conference and Exhibition, San Antonio, Texas, 1997.
105. A. Kumar and N.S. Kumar (1988) A new approach to grid generation based on local optimisation. *Numerical Grid Generation in Computational Fluid Mechanics*, S. Sengupta, J. Häuser, P.R. Eiseman, and J.F. Thompson (eds.), Pineridge Press, Mumbles, Swansea, 177–184.
106. N. Lambeth and R.A. Dawe (1987) Boundary and crossflow behaviour during displacement in nodal systems. SPE 16972. Proceedings of the 62nd Annual Technical Conference of the SPE, Dallas, 1987.
107. T.J. Lasseter (2002) A new approach for the efficient construction of 3D geological models for reservoir applications. Paper E03. Proceedings of the 8th European Conference on the Mathematics of Oil Recovery (ECMOR8), Freiberg, Germany, 3-6 September, 2002.
108. R.C. Laudon (1996) *Principles of Petroleum Development Geology.* Prentice Hall, Upper Saddle River, New Jersey.
109. D.C. Leslie (1973) *Developments in the Theory of Turbulence.* Clarendon Press, Oxford.
110. P.K. Link (1987) *Basic Petroleum Geology.* Oil and Gas Consultants International, Inc., Tulsa.
111. N. Liu and D.S. Oliver (2003) Evaluation of Monte Carlo methods for assessing uncertainty. SPE Journal **8**, 188–195.
112. S.M. Luthi (2001) *Geological Well Logs: Their Use in Reservoir Modeling.* Springer, Berlin.

113. J.-L. Mallet (1989) Discrete smooth interpolation. ACM Transactions on Graphics **8**(2), 121–144.
114. J.-L. Mallet (2002) *Geomodeling.* Oxford University Press, Oxford.
115. K.V. Mardia, J.T. Kent, and J.M. Bibby (1979) *Multivariate Analysis.* Academic Press, London.
116. D. McLaughlin and L.R. Townley (1996) A reassessment of the groundwater inverse problem. Water Resources Research **32**, 1131–1161.
117. A.I. Mees, M.F. Jackson, and L.O. Chua (1992) Device modeling by radial basis functions. IEEE Transactions on Circuits and Systems - I: Fundamental Theory and Applications **39**, 19–27.
118. N. Metropolis, A.W. Rosenbluth, M.N. Rosenbluth, A.H. Teller, and E. Teller (1953) Equations of state calculations by fast computing machine. Jour. Chem. Physics **21**, 1087–1091.
119. W.J. Milliken, A.S. Emmanuel, and A. Chakravarty (2001) Applications of 3d streamline simulation to assist history matching. SPE Reservoir Evaluation and Engineering **4**, 502–508.
120. J.-M. Morel and S. Solimini (1995) *Variational Methods in Image Segmentation.* Birkhauser, Boston.
121. M.E.J. Newman and G.T. Barkema (1999) *Monte Carlo Methods in Statistical Physics.* Oxford University Press, Oxford.
122. A. O'Hagan (1994) *Kendall's Advanced Theory of Statistics, Volume 2B, Bayesian Inference.* Edward Arnold, London.
123. A. Okabe, B. Boots, K. Sugihara, and S.N. Chiu (2000) *Spatial tessellations: Concepts and Applications of Voronoi Diagrams.* J. Wiley and Sons. Ltd., Chichester.
124. D.S. Oliver, A.C. Reynolds, B. Zhuoxin, and Y. Abacioglu (2001) Integration of production data into reservoir models. Petroleum Geoscience **7**, S65–S73.
125. H. Omre (2003) `www.math.ntnu.no/~omre/`.
126. A. Ouenes, R.S. Doddi, Y. Lin, G. Cunningham, and N. Saad (1994) A new approach combining neural networks and simulated annealing for solving petroleum inverse problems. Proceedings of the 4th European Conference on the Mathematics of Oil Recovery, Røros, Norway, 1994.
127. M. Panfilov (2000) *Macroscale Models of Flow Through Highly Heterogeneous Porous Media.* Kluwer Academic.
128. D.W. Peaceman (1978) *Fundamentals of Numerical Reservoir Simulation.* Elsevier, Amsterdam.
129. W.H. Press, S.A. Teukolsky, W.T. Vetterling, and B.P. Flannery (1992) *Numerical Recipes in FORTRAN. The Art of Scientific Computing.* Cambridge University Press, Cambridge.
130. H.H. Rachford (1966) Numerical calculation of immiscible displacement by a moving reference point method. Society of Petroleum Engineers Journal, June 1966, 87–101.
131. C.D. Rodgers (2000) *Inverse Methods for Atmospheric Sounding: Theory and Practice.* World Scientific, Singapore.
132. F. Roggero and L.Y. Hu (1998) Gradual deformation of continuous geostatistical models for history matching. SPE 49004. Proceedings of the SPE Annual Technical Conference and Exhibition, New Orleans, 1998.
133. N. Saad, C.T. Kalkomey, and A. Ouenes (1994) Optimal gridding of stochastic models for scale-up. Proceedings of the 4th European Conference on the Mathematics of Oil Recovery, Røros, Norway, 1994.

134. Y. Saad (1996) *Iterative Methods for Sparse Linear Systems.* RWS Publishing Company, Boston.
135. Schlumberger (1997) *FloGrid^{TM}: A Simulation Grid Generation and Upscaling Software Application.* User Guide, Schlumberger, Abingdon.
136. Schlumberger (1998) *FloGrid^{TM}: A Simulation Grid Generation and Upscaling Software Application.* User Guide, Schlumberger, Abingdon.
137. Schlumberger (2002) *FloGrid^{TM}: A Simulation Grid Generation and Upscaling Software Application.* User Guide, Schlumberger, Abingdon.
138. R.C. Selley (1998) *Elements of Petroleum Geology.* Academic Press, San Diego.
139. R.E. Sheriff and L.P. Geldart (1995) *Exploration Seismology.* Cambridge University Press, Cambridge.
140. Society of Petroleum Engineers (2003) `http://www.spe.org/`.
141. F. Sonier and P. Chaumet (1974) A fully implicit three-dimensional model in curvilinear coordinates. Society of Petroleum Engineers Journal, August 1974, 361–370.
142. A. Tarantola (1987) *Inverse Problem Theory: Methods for Data Fitting and Model Parameter Estimation.* Elsevier, Amsterdam.
143. D. Tearpock and R. Bischke (2002) *Applied Subsurface Geological Mapping.* 2nd edition, Prentice Hall, New Jersey.
144. D.M. Tetzlaff and J.W. Harbaugh (1989) *Simulating Clastic Sedimentation.* Van Nostrand Reinhold, New York.
145. J.F. Thompson, B.K. Soni, and N.P. Weatherill (1999) *Handbook of Grid Generation.* CRC Press, Boca Raton.
146. J.F. Thompson and Z.U.A. Warsi (1982) Boundary-fitted coordinate systems for numerical solution of partial differential equations - a review. J. Comput. Phys. **47**, 1–108.
147. J.F. Thompson, Z.U.A. Warsi, and C.W. Mastin (1974) Automatic numerical generation of body-fitted curvilinear coordinate system for field containing any number of arbitrary two-dimensional bodies. J. Comput. Phys. **15**, 299–319.
148. J.F. Thompson, Z.U.A. Warsi, and C.W. Mastin (1985) *Numerical Grid Generation.* North-Holland, New York.
149. J. Tittman (1986) *Geophysical Well Logging.* Academic Press, London.
150. U. Trottenberg, C.W. Oosterlee, and A. Schüller (2001) *Multigrid.* Academic Press, San Diego.
151. S. Twomey (1996) *Introduction to the Mathematics of Inversion and Remote Sensing and Indirect Measurements.* Dover Publications Incorporated, New York.
152. G. Vincent, B. Corre, and P. Thore (1998) Managing structural uncertainty in a mature field for optimal well placement. SPE 48953. Proceedings of the SPE Annual Technical Conference and Exhibition, New Orleans, 1998.
153. P.L.W. Vinsome (1976) Orthomin, an iterative method for sparse banded sets of simultaneous linear equations. SPE 5729. Proceedings of the 4th SPE Symposium on Reservoir Simulation, Los Angeles, 1976.
154. D.T. Vo, M.I. Sukerin, et al. (2000) Reservoir modelling assists operations to optimize field development: Serang field, East Kalimantan. SPE 59441. Proceedings of the SPE Asia Pacific Conference, Yokahama, 2000.
155. Y. Wang and A.R. Kovscek (2000) Streamline approach for history-matching production data. SPE Journal **5**, 353–362.

156. S. Waryan, D.T. Vo, J. Stites, and M. Swanson (2001) Integrated 3D geological data into fluid flow model improves reservoir management plan: Serang field case study. SPE 68646. Proceedings of the SPE Asia Pacific Conference, Jakarta, 2001.
157. K.G. Wilson (1983) The renormalization group and critical phenomena. Reviews of Modern Physics **55**, 583–600.
158. A. Winslow (1967) Numerical solution of the quasi-linear Poisson equation in a nonuniform triangle mesh. J. Comput. Phys. **2**, 149–172.
159. A.M. Winslow (1981) Adaptive-mesh zoning by the equipotential method. Technical Report UCID-19062, Texas Institute for Computational and Applied Mathematics, Lawrence Livermore National Lab., USA, April 1981.
160. C. Wunsch (1996) *The Ocean Circulation Inverse Problem.* Cambridge University Press, Cambridge.
161. S. Yoon, A.H. Malallah, A. Datta-Gupta, D.W. Vasco, and R.A. Behrens (2001) A multiscale approach to production data integration using streamline models. SPE Journal **6**, 182–199.
162. D. Zhang (2001) *Stochastic Methods for Flow in Porous Media: Coping with Uncertainties.* Academic Press, San Diego.
163. D.A. Zimmerman, G. de Marsily, et al. (1998) A comparison of seven geostatistically based inverse approaches to estimate transmissivities for modeling advective transport by groundwater flow. Water Resources Research **34**, 1373–1413.
164. J. Zinn-Justin (2001) *Quantum Field Theory and Critical Phenomena.* Oxford University Press, Oxford.

Geological Model Building: A Hierarchical Segmentation Approach

Erik Monsen[1], Trygve Randen[1], Lars Sønneland[1], and Jan E. Odegard[2]

[1] Schlumberger Stavanger Research, P.O. Box 8013, N-4068 Stavanger, Norway
[2] Rice University, CITI, Houston, TX, U.S.A.

Summary. A new approach towards model building with the promise of significantly shortening the turnaround time of 3D model building is presented. By introducing a unified framework, efficient representation of models throughout the lifecycle of a reservoir is enabled, all the way from velocity to simulation models. All levels of resolution are maintained and handled simultaneously and information is structured according to geological understanding. This enables the interpreter to work with pre-generated geological objects, rather than spending time creating them in the first place. Consequently, more time is spent being creative, gaining a better understanding of the reservoir. Furthermore, the framework enables use of new semi- or fully automatic interpretation schemes and provides real-time user interaction with large volumes.

1 Introduction

1.1 Model Building

Throughout the lifecycle of a reservoir a variety of models are generated and used as a means of representing the information obtained, and to understand observations, to predict future observations and to plan appraisal strategies. The size and detail of these models depend heavily on tho maturity of the reservoir, which may be in the exploration, appraisal, development or production phase. The various geophysical aspects of the reservoir, including velocities, facies distributions, porosities, permeabilities, and more, are all represented by individual models. Though different in nature and scale, many of these models share several common boundaries, which coincide with the major events in the sub-surface. Viewed across scale, it can be argued that the models relate to each other in a hierarchical fashion, where the more detailed models exist within the coarser models.

Models Used in Hydrocarbon Exploration and Production. To get an appreciation for the wide range of models used in hydrocarbon exploration to production, a small summary is presented here, presenting the purposes of the models and the range of scales covered by them.

- **Velocity Models.** To image the subsurface, acoustic waves are propagated through the earth and recorded at the surface after being reflected from boundaries down below. The two way travel time recorded at the surface (i.e., down and up again) is a direct function of the sound velocities in the different rock types encountered. In order to produce a correct image, it is important to have a good knowledge of the distribution of the velocities, since differences in medium velocities causes ray bending and affects the recorded reflection amplitudes. Velocity models are very coarse, and can have a resolution of up to (a couple of) hundred meters.
- **Structural Models.** As a consequence of gravity, the default orientation and configuration of deposited geological layers are horizontal and parallel. However, due to erosional, depositional and tectonic events, the default configuration is often bent, broken up and rotated, thus, yielding a highly complex configuration. A structural model aims at capturing the major geological boundaries, such as strong reflectors, faults, channel boundaries, etc. The typical resolution of structural models cover a wide range, but is in the order of meters.
- **Geological Models.** The subsurface is a truly heterogeneous medium, composed of many different rock types with highly varying physical properties. Depending on the depositional setting, some rock types are more likely than others. A geological model tries to capture the distribution of rock types, and their associated physical properties, such as porosity, permeability, etc. A geological model represents the most detailed model used in the hydrocarbon exploration and production workflow. It incorporates information all the way down to well log resolution. Hence, the typical resolution is in the range of decimeters vertically and hundred meters horizontally.
- **Reservoir and Simulation Models.** Based on the knowledge obtained about the reservoir, especially from well logs and production data, fluid flow simulations are done to (1) match production data, (2) to update the models, (3) to forecast production behavior, and (4) to plan production strategies. Based on simulations, many different scenarios may be evaluated in terms of optimal reservoir drainage. Mainly due to lack of processing power, simulations are done on models that are scaled up versions of geological models. The typical resolution is in the range of meters vertically and hundred meters horizontally.

There are a number of ways in which these models are interrelated. In this chapter we will consider the geometrical relationships that these models exhibit across scale, and try to exploit this by generating a common framework for model building.

From geology we know that the composition of the subsurface depends on its geological history, and may roughly be separated into units created during different geological time periods, and in different depositional environments. Hence, there exists a time (geological periods) and space (depositional environments) type tessellation of the earth, which possess a hierarchical ordering, describing the relative timing of events. These major geological events are also associated with important geophysical changes. Most of these major geological boundaries exist across several scales and types of models, and motivates the integration into a common modeling framework.

Several methods exist by which the subsurface may be tessellated and labelled. Below we consider *seismic sequence stratigraphy*, which is one of the most commonly used techniques.

1.2 Stratigraphic Analysis and Modeling

Sequence stratigraphy combines logs with fossil data and seismic reflection patterns to explain both the arrangement of rocks and the depositional environment. Understanding how rock layers relate to the sequences of seismic reflections and depositional environments allows more accurate prediction of possible reservoirs, source rocks and caps, even in regions not intersecting the wells.

In general, stratigraphy is the science of describing the vertical and lateral relationships of rocks [17, 21, 30]. These relationships may be based on rock type, called lithostratigraphy, on age, as in chronostratigraphy, on fossil content, labeled biostratigraphy, or on magnetic properties, named magnetostratigraphy. Though different in what they describe, all types of stratigraphy are based on three basic principles - younger layers lie on top of older layers, layers are initially horizontal, and layers continue until they run into a barrier [21].

During exploration seismic data constitutes the dominant part of available data. In the case of no or few exploration wells, we have no direct measurements for the majority of the reservoir. Hence, we have to rely on the subset of sequence stratigraphy that includes seismic data (i.e., seismic sequence stratigraphy) to infer the depositional environments and predict the distribution of different rock types

Sequence stratigraphic analysis is governed by a set of simple principles [21]. A depositional sequence comprises sediments deposited during one cycle of sea-level fluctuation, starting at low sea level, going to high and returning to low (by Exxon convention). One cycle may last a few thousands to millions of years and produce a variety of sediments, such as beach sands, submarine channel and levee deposits, chaotic flows or slumps and deep-water shales (see Figure 1). Sediment type may vary gradually or abruptly, or may be uniform and widespread over the entire basin. Each rock sequence produced by one cycle is bounded by an unconformity at the bottom and top. These sequence boundaries are the main seismic reflections used to identify each depositional sequence, and separate younger from older layers everywhere in the basin.

Sequence stratigraphy was originally developed for explaining sand-shale systems, but it has also been successfully applied to other systems, such as

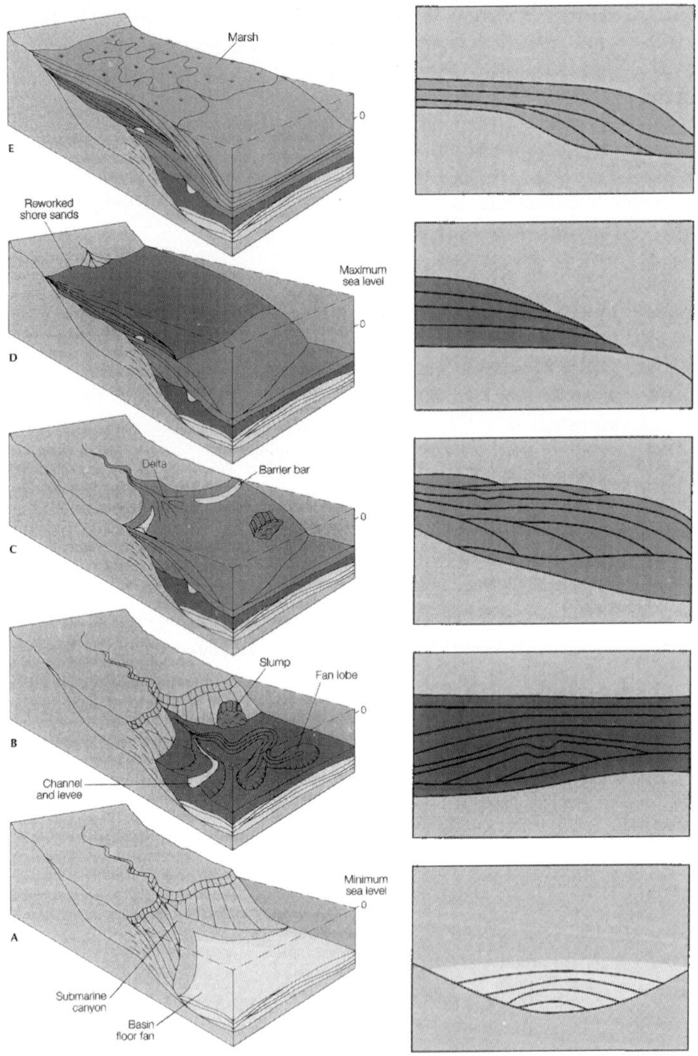

Fig. 1. Sequences in order of deposition [21]. The sequence begins when the sea-level relative to the ocean floor starts to fall (according to Exxon convention). During this period of falling sea-level, sand-rich fans are laid down (A). As the drop in sea-level flattens out and begins to rise again (B), sands and shales are deposited on the continental slope. Slumps and submarine channels often appear in this period. As the sea-level continues to rise (C), sediments are deposited basinward forming wedge-shaped structures, with sands near the shore and shales further out. During a period of rapid rise in sea-level (D), the sand-rich sediments are deposited landward as beaches and sandbars. The following period of lower sea-level rise (E) allows the sediments to build basinward again. Graphic copyright Schlumberger *Oilfield Review*, used with permission.

carbonates. Though being a powerful analysis tool, sequence stratigraphy must be used with care. In environments with low sedimentation rates, the depositional sequences will be very thin, well below the resolving power of the seismic wavelet. With mid to high sedimentation rates, this is not a problem since the layers within a depositional sequence will be tens to hundreds of meters thick. Hence, in regions of thin layering, stratigraphic techniques cannot be used reliably to explain the fine-scale structuring. What can be done, on the other hand, is to concentrate on the larger scale, longer term processes that control the generation of sequences.

As previously explained, depositional sequences have a basic pattern of deposition which results from a relative rise and fall in sea-level. This pattern varies widely, depending on variations in tectonics and sediment supply. However, several levels of sea-level cyclicity may occur in a hierarchy that allows higher-frequency cycles to be superposed or stacked into lower-level cycles (see Figure 2). The first-order cycle, which is the longest one, relates to the breakup of continents and the corresponding creation of new shorelines. The second-order cycle is landward and basinward oscillations of the shoreline that lasts 3 to 50 million years. This oscillation is produced by changes in the rate of tectonic subsidence and uplift, caused by changes in rates of plate motion. The third-order cycle is the sequence cycle, lasting 0.5 to 3 million years. Fourth- and higher order cycles (parasequence cycles) may be correlated with periodic climate changes.

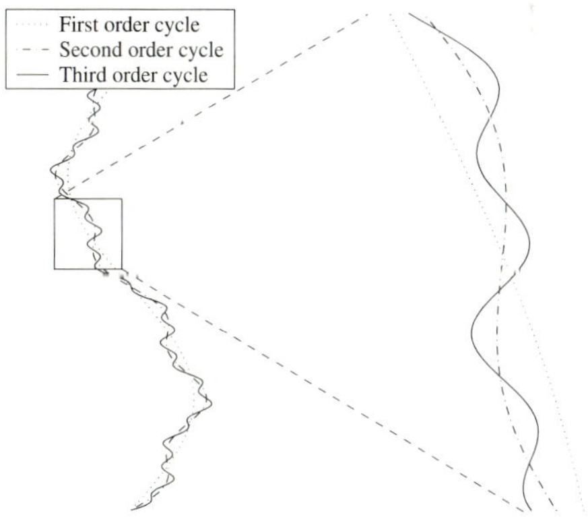

Fig. 2. Hierarchical ordering of depositional cycles.

Remembering how low sedimentation rates often produce depositional sequences below the seismic resolution, and how higher sedimentation rates do not, we see that the third-order cycle represents the lower limit of what we are able to see on seismic data alone. Fourth- and higher order cycles are only visible on well logs.

Soille [42] published an approach based on immersion simulations that the transform really became popular.

Definition. Two of the most popular definitions of watersheds are based on the concepts of rain falling and accumulating in different catchments basins [31], and flooding of the relief by immersing it into water [42]. Though, theoretical treatments of these approaches exists, due to issues of computational speed and special topology handling, the implementations most often used do not adhere to the original definitions. In this section we focus on the algorithmic definition by Vincent and Soille [42].

Considering the grayscale image as a surface, where holes have been created at the positions of regional minima, the algorithm simulates the immersion of it into a lake. The water floods the surface, creating a number of small lakes, or basins (see Figure 5 for an illustration of a 1D analogy). At some point, one or more basins meet and start flowing into each other. To prevent this from happening, dams are erected at these positions, thus, maintaining the separation of the basins. These dams are the sought watershed lines, or simply watersheds. A segmentation of the terrain (i.e., image) now follows directly from the regions covered by the catchment basins.

The flooding process is implemented by the use of sorted sets, geodesic distances and influence zones [27, 42]. Consider a digital grayscale image as a triple $G = (D, E, f)$, where (D, E) is a graph (usually the completely connected digital grid, but may in general be any kind of nodes with arbitrary

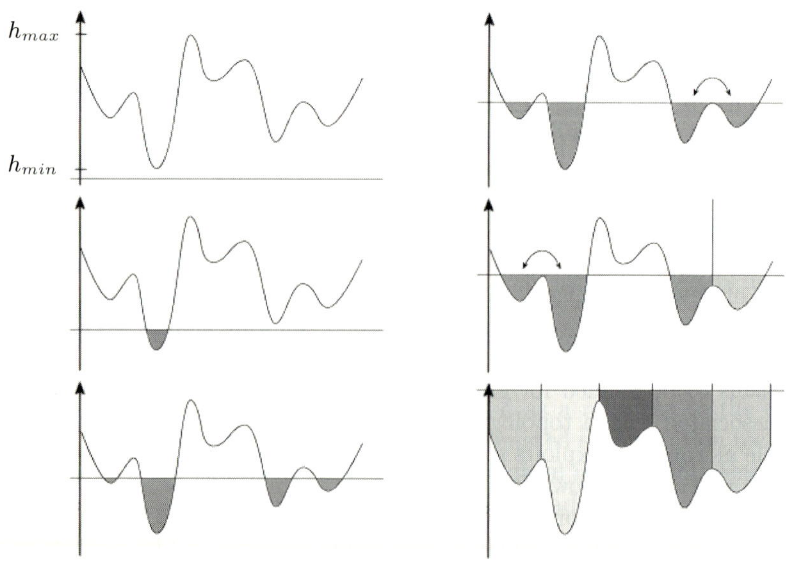

Fig. 5. Watershedding by immersion for a 1D signal.

connectivity), and $f : D \to \mathbb{N}$ is a function assigning integer values to each $p \in D$. The values $f(p)$ represent the gray values, or elevation (when viewing the image as a topological relief).

With $A \subseteq \mathbb{Z}^2$, and $a, b \in A$, the *geodesic distance* $d_A(a, b)$ is defined as the minimum length of any path joining a and b and is totally contained within A. Suppose A contains a set B, $B \subseteq A$, which consists of several connected components $B_1, B_2, \ldots, B_k$. The geodesic distance of any point $a \in A$ to the set B, is then defined as $d_A(a, B) = \min_{b \in B} d_A(a, b)$. Using this definition, we define the geodesic influence zone of the set B_i within A as

$$\mathrm{iz}_A(B_i) = \{p \in A : d_A(p, B_i) < d_A(p, B_j) \text{ for all } j \in [1..k] \setminus \{i\}\}.$$

The union of all the influence zones of the connected components in B within A is defined as,

$$\mathrm{IZ}_A(B) = \bigcup_{i=1}^{k} \mathrm{iz}_A(B_i),$$

and the complement of this set within A defines the SKIZ, i.e., *skeleton of influence zones*,

$$\mathrm{SKIZ}_A(B) = A \setminus \mathrm{IZ}_A(B).$$

The SKIZ represents the set of points within A which are geodesically equidistant from at least two of the nearest connected components of B.

It is based on these concepts that Vincent and Soille [42] algorithmically defines the watershed transform. Considering again the definition of a grayscale image by the valued graph G, having as its minimum and maximum values $h_{\min}$ and $h_{\max}$. A threshold set from this topology is defined as

$$T_h = \{p \in D : f(p) \leq h\}.$$

By the use of a recursion with the gray level rising from $h_{\min}$ to $h_{\max}$, basins associated with minima of f are iteratively expanded. Denote by X_h the connected components detected up to level h. The connected components associated with the threshold set T_{h+1} will either be expansions of already detected basins or new minima. In the former case, one calculates the geodesic influence zones of X_h within T_{h+1} and updates X_{h+1}. The set of new minima is denoted $\mathtt{MIN}_h$.

Definition 1. (Watershed by Immersion)
Starting at level $h_{\min}$, the recursion implementing the flooding becomes:

$$\begin{cases} X_{h_{\min}} = \{p \in D \mid f(p) = h_{\min}\} = T_{h_{\min}} \\ X_{h+1} = \mathtt{MIN}_{h+1} \cup IZ_{T_{h+1}}(X_h), \quad h \in [h_{\min}, h_{\max}) \end{cases}$$

The `watershed` *of f can now be found as the* SKIZ *of $X_{h_{\max}}$ in D,*

$$\mathrm{Wshed}(f) = D \setminus X_{h_{\max}}.$$

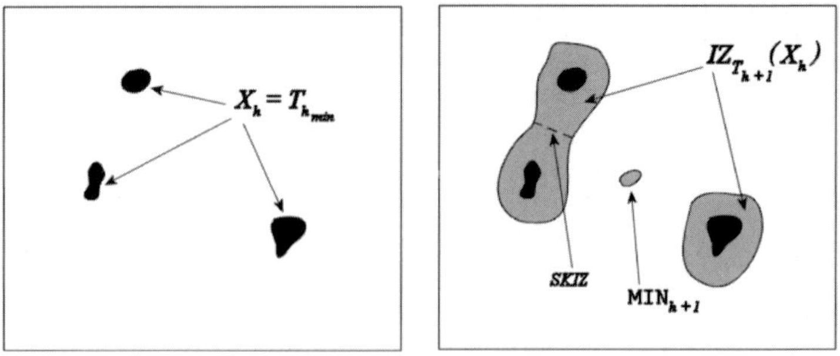

Fig. 6. Watershedding by immersion of the topographic relief corresponding to a grayscale image. This approach extends directly to 3D and higher dimensions.

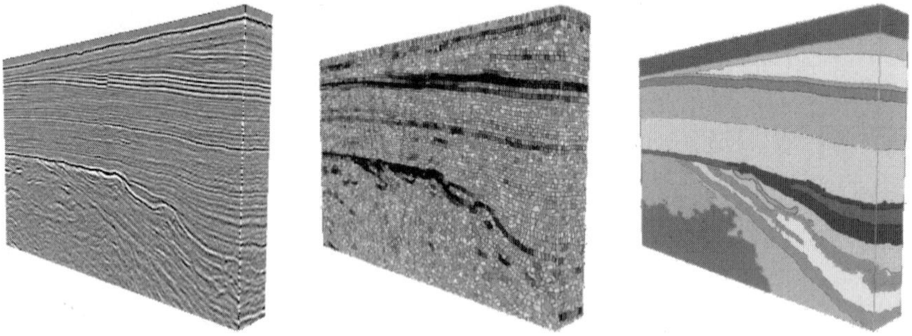

Fig. 7. Direct watershedding of a 3D seismic amplitude volume (left) produces a highly detailed segmentation (middle). This level of segmentation is for most purposes too detailed, and as is shown in Subsection 2.3, the level of detail may be reduced to give the segment size needed (right).

The algorithmic implementation of this definition by Vincent and Soille [42] is linear in its time complexity. However, the implementation has some drawbacks that are discussed and solved in [27] at the expense of a theoretically quadratic time complexity.

An important aspect of this algorithm, is that it *extends directly to 3D and higher dimensions.* The algorithm is based on sets, so all that is required for extending it is an appropriate neighborhood definition. Throughout this chapter, a full 3D version of the algorithm is used, producing 3D volumes.

2.3 The Curse of Oversegmentation

An inherent problem of the watershed segmentation technique is oversegmentation. Much of this oversegmentation is due to small, insignificant minima, which do not contain any relevant information. Removal of these minima prior to applying the watershed transform is one way of reducing the large number of segments.

When imaging the subsurface, noise, improper stacking and migration, are just some of the effects that can make the seismic images deviate from the *true* solution. The result may be distortion of geometry, blurring of the images and degradation of reflector amplitude and continuity. In the context of segmentation, these effects may have severe impact on the final result, particularly if any type of automation is involved. Hence, algorithms extracting information from seismic images will in general benefit from some type of data *preconditioning*. Considering the fact that the watershed transform operates by detecting and propagating regional minima, it is easily seen that any amplitude anomalies can result in increased segmentation. Also, as will be explained later, the order in which the basins flood each other is defined according to the differences in minimum amplitude along the contours of the basins and the amplitudes of the basin minima. This ordering constitutes a hierarchical, multi-resolution view on the segmentation process. Hence, the continuity of the reflectors, in addition to the amplitudes, will necessarily have a big influence on this order.

Data Preconditioning.

Layer-Parallel Smoothing: Various types of image smoothing have been applied in the literature to combat oversegmentation, with signal adaptive smoothing algorithms as the most successful [10, 25]. General lowpass type filtering, such as in scale-space theory, is not successful in preserving contour information, thus, resulting in non-consistent segmentations across scale.

To reduce oversegmentation in our experiments, the layer-parallel smoothing reported in [25] is adopted. The method operates by producing a gradient estimate, that it uses to steer the direction along which a Gaussian filter is applied. Options exist to reduce smoothing in areas of poor directional estimate, e.g., in regions of chaotic texture. A more detailed treatment of the method can be found in the chapter [26] of Randen and Sønneland.

The improvement in segmentation performance, as measured by the degree of oversegmentation, is significant. As an example, consider Figure 8, showing the original seismic data, its smoothed version, and their corresponding segmentations.

Morphological Filtering.

h-Minima Transform: A common technique for simplifying images prior to segmentation is the morphological h-minima operator. This is a so-called connected operator which simplifies the signal by removing small or poorly contrasted regions without destroying the contour information. To understand how this operator works, we will first briefly review some of the fundamental operations involved.

In mathematical morphology the two most basic operators are *erosion* and *dilation*. These operators act on a signal $f(x)$ using a flat structuring element B_n of size n (i.e., a window of size n), in the following way.

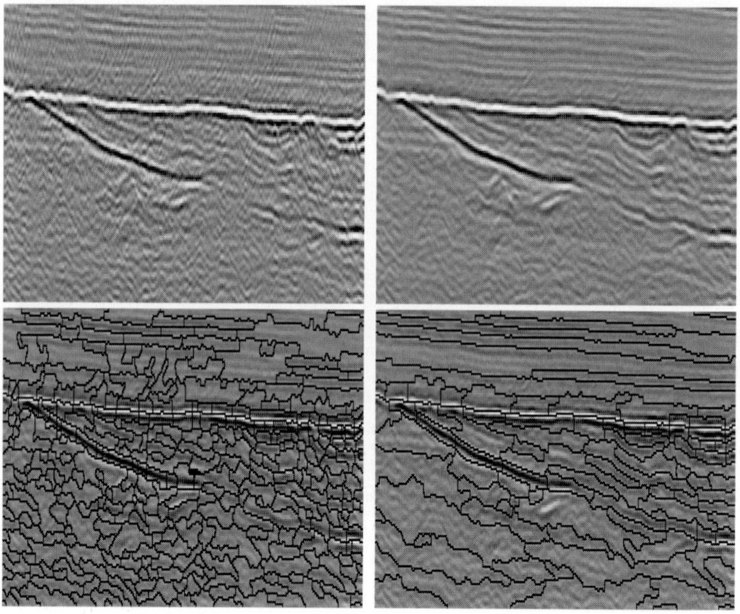

Fig. 8. Layer-parallel smoothing may be used to remove signal noise, and increase reflector continuity prior to feature extraction and segmentation. The original seismic data, along with its segmentation at an intermediate level in the hierarchy is shown on the left hand side. The results after layer-parallel smoothing are seen to the right.

$$\text{Erosion:} \quad \epsilon_n(f)(x) = \min\{f(x+y), y \in B_n\},$$

$$\text{Dilation:} \quad \delta_n(f)(x) = \max\{f(x+y), y \in B_n\}.$$

Yet another set of erosion and dilation operations exist, which are defined using geodesic transforms. These operators are defined using a reference function r, and a structuring element of size one.

$$\text{Geodesic dilation:} \quad \delta^1(f,r) = \min\{\delta_1(f), r\},$$

$$\text{Geodesic erosion:} \quad \epsilon^1(f,r) = -\delta^1(-f,-r),$$

$$= \max(\epsilon_1(f), r).$$

Geodesic erosions or dilations of greater sizes are achieved by iteration of the two basic ones. By iterating until infinity, or stability, a new set of operators emerges, called *reconstruction by dilation (erosion).*

Reconstruction by dilation:

$$\gamma^{\text{rec}}(f,r) = \delta^{\infty}(f,r) = \ldots \delta^{1}(\ldots \delta^{1}(f,r)\ldots,r).$$

Reconstruction by erosion:

$$\varphi^{\text{rec}}(f,r) = \epsilon^{\infty}(f,r) = \ldots \epsilon^{1}(\ldots \epsilon^{1}(f,r)\ldots,r).$$

The h-minima operator is defined using the reconstruction by erosion operator as follows:

$$h - \text{minima}(f) = \varphi^{\text{rec}}(f+h,f).$$

By duality, there also exists a h-maxima transform:

$$h - \text{maxima}(f) = \gamma^{\text{rec}}(f-h,f).$$

Too see the dramatic effect these operators have on the final segmentation result, Figure 9 shows a watershed based segmentation of a seismic volume after applying the h-minima transform with different values for h. We see how larger, more consistent regions emerge, whereas the smaller ones disappear. As a consequence of the contour preserving properties of the operator, we see that the contours of the larger regions are preserved at lower levels of segmentation (i.e., lower h-levels).

2.4 Multiscale Model Building

The suite of models deployed in hydrocarbon exploration through production, displays a progression in granularity from coarse to medium to very detailed, and spans the whole range from kilometers to centimeters. Though different in what they represent in terms of physical quantities, there is clearly a relationship between the boundaries in the models related to changes in these properties. For instance, the boundaries found in a structural model, often impose limitations on the boundaries, and thus distribution of properties, of the more detailed geological, reservoir and simulation models. This behavior agrees with the stratigraphic analysis framework presented in Subsection 1.2.

Among the many multiscale segmentation approaches found in the literature, only a few possess the properties necessary for consistent multiscale model building. Particularly important are the properties of causality of boundaries and their exact positioning across scale. The notion of causality simply implies that if a border exists at a coarse scale, it also exists at the more detailed scales. Imposing these two constraints on the segmentation process, and adopting the ordering dictated by sequence stratigraphy, leads us to consider hierarchical segmentation techniques.

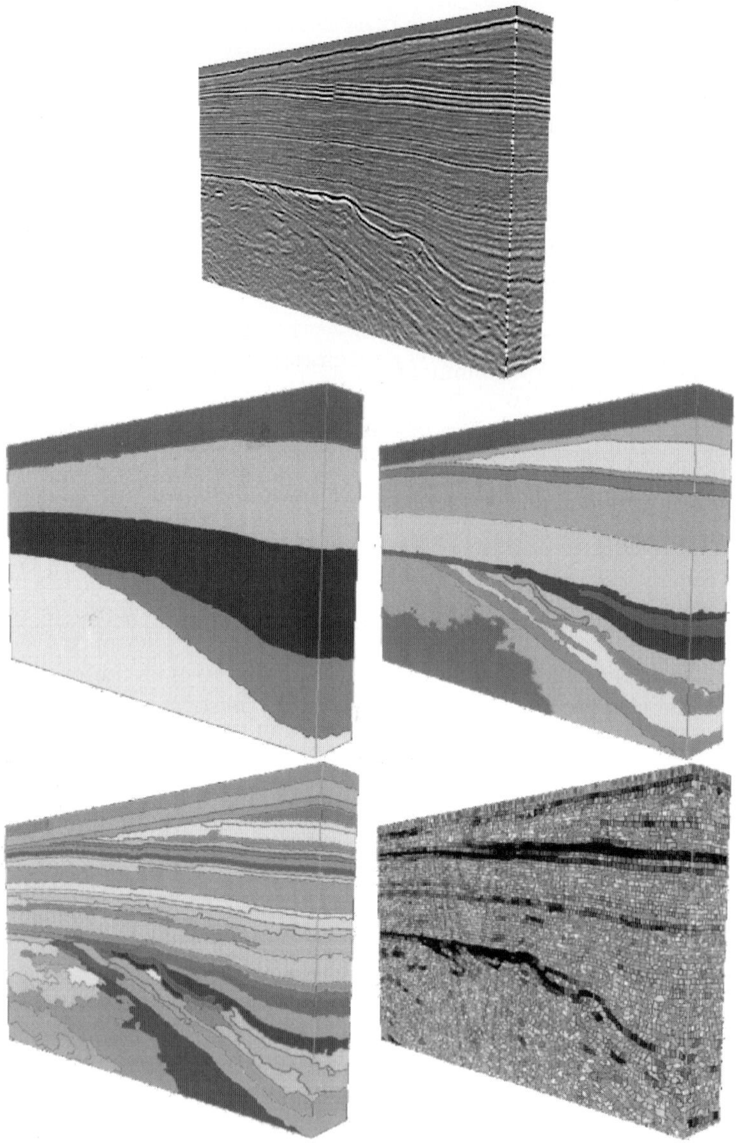

Fig. 9. The h-minima transform removes catchment basins in the images that have a depth smaller than h. Removing the finer details in this manner greatly reduces the oversegmentation resulting from application of the watershed transform directly on the *raw* seismic image.

A possible subset of such techniques include hierarchical clustering and classification, partition trees (herein split and merge techniques), mathematical morphology, scale-space, and more. In practice, however, only the partition tree and mathematical morphology approaches provide viable solutions. The reason being that, clustering and classification approaches tend to produce multiple, unconnected regions for each class or cluster, and the scale-space approaches are very computationally demanding. For our purpose, we will consider the combined use of mathematical morphology and partition trees.

Morphology Goes Multiscale. Geodesic reconstructions and h-minima transforms were shown to produce segmentations of varying granularity when applied to images prior to the watershed transform. Belonging to the set of connected operators, these transforms preserve the positions of contours across scales, so that a contour found at a coarse scale also is present at finer scales. As such, we see that these operators provide us with the tools needed to operate at multiple scales of segmentation. This concept has over the last decade been evolving within the field of mathematical morphology to become a versatile, and powerful way of performing data segmentation [16, 35]. In this section, a more general overview of the techniques available and their possible use is presented.

Consider a segmentation produced by the watershed transform, where each region is a catchment basin corresponding to a regional minima. Without altering the input image in any way, this is the most detailed segmentation the watershed transform is able to generate. Next, consider removing a regional minimum by applying the h-minima transform to the image. Due to the contour preserving properties of this transform, all, but one, of the watershed regions remain as they were in the original segmentation. The last region, corresponding to the removed regional minimum, has now merged with one of its neighboring regions (positioned at the point of lowest amplitude along the contour). Hence, the removal of local minima corresponds to merging regions from the initial, finest segmentation into larger regions. By iteration, even larger regions are produced, so that in the end, the whole image is one region. The results obtained using this methodology on seismic data are very good, bringing the model building process one step closer to automation.

Graph Theoretic Approach. The multiscale segmentation behavior just described may efficiently be posed as a graph problem. As such, we will in this section review some basic graph theory and region definitions.

Partitioning:

Definition 2. (Connected Component)
A `connected component` *is a set* $\mathbf{R}$ *where for all* $x, y \in \mathbf{R}$, *there is a connected path* $\mathbf{p}$ *in* $\mathbf{R}$ *between* x *and* y. *The connected path* $\mathbf{p}$ *is a set of points* $(p_1, p_2, \ldots, p_n)$ *where any two consecutive points,* p_k *and* p_{k+1}, $k \in [1, \ldots, n-1]$, *are connected.*

Arbitrary definitions may be used to decide whether nodes are connected. However, digital grids are by far the most commonly used embedding representations, thus, making 4- or 8-connectivity for 2D images, and 6- and 26-connectivity for 3D images, the most popular choices.

Definition 3. (Neighboring Regions)
Two connected regions $\mathbf{R}_1$ *and* $\mathbf{R}_2$ *are* `neighboring regions`, *if and only if there are elements* $x \in \mathbf{R}_1$ *and* $y \in \mathbf{R}_2$, *that are connected points.*

Definition 4. (Image Partition)
A `partition` *of the image space* $\mathbf{E}$ *is a mapping to a set of connected components* $\mathbf{R}_i$, $i \in [1, \ldots, n]$, *satisfying* $\mathbf{R}_i \cap \mathbf{R}_j = \emptyset$ *for all* $i \neq j$, *and* $\bigcup_{i=1}^{n} \mathbf{R}_i = \mathbf{E}$.

The connected regions produced by segmentation and their interrelationships are often represented as graphs. In doing so, each connected component $\mathbf{R}_i$ is represented as a node $\mathbf{n}_i$, and edges between the nodes are derived from the neighborhood structure dictated by Definition 3. The resulting graph structure is referred to as a *Region Adjacency Graph (RAG).*

Segmentation: The extreme cases of image partitioning are (1) when each connected region represents a single pixel, and (2) when only one connected region is produced, representing the whole image. These two extremes also represent the outer limits for multiscale segmentation, where scale refers to the number of regions produced by the segmentation.

Generally speaking, multiple scale segmentations are essentially a collection of single scale segmentations. In theory, each of these single scale segmentations may be produced independently of the others. However, this produces segments that have no apparent relationship across scale. This does not fit with the requirements set out by analysis of the hierarchical nature of the geological models. As such, among the many possible single scale segmentations, only those displaying a nesting of the regions are allowed. That is, stepping down[3] from scale k to scale $k+1$, all but one region is kept exactly the same (see Figure 10). The only change is that this region is split into two new connected regions according to Definition 4. This defines a *nested partitioning* scheme that agrees with the hierarchical ordering dictated by seismic sequence stratigraphy (see Subsection 1.2).

One approach for generating nested partitions have already been introduced, i.e., the h-minima transform. The set of segmentations produced when applying this transform for increasing levels prior to the watershed transform, induces a set of nested partitions. In each step, the transform simplifies the image, producing larger, and fewer, connected regions[4]. Repeated application of the transform, produces a nested, hierarchical ordering of the regions

[3] In the hierarchy notation adopted here, the coarsest segmentation (the whole image), is represented as the top node (scale 1), whereas the regions of the finest segmentation constitute the leaf nodes.

[4] A general property of connected operators.

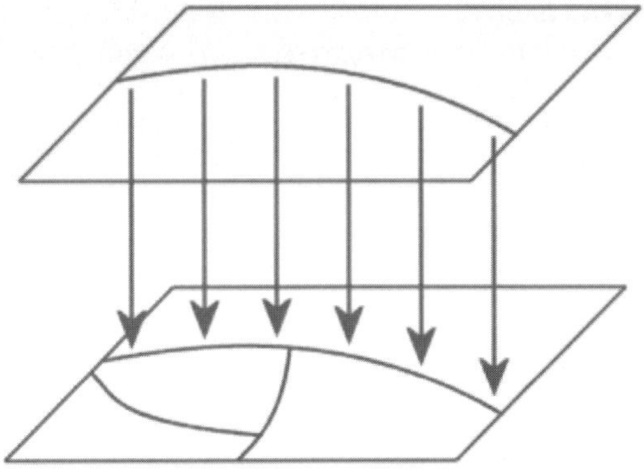

Fig. 10. Causality and persistence of contours across scale follows directly from nesting of partitions.

in a bottom-up fashion. We see that, starting with the most detailed segmentation, and its corresponding region adjacency graph, all subsequent segmentations may be generated purely by operations on this underlying graph structure.

Trees and Hierarchies: In the case of increasing removal of detail by the h-minima transform, the incremental change from one scale to another is the merging of two regions. On the underlying RAG, this corresponds to substituting the two nodes corresponding to the two regions and their common edge by a new node, thus reducing both the number of nodes and edges by one. Optionally, we may generate a hierarchy by introducing a new node having the two nodes to be merged as its children (see Figure 11).

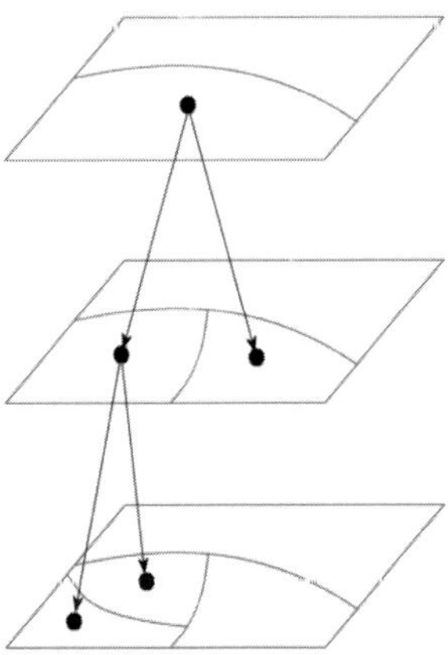

Fig. 11. Hierarchy representing a binary merging of nested regions.

Definition 5. (Hierarchy)
A partition A is said to be a `hierarchy`*, iff it satisfies the following two properties.*

- **Intersection Axiom:** *Two elements of A which are not comparable by the inclusion relation have an empty intersection, i.e.,*

$$a \cap b \in \{a, b, \emptyset\}, \qquad \textit{for all } a, b \in A.$$

- **Union Axiom:** *Any element $a \in A$ is the union of all other elements in A contained in a, i.e.,*

$$\bigcup \{b : b \in A; b \neq a; b \subset a\} = \{a, \emptyset\}, \qquad \textit{for all } a \in A.$$

A particularly interesting construction, which we will make extensive use of, is the *stratified hierarchy* [16].

Definition 6. (Stratified Hierarchy)
A hierarchy is said to be `stratified`*, iff it is associated with an index function $f : A \to \mathbb{R}$ which is strictly increasing with the inclusion order, i.e.,*

$$a \subset b \textit{ and } b \neq a \quad \Rightarrow \quad f(a) < f(b), \qquad \textit{for all } a, b \in A.$$

The h-minima transform, for instance, can in some instances[5] produce a stratified hierarchy where the level of detail removed, h, is used as the index function. Since all connected components associated with a level h_1 is included in the connected components associated with a level $h_2 > h_1$, the index function strictly increases with the inclusion order, and produces a stratified hierarchy.

The reason stratified hierarchies are so interesting is the flexibility they offer in hierarchy generation and segmentation. In particular, the combined use of stratified hierarchies and *ultrametric distances*, enables efficient generation of hierarchies based on any properties of interest.

Definition 7. (Ultrametric Distance)
For a stratified hierarchy A, with stratification index value 0 for its finest partition, the following represents an ultrametric distance index.

$$d(i, j) = \inf\{f(a) : a \in A \textit{ and } i, j \in a\}, \qquad \textit{for } i, j \in \wp(E),$$

where $\wp(E)$ denotes the set of all subsets of partitions of the image E. An `ultrametric distance` *satisfies the following three properties.*

(a) $d(i, j) = 0$ *implies* $i = j$, *for* $i, j \in A$;
(b) $d(i, j) = d(j, i)$ *for all* $i, j \in \wp(E)$;
(c) $d(i, j) \leq \max\{d(i, k), d(k, j)\}$ *for all* $i, j, k \in \wp(E)$.

[5] By allowing the nodes to have multiple children.

The ultrametric distance (DU) is basically a dissimilarity measure between the tiles/nodes constituting the hierarchy. The more dissimilar two tiles are, the later they are merged and the higher their ultrametric distance is. Hence, if we have an arbitrary dissimilarity measure between the tiles, and merge tiles below a dissimilarity index (DI), taking as the new DU the maximum among all DU's below DI, we are able to construct a stratified hierarchy on which the ultrametric distance is defined. The resulting hierarchy (see Figure 12), in this case, is equivalent to a *Minimum Spanning Tree (MST)* of the region adjacency graph.

2.5 Attributes and Region Merging

Nice segmentation results have been obtained on nicely behaving seismic data. However, it is seldom the case that data are well behaved, and furthermore, that the ordering of the objects in the image corresponds to the semantic objects that we want to extract. A simple example is shown in Figure 13, where we want to segment a signal into two regions having different peak spacings (or frequencies). However, due to the uniform amplitudes, a segmentation of this signal will be determined by the order in which the regions are visited. Thus, the result will inevitably be different from what we seek.

Generally, segmenting textured images is a challenging task. Starting with the initial watershed segmentation as leaf nodes, and then constructing the hierarchy based on the h-minima transform does no longer make sense. The detail removed by this transform does not necessarily correspond to discriminating features of the textures. Hence, there is a need for also using attributes (e.g. texture attributes) in the hierarchy building process.

This is where the stratified hierarchy, with its ultrametric distance comes into play. If we are able to define a dissimilarity measure based on the attributes of interest, that respects the features of interest, we will be able to generate a hierarchy. Hence, we now have the means to generate tailor made hierarchies, and thus segmentations, respecting the important properties of the data.

As a general rule of thumb, *(1) watershedding of the seismic data is used to generate the initial segmentation, or building blocks, whereafter (2) a dissimilarity measure between the aggregated attributes within the regions is used to build the hierarchy.*

Seismic Texture Attributes. Subsection 1.2 introduced seismic sequence stratigraphy as a means of explaining the structure of the subsurface, helping determine the depositional environments and possible rock type distributions. Essentially, this process boils down to an analysis of seismic bodies defined by their internal textures and external shape, often referred to as seismic facies analysis. This type of analysis is a must in seismic interpretation to locate potential reservoirs, especially in complex oilfields.

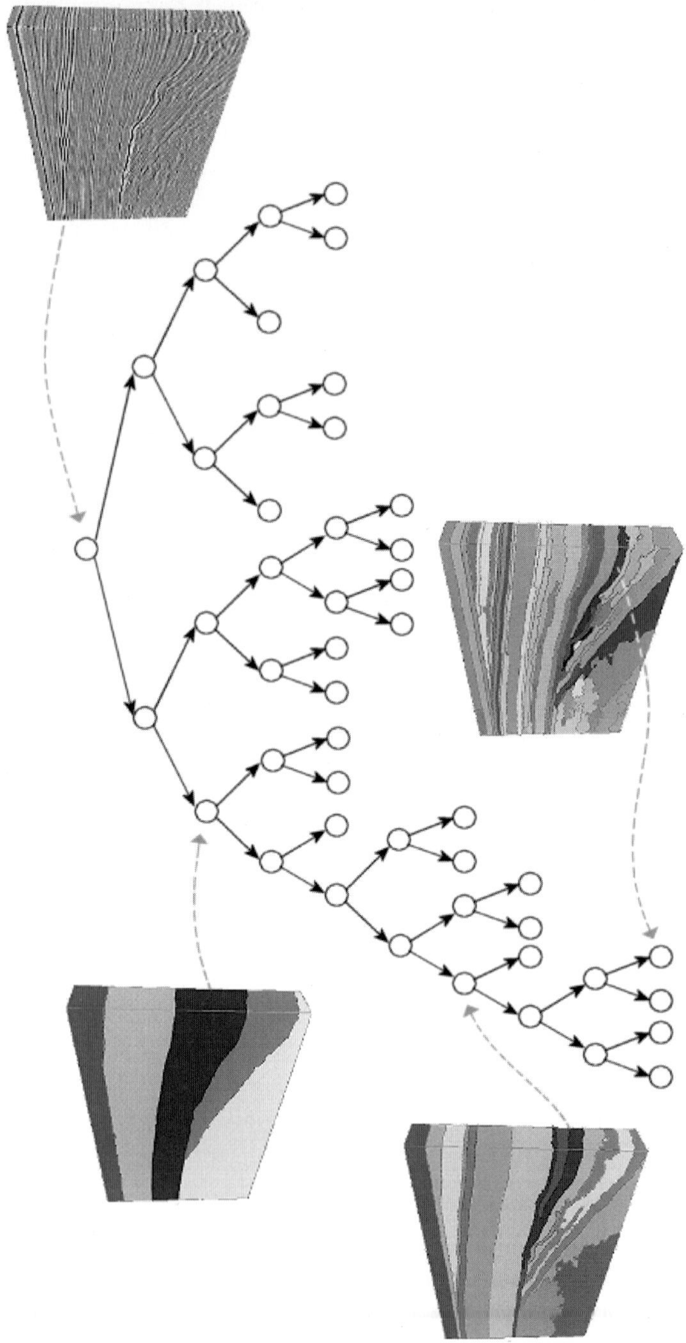

Fig. 12. Hierarchical representation of a multiscale segmentation.

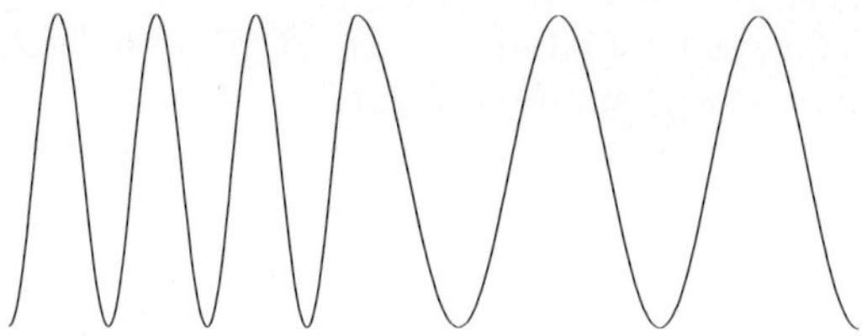

Fig. 13. Textured regions provide a challenge when using the h-minima transform to build the segmentation hierarchy. Seen above is a 1D texture, composed of two time-frequency disjoint [22] signals with identical amplitudes. In this case the h-minima transform has no discriminating effect, thus, rendering the order of grouping (defining the hierarchy) a function of the order in which the regions are considered. To remedy this, signal and image attributes must also be included in the hierarchy building process. In the above example, measuring the dissimilarity of peak spacings would suffice.

The choice of seismic texture attributes for hierarchy building heavily depends on the data, on what kind of model is being built and on the characteristics of the objects comprising the model. If the objects to be modelled are outlined by strong reflectors, the traditional h-minima transform may actually suffice. In other instances, where the reflectors outlining the objects are vague or non-existing, attributes are needed to guide the segmentation process.

For further information, extensive overviews of the interplay between seismic facies analysis and seismic texture attributes are given in the chapter [8] of Carrillat and Vallès.

Attribute Modeling and Dissimilarity Measuring. The initial segmentation generated by the watershed transform consists of a set of connected regions, with a corresponding region adjacency graph. Each region has a finite extent, within which attributes may be calculated. As such, each region contains a set of attributes with some overall behavior. It is not tractable to keep track of all the individual attribute responses, so some means of overall attribute behavior is needed.

There are many ways in which we can represent the overall attribute behavior of the regions. Most popular, however, are the two lower order statistical moments, mean and variance. Calculating these measures within the regions is very simple and fast, and when merging regions, it is straightforward to calculate the mean and variance of the new region based those of the regions merged.

Using the two lower order statistical moments also allows for simple dissimilarity calculations, such as the Fisher metric. For any two regions $\mathbf{R}_i$ and $\mathbf{R}_j$, this measure has the form

$$J_F = \frac{\left(\mu_{\mathbf{R}_i} - \mu_{\mathbf{R}_j}\right)^2}{\sigma^2_{\mathbf{R}_i} + \sigma^2_{\mathbf{R}_j}}.$$

Evaluating this expression for all edges in the RAG, a valuated graph emerges on which we build the segmentation hierarchy.

By keeping track of more information, other dissimilarity (distance) measures may also be applied. Examples are the Bhattacharya and Mahalanobis metrics.

Intelligent Merging. In order to obtain objects that exhibit geologically meaningful shapes, a set of constraints must be put on the hierarchy building process. For instance, a natural assumption when building a model, is that the objects, at a particular scale, should be roughly the same size. This implies that it is not wise to merge a very big region with a very small one. An easy, and effective way to implement such a behavior, is to increase the weights of those edges that exist between regions of radically different sizes. The same argument holds for region shape, where an example of a very simple weight function is the absolute difference of region compactness[6].

The edges of a RAG can in general take on values representing any type of information. As such, the edges may also be valued according to attribute values existing at the border corresponding to the edge. In seismic model building, this is of interest for extracting continuous layers. By imposing the constraint that flow lines (normal to the amplitude gradient field) at the boundary between merged regions should conform to the flow field within the two regions, the hierarchy will favor merging of volume primitives along reflectors, rather than across.

The above mentioned examples goes to show how constraints may be put on the hierarchy building process in order to tailor make the hierarchies for specific purposes. In the end, one may want to generate a suite of hierarchies, each modeling different aspects of the data set, and combine these for generating the final model.

Furthermore, there also exist other ways by which to build hierarchies, where we also take into account properties of the RAG nodes themselves [15]. As such, we can combine both edge and node-based techniques to increase our flexibility in automating the process.

3 Application: Velocity Model Building

Background. To appreciate the power of the model building approach presented in this chapter, we here present an application of the new approach for geometric mapping of salt domes for velocity model building. To understand why this is a difficult, but import task, we need to understand what a salt dome is, its properties, how to image it and the problems associated with that.

[6]Compactness measures how sphere-like an object is by relating the area of the object surface to its internal volume. Normalized by 4π, a perfect sphere would give a value of 1, whereas all other shapes give values less than this.

What is a salt dome? A salt dome [2, 32] is a teardrop-shaped diapir of salt that migrates up through the subsurface. Below a certain depth, the density of salt becomes much less than its surrounding sediments, causing a buoyancy difference. As the surrounding sediments subside, the salt will therefore maintain its position in depth causing the salt to push through the overlying sediments. After some time of upward migration, drops of salt tend to separate from the main salt and propagate further by themselves (see Figure 14). Hydrocarbons are commonly found around salt domes because of the abundance and variety of traps created by the salt movement. That is, the salt dome bows up sedimentary beds, produces faulting, and affects the nature of the beds being deposited. Consequently, traps may be produced over or around the sides of the dome. Also, due to the impermeability of salt, hydrocarbons may be trapped underneath the salt.

Two examples displaying typical salt dome behavior are depicted in Figure 15. Here we see bodies having chaotic internal structure (explicitly measured in Figure 16), and the effect of upward flow causing sedimentary beds to bend up, fault and terminate against the sides of the salts. On the salt dome to the right in the figure we also see a clearly defined cap rock overlying the internal chaotic regions of the salt. The figure also serves as a good example of good and poorly imaged salt domes. The domes are commonly associated with complex geological structures and velocity fields, thus, making imaging them a difficult task. Salt has a medium velocity of 4500-5000 m/s, which is significantly different from typical sediment background velocities of 2000-4000 m/s. This contrast in velocity causes strong reflections and displaces and distorts the events we try to image significantly. In addition to being a velocity anomaly, salt domes typically have highly irregular boundaries. Hence, the path of any acoustic energy entering such a region is highly unpredictable.

Consequently, high demands must be put on the inversion process to avoid the images appearing unfocussed (as in the case of the left salt dome in Figure 15).

To properly image salt domes advanced imaging techniques must be utilized. However, the operations and processing time involved in these types of techniques have previously rendered their use impractical. As a consequence of better acquisition equipment and faster computers, salt imaging have be-

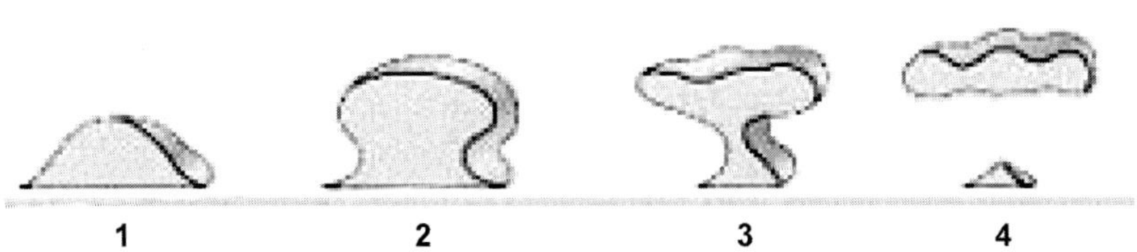

Fig. 14. Illustration of the evolution of a salt diapir.

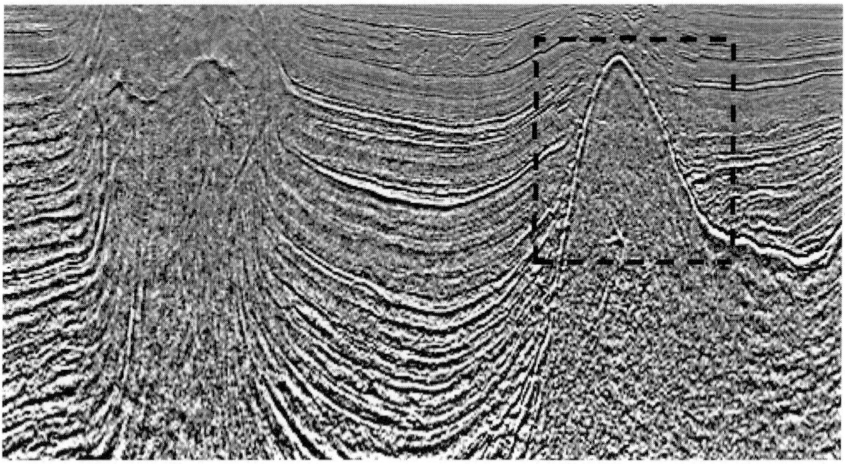

Fig. 15. Seismic section containing two salt domes. The framed region is shown in greater detail in Figure 17.

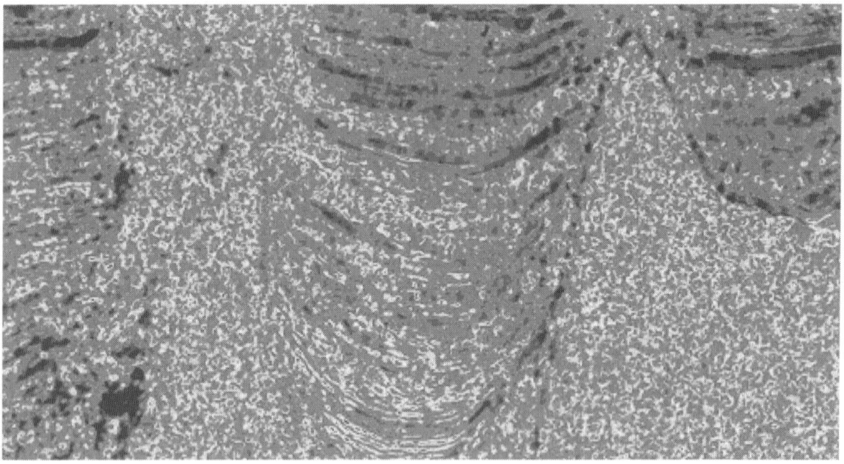

Fig. 16. By measuring the amount of *chaosness* present in Figure 15 we see how the highly chaotic areas corresponds to the internal regions of the salt domes.

come more tractable over the last decade. In fact, more and more discoveries are made in or around salt domes. These structures are often encountered in offshore West Africa, the Gulf of Mexico, the Red Sea, and the North Sea.

Seismic imaging consists of two main components: stacking and migration. Stacking is the process of enhancing the signal-to-noise ratio by summing signals from several angles reflecting at the same point. The process implicitly assumes that the signal energy comes from reflections at the midpoint between source and receiver pairs. This assumption is only fulfilled with an earth having flat layers and stepwise uniform velocity distributions (i.e., layer-cake model). As this is usually not the case, migration is needed to redistribute the assumed midpoint reflection energy to its correct position [3]. By using a velocity model, the migration process re-focuses the image by taking into account the *true* propagation paths of the acoustic energy as defined by the velocities.

Depending on the complexity of the structural framework and velocity distribution, migration may take on a number of forms. Post-stack time, pre-stack time, post-stack depth and pre-stack depth migration are the main choices, with the latter being the most accurate and sophisticated.

Pre-stack depth migration is used for imaging in areas of high risk, such as thin reservoirs, structurally complex geological settings and areas having large lateral velocity gradients.

In order to correct for the velocity anomaly that the salt represents, we need a correct geometric model of the salt body. However, due to the complex nature of the body and its boundaries, it is not straightforward to use the traditional interpretation tools. The question is now, can we use the approach presented in this chapter to automate or improve on the interpretation of salt structures?

Salt Body Extraction. Applying the watershed transform directly to the data set, and building a hierarchy based on the basin depths (equivalent to the h-minima transform) does in this case not make sense. There is in general no continuous cap rock that will give us the correct segmentation according to this approach. However, if we consider Figure 16, we see that there is a strong evidence of the salt in the measure of chaotic structure. Hence, an approach might be to use the chaos measure as an attribute, and build the hierarchy based on this. In testing this approach, the algorithms are run on the sub-volumes of Figure 15 and 16, shown in Figure 17.

The segmentation hierarchy is built as follows:

- The basic volume primitives are produced by applying the watershed transform directly to the seismic data (Figure 18, top), producing a very detailed segmentation.
- Using the chaos attribute, each volume primitive calculates the two first statistical moments, mean and variance, of the attribute values falling inside the sub-volumes they cover.
- The edges of the region adjacency graph are valued with the Fisher metric as the dissimilarity measure.
- A stratified hierarchy is built using the edge weights to determine which regions to merge. Every time two sub-volumes are merged, the new volume attributes are calculated, and the associated edge weights updated.
- Using the stratification index, the multiscale segmentations can now be browsed, as in Figure 18.

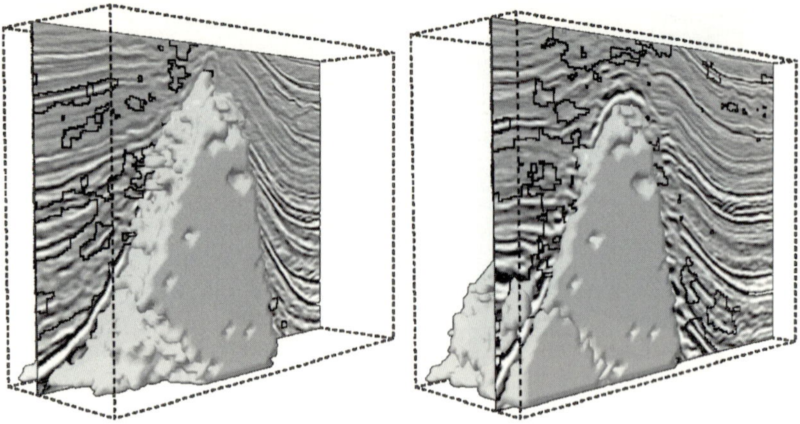

Fig. 19. Instead of working with the objects projected onto image slices (as in Figure 18), the user may also work with the corresponding segments in 3D. The hierarchy is used real-time, and is highly interactive. In the case seen here, only two segments need to be extracted to build a complete model of the salt.

spending time creating them in the first place. Consequently, more time is spent being creative, gaining a better understanding of the reservoir.

This is in stark contrast to the current model building workflow, where the user spends a lot of time doing tedious tasks that does not really add value. Moreover, as more information is gained over the lifecycle of the reservoir, the models are not necessarily updated due to the amount of model reworking that needs to be done. Using the framework presented here, new information is visible across all scales and is directly ready for utilization in all models once it becomes available. As such, fast model updates are now possible.

Furthermore, the hierarchical framework enables use of new semi- or fully automatic interpretation schemes and provides real-time user interaction with large volumes.

Acknowledgement

The work was partly supported by the European Union through the research and training network NetAGES, contract no. IST-1999-29034. We also wish to thank Dave Hunt, Kai Hogstad, and Geir Elvebakk and the Barents Sea Seismic Area C partners Norsk Hydro, Statoil, AGIP, and Fortum for collaboration and permission to publish data.

References

1. The MPEG Home Page. `http://www.chiariglione.org/mpeg/index.htm`
2. Schlumberger Oilfield Glossary. `http://www.glossary.oilfield.slb.com/`
3. U. Albertin, J. Kapoor, R. Randall, M. Smith, G. Brown, C. Soufleris, P. Whitfield, R. Dewey, J. Farnsworth, G. Grubitz, and M. Kemme (2002) The time for depth imaging. Oilfield Review.
4. M. Bahorich and S. Farmer (1995) 3-D seismic discontinuity for faults and stratigraphic features: The coherence cube. The Leading Edge **14**(10), 1053–1058.
5. S. Beucher (1999) Mathematical morphology and geology: when image analysis uses the vocabulary of earth science: a review of some applications. Proceedings of GeoVision, 13–16.
6. S. Beucher and C. Lantuéjoul (1979) Use of watersheds in contour detection. Proc. Int. Workshop on Image Processing, Real-Time Edge and Motion Detection/Estimation.
7. A. Carrillat, T. Randen, L. Sønneland, and G. Elvebakk (2002) Automated mapping of carbonate mounds using 3D seismic texture attributes. Proceedings of the Society of Exploration Geophysicists, Annual Meeting.
8. A. Carrillat and B. Vallès (2004) From 3d seismic facies to reservoir simulation: an example from the Grane field. This volume.
9. H. Digabel and C. Lantuéjoul (1977) Iterative algorithms. *Proc. 2nd European Symp. on Quantitative Anal. Microstructures in Material Science, Biology and Medicine*, J.L. Chermant (ed.), 85–99.
10. G.C. Fehmers and C.F.W. Höcker (2003) Fast structural interpretation with structure-oriented filtering. Geophysics **68**(4), 1286–1293.
11. D. Gao (2003) Volume texture extraction for 3D seismic visualization and interpretation. Geophysics **68**(4), 1294–1302.
12. J. Grötsch and C. Mercadier (1999) Integrated 3-D reservoir modeling based on 3-D seismic: the tertiary Malampaya and Camago buildups, offshore Palawan, Philippines. AAPG Bulletin **83**(11), 1703–1728.
13. H.K. Hahn and H.-O. Peitgen (2003) IWT - Interactive Watershed Transform: a hierarchical method for efficient interactive and automated segmentation of multidimensional grayscale images. Proceedings of Medical Imaging, SPIE.
14. B.S. Manjunath, P. Salembier, and T. Sikora (2002) *Introduction to MPEG-7: Multimedia Content Description Interface*. John Wiley & Sons, Ltd.
15. F. Meyer (2001) An overview of morphological segmentation. Int. Journal of Pattern Recognition and Artificial Intelligence **15**(7), 1089–1117.
16. F. Meyer (2003) Morphological segmentation: recent advances - part 2. Tutorial at the International Conference on Image Processing.
17. R.M. Mitchum, P.R. Vail, and J.B. Sangree (1977) Seismic stratigraphy: applications to hydrocarbon exploration. Seismic stratigraphy and global changes of sea level, AAPG Memoir., part 6-7.
18. E. Monsen and J.E. Ødegård (2002) Segmentation of seismic data with complex stratigraphy using watershedding. Proceedings of the 10th Digital Signal Processing Workshop, 13-16 October, 2002, 68–71.
19. E. Monsen, J.E. Ødegård, J. Romberg, H. Choi, and R.G. Baraniuk (2001) Seismic texture classification by hidden Markov tree modeling of the complex wavelet transform. Proceedings of Norwegian Signal Processing Society.

20. E. Monsen and T. Randen (2001) Local 3d texture anisotropy analysis of seismic reflection data. Proceedings of 12th Scandinavian Conference on Image Analysis.
21. J. Neal, D. Risch, and P. Vail (1993) Sequence stratigraphy - A global theory for local success. Oilfield Review, 51–62, January 1993.
22. P. Steeghs, R. Baraniuk and J.E. Odegard (2003) Time-frequency analysis of seismic reflection data. *Applications in Time-Frequency Signal Processing*, A. Papandreou-Suppappola (ed.), CRC Press.
23. H.W. Posamentier and D.P. James (1993) An overview of sequence-stratigraphic concepts: uses and abuses. *Sequence stratigraphy and facies associations.* Special Publication of the International Association of Sedimentologists, no. 18. Blackwell Scientific Publications, 3–18.
24. T. Randen, E. Monsen, C. Signer, A. Abrahamsen, J.O. Hansen, T. Sæter, and J. Schlaf (2000) Three-dimensional texture attributes for seismic data analysis. Proceedings of the Society of Exploration Geophysicists Annual Meeting.
25. T. Randen, L. Sønneland, A. Carrillat, T.S. Valen, T. Skov, S.I. Pedersen, B. Rafaelsen, and G. Elvebakk (2003) Preconditioning for optimal 3D stratigraphical and structural inversion. Proc. 65th EAGE Conference & Exhibition.
26. T. Randen and L. Sønneland (2004) Atlas of 3d seismic attributes. This volume.
27. T.M. Roerdink and A. Meijster (2001) The watershed transform: definitions, algorithms and parallelization strategies. Fundamenta Informaticae **41**, 187–228.
28. P. Salembier and L. Garrido (2000) Binary partition tree as an efficient representation for image processing, segmentation, and information retrieval. IEEE Trans. Image Proc **9**(4), 561–576.
29. P. Salembier, J. Llach, and L. Garrido (2002) Visual segment tree creation for MPEG-7 description schemes. Pattern Recognition **35**(3), 563–579.
30. R.M. Schoch (1989) *Stratigraphy: Principles and Methods.* Van Nostrand Reinhold, New York.
31. J. Serra (1982) *Image Analysis and Mathematical Morphology.* Academic Press, New York.
32. R.E. Sheriff and L.P. Geldart (1999) *Exploration Seismology.* Cambridge University Press, 2nd edition.
33. J. Shi and J. Malik (2000) Normalized cuts and image segmentation. IEEE Trans. Pattern Anal. and Machine Intell. **22**(8), 888–905.
34. J. Sijbers, P. Scheunders, M. Verhoye, A. Van der Linden, D. Van Dyck, and E. Raman (1997) Watershed-based segmentation of 3D MR data for volume quantization. Magnetic Resonance Imaging **15**(4).
35. P. Soille (2003) *Morphological Image Analysis - Principles and Applications.* Springer-Verlag, 2nd edition.
36. P. Steeghs and G. Drijkoningen (2001) Seismic sequence analysis and attribute extraction using quadratic time-frequency representations. Geophysics **66**(6), 1947–1959.
37. T.P.H. Steeghs (1997) *Local power spectra and seismic interpretation.* PhD thesis, Tech. Univ. Delft, The Netherlands.
38. M.T. Taner, F. Koehler, and R.E. Sheriff (1979) Complex seismic trace analysis. Geophysics **44**, 1041–1063.

39. M.T. Taner, J.S. Schuelke, R. O'Doherty, and E. Baysal (1994) Seismic attributes revisited. Proceedings of the 64th Annual International Meeting, vol. 94, Society of Exploration Geophysisists, 1104–1106.
40. J. Tveiten (1992) *Automatisk horisontfølging i tredimensjonale seismiske volumer.* Master's thesis, Stavanger University College.
41. P.R. Vail, J. Hardenbol, and R.G. Todd (1984) Jurassic unconformities, chronostratigraphy, and sea-level changes from seismic stratigraphy and biostratigraphy. Interregional Unconformities and Hydrocarbon Accumulation, AAPG Memoir., no. 36, 129–144.
42. L. Vincent and P. Soille (1991) Watersheds in digital spaces: An efficient algorithm based on immersion simulations. IEEE Trans. Pattern Analysis and Machine Intelligence **13**(6), 583–598.
43. B.P. West, S.R. May, J.E. Eastwood, and C. Rossen (2002) Interactive seismic facies classification using textural attributes and neural networks. The Leading Edge **10**, 1042–1049.
44. F. Zanoguera, B. Marcotegui, and F. Meyer (2000) A segmentation pyramid for the interactive segmentation of 3-D images and video sequences. Proceedings of the International Symposium of Mathematical Morphology, Palo Alto, California, USA, Kluwer Academic Publishers, 263–272.

Mapping 3D Geo-Bodies Based on Level Set and Marching Methods

Stine Kjersti Richardsen[1] and Trygve Randen[2]

[1] Stavanger University College, TekNat, N-4068 Stavanger, Norway
[2] Schlumberger Stavanger Research, P.O. Box 8013, N-4068 Stavanger, Norway

Summary. In this chapter, a simplified method for mapping objects based on the level set method is introduced. Level set and marching methods are used to map connected volumes within 3D seismic data. The simpler marching method solves the stationary problem stated by the level set formulation. The evolution of the object, from a seed point to the boundary, is described by a differential equation.

1 Introduction

This chapter presents a procedure for mapping connected sets of voxels in seismic data. These connected sets make up geo-bodies, holding the same geological characteristics, e.g. impedance, and intrinsic velocity. The request for algorithms extracting areas and volumes of interest from a larger sized data set has been the main motivation for this work.

Mapping of objects is a topic within the more general theory of segmentation. Segmentation includes other topics known from pattern recognition such as classification, clustering, extraction, and more. In addition to pattern recognition, the procedure is based on work within applied mathematics and physics. Here we apply the mapping to both 2D and 3D seismic data. In Figure 1 we see a 2D example of a segmented seismic survey. The data might often be of a more disrupt and chaotic nature than seen in the example, and it is not always easy to detect connectivity by visual inspection. An algorithm would generate decisions of a higher degree of objectivity.

Some earlier segmentation methods are characterized as *explicit*. This means the front representing the boundary of the preliminary object is moved explicitly, point by point. A popular explicit segmentation method is the active contours, or *snake method*. As both these names indicate, the front (the contour) is active in *crawling* to find the true boundary. The points indicating the front are updated by a minimization of the energy formulation. More on theory and applications can be found in [2] and [8].

Experience tells us that the explicitness often leads to numerical instability. This can be improved by *implicitly* moving the front, as will be explained in the level set formulation in Subsection 2.1. The underlying theory presented is based on the implicit level set and marching methods, further introduced in Section 2. The work on level sets was initiated by Sethian and Osher in

Second, the literature [1, 6] points out that for the method's performance, the level set function must be kept a distance function at all time. The distance function gives the value indicating how far a node is from the temporary object's front. When applying the narrow band approach, errors are introduced. Hence we need to re-initialize the level set distance function from time to time. The narrower the band in the narrow band approach is chosen, the more often the updating is performed, but the more frequent the re-initialization is required. Some algorithms use a marching method as a re-initialization step. We want to use the marching method for initialization purposes. The basic idea is to produce a preliminary solution by the marching method, which is discussed next. With *well-behaving* data, the marching method can process all the way to the wanted boundary. When the data is more *defiant*, the situation changes and the solution will be further elaborated by the level set method to produce a satisfactory result.

2.2 The Marching Method

The marching method, Figure 6, is designed to find the solution to the mapping problem through solving a differential equation in spatial dimensions. The basic principles are similar to the case of the level set method, and therefore we also work with a function containing information about the front, denoted as $\gamma(t)$.

Fig. 6. Flowchart of the mapping algorithm, from raw seismic data to segmented geo-body output. The enhanced box incorporates a solver for a spatial differential equation.

The Stationary Case. The marching method is used to solve the stationary case of the level set formulation. When moving from the original level set formulation to the stationary formulation, we are able to speed up the computations. The price to pay is that the velocity function F we work with has to be strictly monotone when solving the simplified scenario.

This stationary solution is obtained through the differential equation in (5), called *eikonal equation* [18]. The eikonal equation contains the spatial differential of a function $T(\mathbf{x})$, given the velocity function F,

$$F|\nabla T(\mathbf{x})| = 1 \ , \tag{5}$$

where the gradient norm in n dimensions is defined as

$$|\nabla T(\mathbf{x})| = \sqrt{T_{x_1}^2 + \ldots + T_{x_n}^2} .$$

We know the velocity function F from the level set theory, and it will be further elaborated in Subsection 2.3. The only difference from the level set version is the additional monotonicity requirement.

Let us elaborate a little on the $T(\mathbf{x})$. This is the *travel-time function*, and indicates at each node $\mathbf{x}$ the time the front has spent on evolving from the initial seed point to the current node. The time of the moving front at every node is recorded. This means that only grid nodes swept by the front are assigned values. The front $\gamma(t)$ at any instant in time is given by

$$\gamma(t) = \{\mathbf{x} : T(\mathbf{x}) = t\}. \tag{6}$$

In Figure 7 we view the concept of the growing circle again, which we remember from Figures 3 and 4. The front $\gamma(t)$ is represented at the times $t = 0$, $t = 1$, and $t = 2$, through (6), where we no longer use the zero level, but rather the t-level.

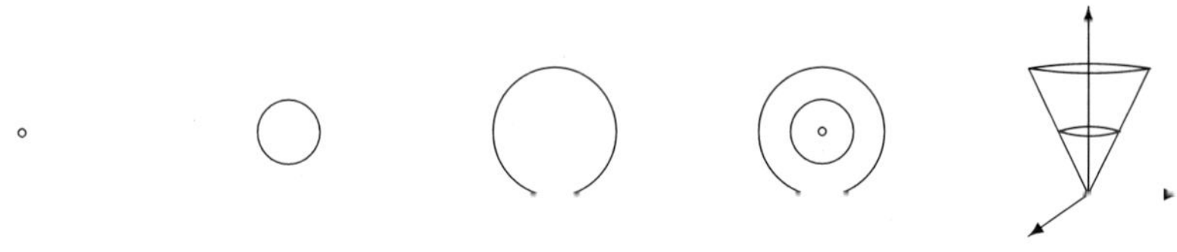

(a) $t = 0$ (b) $t = 1$ (c) $t = 2$ (d) above (e) sideways

Fig. 7. The evolving front $\gamma(t)$ at different times. It is indicated through the t-level of the travel-time function and defines a three-dimensional cone.

Sethian makes the concept of travel-time easily understood on his webpage [17]. Suppose the computational domain divided into cells by a suitable grid, see Figure 8. In each of the nodes we place a watch (w). All the watches are started simultaneously as the seed point is picked. The watches are stopped when the front passes the particular node. Nodes, which are never passed by the front, are assigned to travel-time *infinity*.

In consequence, the velocity function is monotonic, either positive or negative, so that the front propagates strictly outwards or inwards. In this way

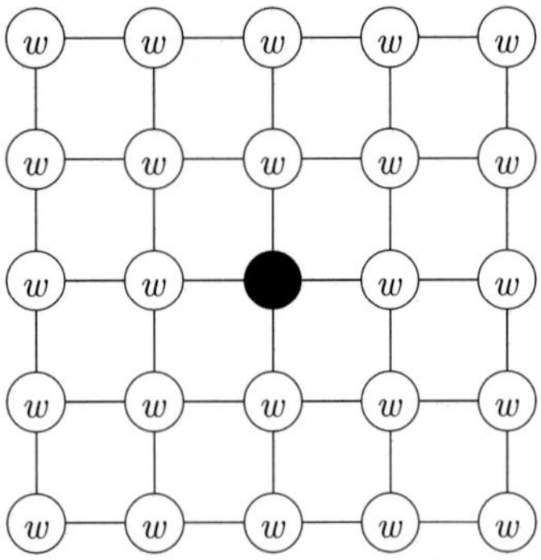

Fig. 8. The front sweeps the grid. At time equal zero the watches at all nodes starts, and are stopped as the front passes that node. The watches at the nodes are indicated with a w encircled, while the black circle in the middle represents the seed point.

only one value is recorded at each node, and so no information is lost, due to a second pass of the front.

The evolving front moves according to the velocity function F. It may be difficult to understand this function, since it represents a motion in a stationary data set. In order to get a better insight into the method, let us further explain the velocity function. If we use a 2D intensity image as input data, we may interpret the 2D velocity function as a topological map based on the original data. Imagine moving through the intensity landscape; when changes are fast, the velocity goes down. Hence, we may be close to a boundary and want to stop. In areas of little change in intensity the velocity should be constant and relatively high valued, so we easily can move through. This gives a first idea on the features of the velocity function, a more comprehensive mathematical discussion on this is provided in Subsection 2.3.

The spatial differentials T_{x_i} included in (5) are approximated by a nearest neighbor differential scheme. The scheme is shown in Figure 9, for the sake of simplicity merely in two dimensions, $d = 2$. From the current source node (i, j) there are $2^d = 4$ possible directions for the front to propagate. The minimum of the travel-time solutions indicates where the front passes first, and hence the neighbor in this direction is selected as the next source node.

The derivatives are approximated by neighboring differential schemes, using the information available at the current node (i, j). In (7) the formula for the computations made in the x_1- and x_2-direction are shown,

$$T_{x_1}^- = \frac{T(i,j) - T(i-1,j)}{\Delta x_1} \qquad T_{x_1}^+ = \frac{T(i+1,j) - T(i,j)}{\Delta x_1}$$
$$T_{x_2}^- = \frac{T(i,j) - T(i,j-1)}{\Delta x_2} \qquad T_{x_2}^+ = \frac{T(i,j+1) - T(i,j)}{\Delta x_2}, \tag{7}$$

where we are using an equidistant Cartesian grid, so that $\Delta x_i = h$, for all i.

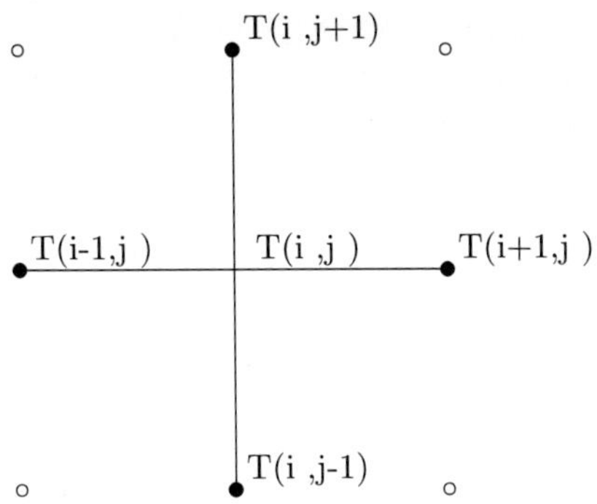

Fig. 9. The neighboring computational scheme of travel-time T for approximating the directional derivatives T_{x_1} and T_{x_2}. The grid nodes are indicated by circles, the dark ones are used in the computation.

The Marching Algorithm. The main motivation for solving the eikonal equation (rather than regarding the level set scenario) is the availability of several fast solvers: the celebrated *fast marching method* [18, 19], based on Dijkstra's method [5], is easy to understand. However, we have so far concentrated more on the *group marching method* [11], based on Dial's algorithm [4]. The reason for working with the group marching is mainly because the workload of this method is of the same order as number of grid nodes, ($\mathcal{O}(N)$), while the fast marching approach takes $\mathcal{O}(N \log_2 N)$. We present a slightly modified version of the group marching method, which finds a stationary solution rather quickly.

The marching algorithm mainly takes care of two tasks. Firstly, it updates the index matrix; whether a node is outside ($'0'$), at the front ($'1'$), or inside ($'2'$) the object. The other task is to update the travel-time value for the current node and its neighboring nodes. In addition, we want the algorithm to do this fast as well. Two important reasons for why a group marching-based method works fast are:

- Each iteration advances a *group* of nodes.
- *Direction of propagation* is incorporated through a travel-time constraint.

One way to speed up the algorithm is advancing more nodes for each iteration. In order to do this, information about the direction of front propagation is explicitly included. The group marching method selects a group of nodes from the neighbors of the front. Hence, instead of marching a single node forward, a whole group is advanced in the direction of evolution. Naturally, the question on how to choose these nodes arises.

For nodes belonging to the front, hence indexed ($'1'$), we choose a set of nodes to be changed into processed nodes (indexed ($'2'$)), denoted $\mathcal{G}$. The decision is made based on the travel-time values. Nodes whose travel-time value is below a certain threshold are included as inner nodes.

The example in Figure 10 illustrates the performance of marching. Four different grid images are shown; from iteration step (t) in (a), through an intermediate step in (b), to iterations $(t+1)$ and $(t+2)$ in (c) and (d), respectively.

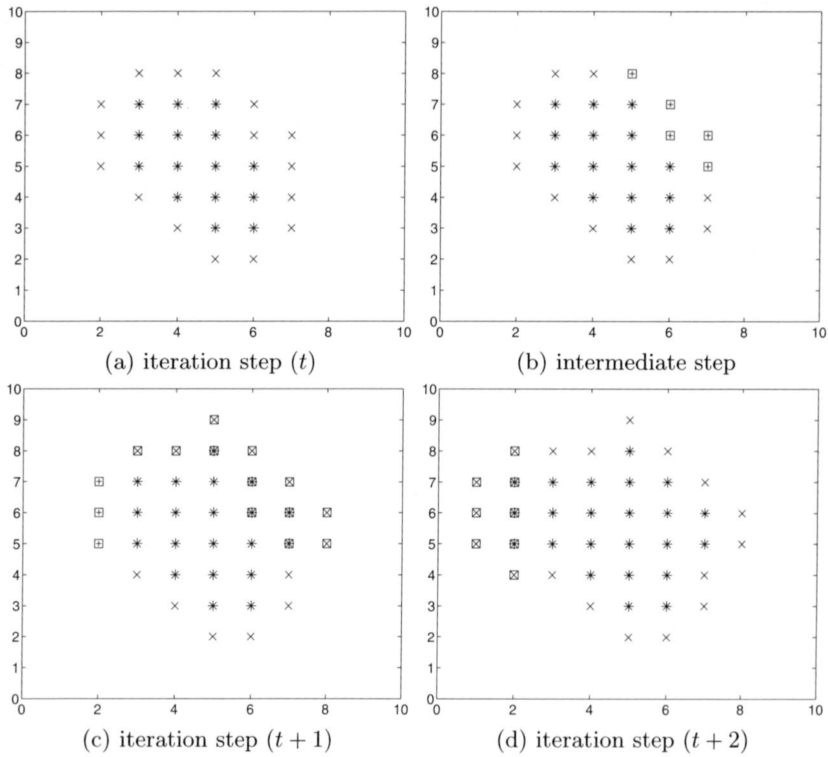

Fig. 10. $'*'$ means source. $'\times'$ means neighbor. $'+'$ means group node. The squares indicate nodes having changed condition from one image to the next. In (a) we see source nodes and neighbors. The group nodes are sorted out, and indicated, in (b). New neighbors are included as the group becomes source nodes, in (c). New group members are again indicated by squares in (c). Yet, another step is shown in (d), where the front now has moved in the direction of group nodes indicated in (c).

The explanation is based on iterations in Figure 10, executed at some time into the marching algorithm. Sources and neighboring nodes are labelled with $'*'$-es and $'\times'$-es, respectively. From the set of neighboring nodes in Figure 10 (a) a group is chosen, utilizing information on the dominant propagation direction.

Figure 10 (b) indicates the chosen group nodes as $'+'$-es wrapped in squares. The square indicates a change in the status of the node. The following iteration step $(t+1)$, shown in Figure 10 (c), displays the group nodes as new sources. The algorithm chooses yet another group of advancing nodes, this time the front moves to the left. In Figure 10 (d) these group nodes are included as sources, and new neighbors are labelled.

The travel-time values are computed according to the eikonal differential equation in (5), comprising information from the velocity function. The velocity function indicates how the front will evolve in the different directions. If we follow the front in the dominant propagation direction, we will end up at nodes where the travel-time values are within a bounded interval. Hence, the propagation direction is respected by choosing to advance nodes within a certain travel-time limit. In Figure 9 possible ray-paths in one of the quadrants are shown. The wave propagates from node (i, j) into the quadrant, represented by minimum travel-time, thus closest in time.

First, one marching sweep is performed in reverse order through the accepted nodes in $\mathcal{G}$ to update the travel-times of the corresponding 2D neighbors. Thereafter one forward marching sweep re-computes the travel-times. Then by using the newly computed information, we find suitable nodes for $\mathcal{G}$. The double computation of the travel-times improves the stability. The status indicator changes as the set of group node $\mathcal{G}$ is updated. The processed node (i, j) becomes a source node, labelled ('2'). Those of the neighbors not already labelled as possible source nodes ('1'), will be labelled now.

The selection of group nodes is taken from a constrained travel-time neighborhood. Hence, instead of just choosing one node to be included, the idea is to convert all neighboring nodes with a travel-time range below a certain threshold into inner nodes.

The threshold is set to the minimum of travel-times among the front nodes, plus an additional term. The limiting value $T_{\min}$ is the minimum travel-time among the first neighboring nodes. Then at each iteration the limit was updated by a certain amount of time, $\Delta\tau$, since we have advanced further away from the initial seed point. This additional term depends on both the spacing (h), the maximum velocity of the front, and dimension d, see Figure 11.

$$\Delta\tau = \frac{h}{\sqrt{d}\max_{\mathbf{x}\in\gamma} v}$$

$$T(\mathbf{x}) \leq T_{\min}(\mathbf{x}) + \Delta\tau \ .$$

Thus the group of nodes $\mathcal{G}$ is given by

$$\mathcal{G} = \{\mathbf{x} \in \Gamma : T(\mathbf{x}) \leq T_{\min}(\mathbf{x}) + \Delta\tau\},$$

$$\Gamma = \{\mathbf{x} \colon \operatorname{index}(\mathbf{x}) = 1\},$$

where *index* is the operator giving the status of the node.

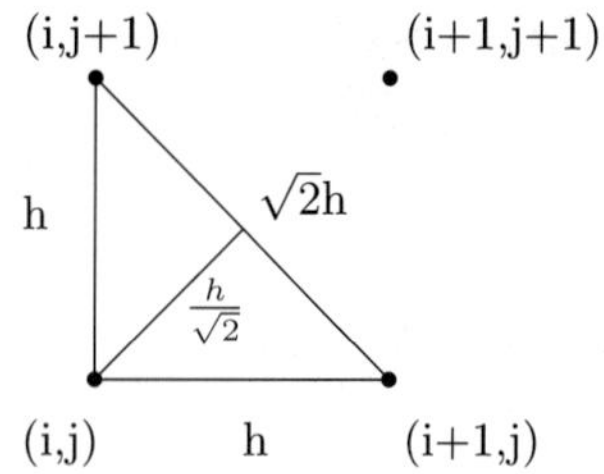

Fig. 11. The $\Delta\tau$ based on half nodes, updates the travel-time limit at every iteration.

2.3 The Velocity Function

The velocity function F is involved in both of the differential equations (4) and (5). In one sense, the function is the same in the two cases; it contains the available information from the data-set. This is why the choice of the velocity function is the salient point of the algorithm. The rate of success for the overall result depends on a well-defined velocity function.

In general the two velocity functions need not necessarily coincide. For the level set formulation, the velocity function is adequately defined without constraints, whereas the velocity function applied in the stationary formulation is supposed to be monotonic. The travel-time value, which a node acquires as the front sweeps by, is retained because the monotonicity prohibits a second sweep and and so an update of the already computed value.

The velocity function is either defined locally or globally, or it is of composite structure [22]. In either case, we want the front to evolve fast in homogeneous areas and to brake down near edges. This leads us to the following requirements for the velocity function.

- Zero at the boundary.
- High values (fast moving) in areas of little relevant information, hence without significant edges.

These requirements can be satisfied in various ways. For instance we know that the gradient operator will indicate edges. Abrupt behavior causes large gradient values, hence an inverse version of the gradient operator leads to small values at the boundaries. Using the gradient concerns global behavior, whereas locally dependent characteristics, such as curvature, may also be included. Using morphologic theory [7, 21] to define a suitable global velocity function is another alternative.

To exemplify more clearly the relation between an input intensity image and its corresponding velocity function image, see Figure 12. A horizontal line is drawn in the lower plots in order to visualize how the large intensity changes (boundaries) to the left lead to small velocity values in the right plot.

When defining a velocity function, it is important to recall that in addition to drive the front towards a boundary, it also acts as a stopping term when the

(a) input intensity image (b) velocity function image

Fig. 12. The upper figures show the intensity image (a) and the corresponding velocity image (b). The lower figures show an intersection at a horizontal line. The changes in intensity lead to small velocity values.

true boundary is reached. For computational reasons, a maximum number of iterations is given. However, that stopping criterion is only invoked in very complicated cases. We have included a Canny edge detector [3] in our velocity function, in order to guarantee a zero velocity at the boundaries. The output from the edge detector indicates where all possible boundaries can be found. In addition to the binary edge indicator $\mathcal{C}(\mathbf{x})$, our velocity function is gradient-based,

$$F(\mathbf{x}) = \frac{\mathcal{C}(\mathbf{x})}{1 + \alpha|\nabla \mathcal{I}(\mathbf{x})|},$$

where $\mathcal{I}$ is the available data-set and α a shift-parameter shifting the velocity distribution.

3 Mapping Geo-Bodies from Barents Sea Seismic Data

Let us finally show some numerical results obtained by using the method explained in this chapter. Recall that the input data determines the velocity function, and so this relies on the underlying application. In our particular application, we first specify the procedure and show examples of mapping bodies within seismic data.

The mapping is initialized by placing a seed point within the body of interest. As the objects often are large and complicated in seismic data, it is often necessary to give more seed points. The multiple seed points are either given as scattered points at the initialization step or after some iterations.

We have worked on seismic data from the Barents Sea, using both the seismic data and computed attribute data.

3.1 Two-Dimensional Example

In the 2D case, a cross-section from an azimuth attribute cube [16] can be used as input data. An attribute is defined as a mathematical operator, or a composition of operators, capturing properties from seismic data. The azimuth attribute has, together with other attributes, such as dip, chaotic texture and continuity, been successfully applied to mapping carbonate reef structures, gas chimneys, channels, fans, faults etc.

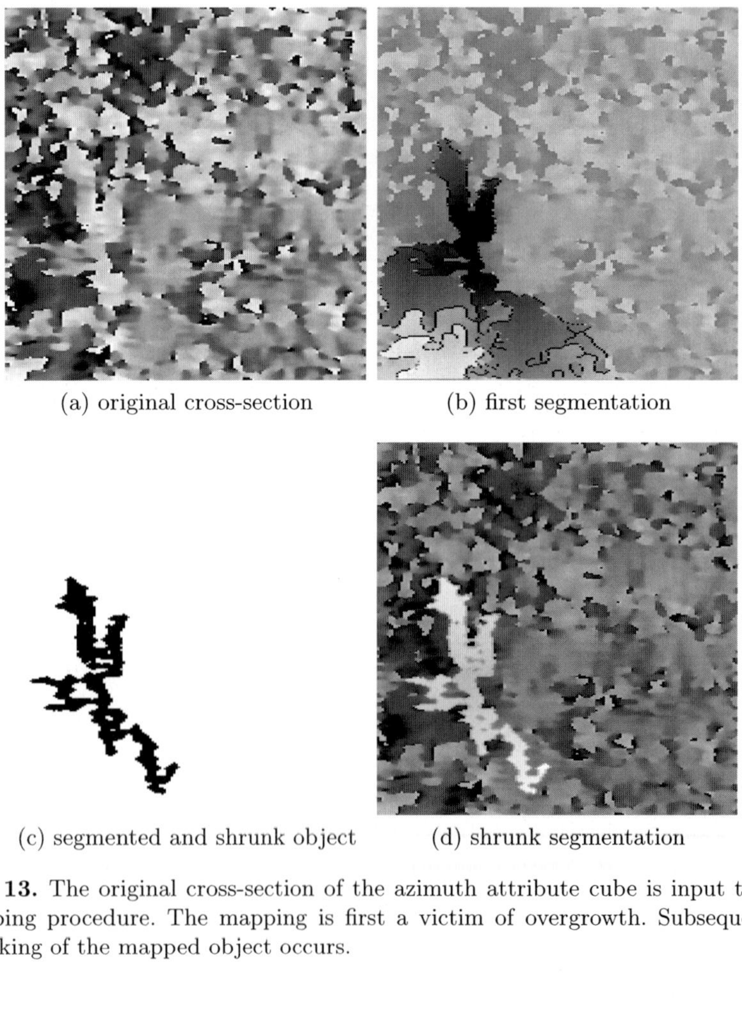

(a) original cross-section (b) first segmentation

(c) segmented and shrunk object (d) shrunk segmentation

Fig. 13. The original cross-section of the azimuth attribute cube is input to the mapping procedure. The mapping is first a victim of overgrowth. Subsequently, shrinking of the mapped object occurs.

Figure 13 shows some result from the 2D mapping procedure. This exemplifies an overgrowth, which may occur if the stopping criteria set by the velocity function is not strong enough. We then shrink the mapped area by limiting the travel-time by a maximum value. The final mapped result is shown in Figures 13 (c) and (d).

3.2 Three-Dimensional Example

For the 3D example we have used the seismic cube as input data. In this case, we are mapping a horizontally oriented geo-body. Figure 14 (a) shows a cross-section of the input seismic cube. The two darker line segments (below the middle) indicate the extent of the geo-body.

The obtained result is shown in Figure 14 (b). We see how the body is mapped in the foreground, on the left side we can see the background extraction.

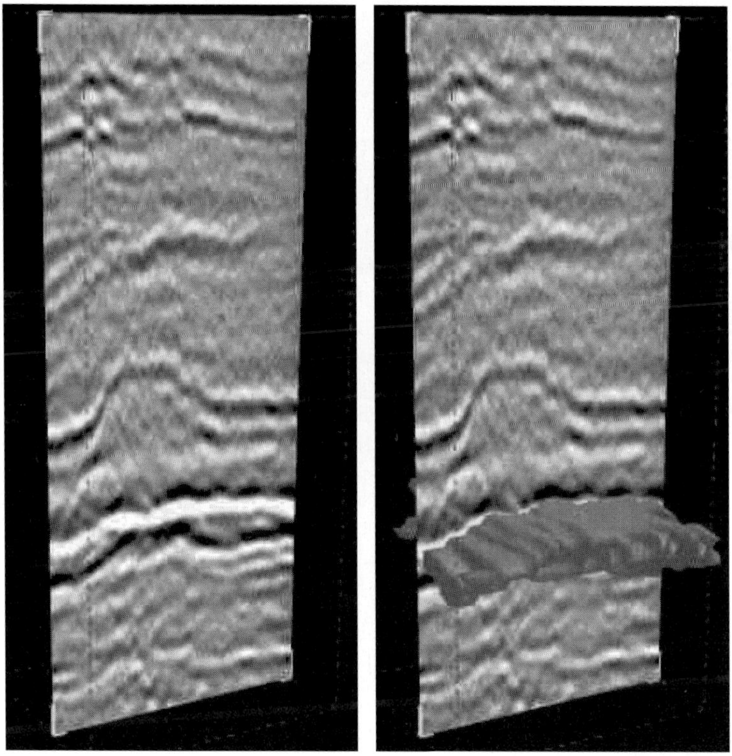

(a) input seismic (b) segmented seismic

Fig. 14. [Reproduced in colour in Plate 15 on page 432.] From the input seismic a horizontal geo-body is segmented. The resulting body is seen on the right.

4 Conclusions and Future Directions

This chapter presents a procedure for mapping geological bodies from seismic data. Theoretical aspects of the mapping procedure, concerning the level set method and marching method, are discussed. The overall method is adapted to applications from seismic data interpretation, where particular attention is placed on the design of the velocity function. Numerical examples, involving real-world seismic data taken from Barents Sea, support the utility of our method.

Altogether, this chapter provides a complementary step in the workflow for visualization and connectivity study. Moreover, this leads to application specification for seismic data, developing velocity function enhancing seismic characteristics and properties.

The tool, presented in this chapter, is intended to be a part of new and improved workflows for seismic data interpretation. One of motivations is to automate the steps in order to save manpower and reduce subjectivity.

Acknowledgement

The work was partly supported by the European Union through the research and training network NetAGES, contract no. IST-1999-29034.

References

1. A. Balabel (2001) *Numerical simulation of gas-liquid interface dynamics using the level set method.* PhD thesis, RWTH Aachen.
2. A. Blake and M. Isard (1998) *Active Contours.* Springer.
3. J. Canny (1986) A computational approach to edge detection. IEEE Transactions on Pattern Analysis and Machine Intelligence **8**(6), 679–698.
4. R.B. Dial (1971) A Probabilistic Multipath Traffic Assignment Model which Obviates Path Enumeration. Transportation Research **5**, 83–111.
5. E.W. Dijkstra (1959) A Note on Two Problems in Connection with Graphs. Numerische Mathematik **1**, 269–271.
6. J. Gomes and O. Faugeras (1999) Reconciling distance functions and level sets. Technical report, INRIA Sophia Antipolis.
7. R.C. Gonzalez and R.E. Woods (1992) *Digital image processing.* Addison-Wesley.
8. V. Hadziavdiz (2000) A comparative study of active contour models for boundary detection in brain images. Master's thesis, University of Tromsø.
9. H. Holden and N.H. Risebro (2002) *Front tracking for hyperbolic conservation laws.* Springer.
10. K.H. Karlsen, K.A. Lie, and N.H. Risebro (2000) A fast level set method for reservoir simulation. Computational Geosciences **4**(2), 185–206.
11. S. Kim (2001) An $\mathcal{O}(N)$ level set method for eikonal equations (2001) SIAM Journal of Scientific Computation **22**(6), 2178–2193.

12. R. Malladi and J.A. Sethian (1997) Level set methods for curvature flow, image enchancement, and shape recovery in medical images. Visualization and Mathematics, Springer-Verlag, 329–345.
13. S. Osher and R. Fedkiw (2003) *Level Set Methods and Dynamic Implicit Surfaces.* Springer, 2003.
14. S. Osher and J.A. Sethian (1988) Fronts propagating with curvature-dependent speed: algorithms based on Hamilton-Jacobi formulation. Journal of Computational Physics **79**(12–49), 898–903.
15. C. Pozrikidis (1997) *Introduction to Theoretical and Computational Fluid Dynamics.* Oxford University Press.
16. T. Randen, E. Monsen, C. Signer, A. Abrahamsen, J.O. Hansen, T. Sæter, J. Schlaf, and L. Sønneland (2000) Three-dimensional texture attributes for seismic data analysis. 70th Ann. Internat. Mtg: Soc. of Expl. Geophys., 668–671.
17. J.A. Sethian. `http://math.berkeley.edu/~sethian/level set.html`.
18. J.A. Sethian (1996) *Level Set Methods and Fast Marching Methods.* Cambridge University Press.
19. J.A. Sethian (1996) A Fast Marching Level Set Method for Monotonically Advancing Fronts. Proceedings of the National Academy of Science **93**(4), 1591–1595.
20. I.N. Sneddon (1957) *Elements of Partial Differential Equations. International Series in Pure and Applied Mathematics*, volume IX, McGraw-Hill.
21. P. Soille (2003) *Morphological Image Processing.* Springer, 2nd edition.
22. J. Suri, K. Liu, S. Singh, S.N. Laxminarayan, X. Zeng, and L. Reden (2002) Shape recovery algorithms using level sets in 2D/3D medical imagery: a state-of-the-art review. IEEE Transactions on Information Technology in Biomedicine **6**(1).
23. W.G. Vincenti and C.H. Kruger (1975) *Introduction to Physical Gas Dynamics.* Huntington.
24. O. Yilmaz (1987) *Seismic Data Processing.* Society of Exploration Geophysicists.

Modern Techniques in Seismic Tomography

Alexander A. Boukhgueim

University of Vienna, Department of Mathematics, Austria

Summary. This chapter is focussed on two local inverse kinematic problems of seismology, concerning *reflected rays* and *refracted rays*. Both model problems are reduced to a sequence of 2D problems, where theoretical and numerical results are offered. In the case of reflected rays, it is shown how to select a stable problem of recovering a velocity distribution in a layer, by using travel time measurements along rays with one reflection on the boundary. This way, a simple inversion algorithm is obtained for the linearized near a constant velocity case. In the case of refracted rays, a Newton-type algorithm for finding the 3D velocity distribution from 3D travel time measurements is constructed for the local inverse kinematic problem. To this end, a sound velocity that increases linearly with depth is chosen as a first approximation. With this particular choice for the linearization, the underlying problem reduces to a sequence of 2D Radon transforms in discs.

1 Introduction

One-dimensional inversion of travel time data goes back to G. Herglotz [13] and E. Wiechert [22]. The mathematical background of their works relies on the inversion formula for Abel's integral equation which was found by Abel in 1826. Two and three dimensional inversions of travel time data take their roots in works of J. Radon (1917) [17] (recovering a function through its integrals along all straight lines) and P. Funk (1916) [12] (recovering a function on the unit sphere through its integrals along all great circles). G. E. Backus (1964) [1] reduced the problem of determining upper mantle heterogeneity from phase-speed measurements to the Funk problem, where it was discovered that only the even part of the unknown function could be determined. This observation is rather obvious, since in the Funk problem every odd function on a sphere has zero great-circular average data.

In reality, rays are usually neither straight lines nor great circles. Moreover, the arising problems are typically nonlinear, since the rays depend on the unknown velocity distribution, and so there is high need for suitable mathematical models. On the one hand, important theoretical steps in this direction were made in Russia, especially in the Novosibisk school, where

- M.M. Lavrentiev, V.G. Romanov, and V.G. Vasiliev (1970) [15] considered the linearized inverse kinematic problem;
- R.G. Mukhometov (1975, 1977) [16] proved global uniqueness results for 2D inverse kinematic problem with full data;
- V.G. Romanov (1979) and R.G. Mukhometov (1979) showed the multidimensional global uniqueness for inverse kinematic problem;
- Yu. E. Anikonov (1978) proved uniqueness in the class of real analytic functions;
- N.N. Bernstein and M.L. Gerver (1980) [2] derived conditions on distinguishability of metrics by hodographs;
- A.L. Bukhgeim (1983) [4] proved necessary and sufficient conditions for the solvability of the two-dimensional inverse kinematic problem with partial local data in class of real analytic functions, and
- V. A. Sharafutdinov (1993) [20] studied integral geometry problems of tensor fields.

On the other hand, this problem was *solved* in practice by seismologists, by using real-world data, and this way first images of the Earth's interior were obtained by A.M. Dziewonski and others [6, 7, 8, 9, 10, 11, 18, 19, 23, 24].

In this chapter, we consider two *local* (i.e., in a layer) 3D seismic tomography problems. The first case, concerning *reflected rays*, is considered in the following Section 2, whereas a second case, being subject of the discussion in Section 3, is concerning *refracted rays*. In either case, the arising problem is reduced to a sequence of 2D problems (i.e., in sections). New theoretical results are obtained, and illustrative numerical examples are presented, respectively.

More precise, in Section 2 we show how to select a stable problem of recovering a velocity distribution in the layer. This is done by using travel time measurements along rays with one reflection on the boundary. We so obtain simple inversion algorithms for the linearized near a constant velocity case. We also consider the case of a dipping reflector.

Moreover, in Section 3 we construct a Newton-type algorithm for finding the 3D velocity distribution from 3D travel time measurements for the local inverse kinematic problem. Initially, as a first approximation, we choose a sound velocity that increases linearly with the depth. This is since it was shown in [5] that with this choice of linearization our problem reduces to a sequence of 2D Radon transforms in discs. Our case is much harder, since we consider solving a *nonlinear* problem, and therefore we need to solve a direct 3D problem on each iteration. However, we can show that, in our case, already the second iteration is often much better than the solution from the linearized approximation.

2 Inversion of Travel Time in a Layer

In order to find the most stable three-dimensional (local) inverse kinematic problem it is appropriate to start from the simpler one-dimensional case. The following considerations are made along the lines of [21, pp. 102–103]. Let us consider a ray propagation in a layer of thickness H with a sound velocity V over a half-space of velocity $V^* > V$, see Figure 1.

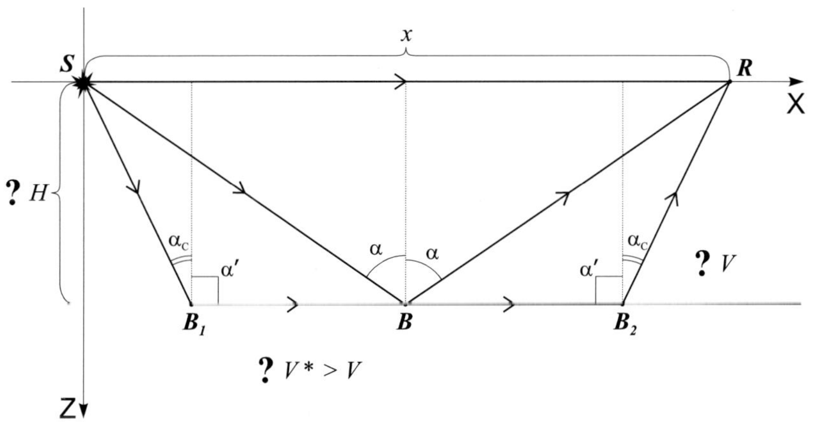

Fig. 1. A layer of thickness H with sound velocity V over a half-space with sound velocity $V^* > V$. All these three quantities (V, V^*, H) are unknown constants.

This example represents, in a most simplified way, the situation of the sediments over a rock base or the Earth's crust over its upper mantle for small distances, for which a *flat* Earth approximation is reasonable.

At a receiver point R, at a distance x from the source S, we encounter three different types of rays:

- direct rays (SR), traveling from S to R with travel time

$$t_1(x) = x/V,$$

- rays (SBR) reflected on the boundary $z = H$ with travel time

$$t_2(x) = \frac{2(H^2 + x^2/4)^{1/2}}{V}, \tag{1}$$

- critically refracted rays (SB_1B_2R) or head waves, i.e., rays with the critical angle of incidence α_c at the contact surface $z = H$ that propagate horizontally through the half-space and return to the free boundary $z = 0$ with the same angle α_c at travel time

$$t_3(x) = \frac{2H}{V \cos \alpha_c} + \frac{x - 2H \tan \alpha_c}{V^*}. \tag{2}$$

By *Snell's law*

$$\frac{\sin\alpha_c}{V} = \frac{\sin\alpha'}{V^*},$$

for $\alpha' = \pi/2$, we find $\sin\alpha_c = V/V^*$, and so (2) can be rewritten as

$$t_3(x) = \frac{x}{V^*} + \frac{2H(V^{*2} - V^2)^{1/2}}{VV^*}. \tag{3}$$

For sufficiently small x we have

$$t_1(x) < t_3(x) < t_2(x),$$

and hence we can measure t_1, t_2, t_3 at the point R separately, so that we obtain $V = x/t_1$, before we find H from equation (1) and V^* from (3). This way, we obtain the three constant parameters V, V^*, and H from t_1, t_2, and t_3, which are known for any fixed x.

Now let us turn to a more complicated situation. Suppose that we know V^*, H (from the previous model, for example) and wish to find $V = V(z)$ in a layer, provided that $V_z' > 0$ for all $z \in [0, H]$.

Then, for *direct* rays, continuously refracted in the layer, $t_1(X)$ reduces to

$$t(X) = \int\limits_{\Gamma(X)} \frac{ds}{V} = \int\limits_{\Gamma(X)} n ds, \tag{4}$$

see Figure 2, where $n = 1/V$ is the *slowness*, and $\Gamma(X)$ is the ray passing through the points $(0,0)$ and $(X,0)$.

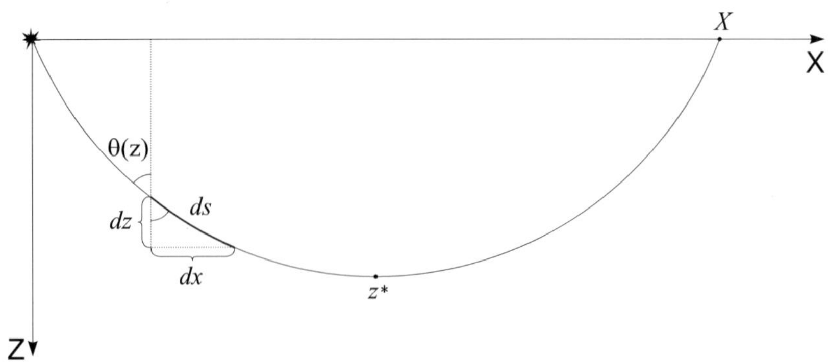

Fig. 2. Illustration of the one dimensional kinematic problem. Direct ray, continuously refracted in the medium with the slowness being the function of the variable z.

As regards the slowness $n(z)$, Snell's law takes the form

$$n(z)\sin\theta(z) = p,$$

where the ray parameter p is constant on $\Gamma(X)$. On the ray Γ, see Figure 2, we have

$$\frac{dx}{ds} = \sin\theta = \frac{p}{n},$$

$$\frac{dz}{ds} = \cos\theta = (1-\sin^2\theta)^{1/2} = \frac{(n^2-p^2)^{1/2}}{n}.$$

Hence, using the chain rule we find

$$\frac{dx}{dz} = \frac{p}{\sqrt{n^2(z)-p^2}},$$

$$\frac{dt}{dz} = \underbrace{\frac{dt}{ds}}_{n}\frac{ds}{dz} = \frac{n^2(z)}{\sqrt{n^2(z)-p^2}}.$$

Moreover, integrating from 0 to the turning point z^* (w.r.t. z), and taking into account the symmetry of Γ, we obtain

$$X(p) = 2p\int_0^{z^*}\frac{dz}{\sqrt{n^2(z)-p^2}}, \tag{5}$$

$$T(p) = 2\int_0^{z^*}\frac{n^2(z)dz}{\sqrt{n^2(z)-p^2}},$$

where

$$T(p) = t(X(p)).$$

For the ray parameter p, we find

$$p = \frac{dt}{dX},$$

see Figure 3. This is due to Snell's law, since

$$ds = dX\cos\left(\frac{\pi}{2}-\theta_0\right) = dX\sin\theta(0),$$

and

$$\frac{dt}{dX} = \underbrace{\frac{dt}{ds}}_{n(0)}\frac{ds}{dX} = n(0)\sin\theta(0) = p.$$

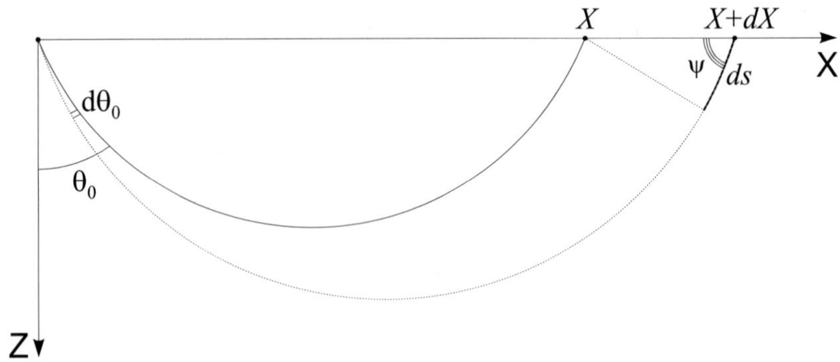

Fig. 3. The angle $\psi = \frac{\pi}{2} - (\theta_0 - d\theta_0) \approx \frac{\pi}{2} - \theta_0, \quad \theta_0 = \theta(0)$.

So if we know $t(X)$, then we can find $p = t'(X)$ and hence the inverse function $X = X(p)$, and $T(p) = t(X(p))$. For $X \to 0$, $\theta(0) \to \frac{\pi}{2}$, with using Snell's law, we find

$$\left.\frac{dt}{dX}\right|_{X=0} = p_0 = n(0) \geq p.$$

By using the inverse function $z = z(n)$, we can rewrite (5) as a linear equation of Abel's type (note also, that $V'_z > 0$ implies $n'_z < 0 \sim z'_n < 0$), so that

$$\frac{X(p)}{2p} = \int_{n(0)}^{p} \frac{z'(n)dn}{\sqrt{n^2 - p^2}} = -\int_{p}^{n(0)} \frac{z'(n)dn}{\sqrt{n^2 - p^2}}$$

and hence we obtain (by using Abel's inversion formula) the solution

$$z(n) = \frac{1}{\pi} \int_{n}^{n_0} \frac{X(p)dp}{\sqrt{p^2 - n^2}}.$$

Note that this formula works only locally. However, we can apply it step by step, in which case it is possible to show that in the absence of a low velocity zone $V'_z \leq 0$, we can find $V(z)$ in the whole layer $[0, H]$. This result was obtained by G. Herglotz [13] and E. Wiechert [22] about hundred years ago.

An analogous equation for rays reflected on $z = H$ (at least in the linearized near the constant velocity case) contains only the information about the average velocity $\bar{V}$ in a layer, since it reduces either to

$$t_2(x) = \frac{2}{\cos\alpha} \int_{0}^{H} \frac{dz}{V(z)},$$

see Figure 1, or, since $\cos\alpha = \frac{H}{\sqrt{H^2+x^2/4}}$, to

$$t_2(x) = \frac{2(H^2 + x^2/4)^{1/2}}{H} \int\limits_0^H \frac{dz}{V(z)}. \tag{6}$$

So all we can find from this equation is

$$\frac{1}{\bar{V}} = \frac{1}{H} \int\limits_0^H \frac{dz}{V(z)}.$$

This situation changes dramatically when we go to the two-dimensional case $V = V(x, z),\ x \in \mathbb{R},\ z \in [0, H]$. Analog of the equation (4),

$$\int\limits_{\Gamma(x_0,x_1)} n ds = f(x_0, x_1), \quad n = \frac{1}{V}$$

becomes *exponentially* unstable in terms of the Fourier transform with respect to a lateral variable x in the linearized case, as this is shown in [15].

At the same time, an analog of the equation (6) is sufficiently stable, at least in the case when we apply linearization near a constant velocity. Indeed, suppose that we consider a half-space $\mathbb{R}^3_+ = \{x, z|\ x \in \mathbb{R}^2,\ z \in \mathbb{R},\ z > 0\}$ (we consider a more general 3D case here, so that this linearization scheme remains valid for the next section as well). The task is to find the slowness $n(x, z)$ (here it is more convenient to work in terms of slowness rather than sound velocity) from the conditions

$$|\nabla\tau(x, z, x_0)|^2 = n^2(x, z), \tag{7}$$

$$\tau\big|_S = g(\alpha, \beta, r), \quad n'_z < 0,$$

where $S = \{x, z, x_0|\ z = 0,\ |x| = |x_0| = r\}$, $\alpha, \beta \in [0, 2\pi)$, $r \in [0, \rho]$, and where (α, r) and (β, r) are the polar coordinates of the points x and x_0, respectively. Moreover, ∇ is a gradient vector with respect to the variables x and z. Here we assume that

$$n(x, z) = n_0(z) + u(x, z), \quad |u| \ll n_0, \tag{8}$$

where $n_0(z)$ is known and $u(x, z)$ is an unknown part of the slowness (smaller in magnitude than n_0). In this case,

$$\tau(x, z, x_0) = \tau_0(x, z, x_0) + t(x, z, x_0), \tag{9}$$

where the component τ_0 of the travel time τ corresponds to the slowness n_0:

$$|\nabla\tau_0|^2 = n_0^2. \tag{10}$$

Substituting (8), (9) into (7) and taking into account (10) we get

$$\langle \nabla\tau_0 + \nabla t, \nabla\tau_0 + \nabla t\rangle = (n_0 + u)^2,$$

$$|\nabla\tau_0|^2 + 2\langle \nabla\tau_0, \nabla t\rangle + |\nabla t|^2 = n_0^2 + 2n_0 u + u^2,$$

$$\langle \nu_0, \nabla t\rangle = u + \frac{1}{2n_0}\left(u^2 - |\nabla t|^2\right), \quad \nu_0 = \frac{\nabla\tau_0}{n_0},$$

where ν_0 is a unit vector, tangent to the geodesic Γ that corresponds to the slowness n_0.

The solution of this equation can be found by the method of successive approximations

$$\langle \nu_0, \nabla t_k\rangle = u_k + \frac{1}{2n_0}\left(u_{k-1}^2 - |\nabla t_{k-1}|^2\right),$$

$$t_k\big|_S = f(\alpha, \beta, r), \quad k = 1, 2, \ldots,$$

where $u_0 = n_0$, $t_0 = \tau_0$, $f = g - \tau_0\big|_S = t\big|_S$. So, on each step one needs to solve a linear problem, that can be written in an integral form as

$$Pu_k \equiv \int\limits_{\Gamma(\alpha,\beta,r)} u_k d\sigma = f_k(\alpha, \beta, r), \tag{11}$$

where $\Gamma(\alpha, \beta, r)$ is a geodesic, that corresponds to the slowness $n_0(z)$, and f_k is a given function that can be computed from f, u_{k-1}, and $|\nabla t_{k-1}|$, so that in particular $f_1 = f$.

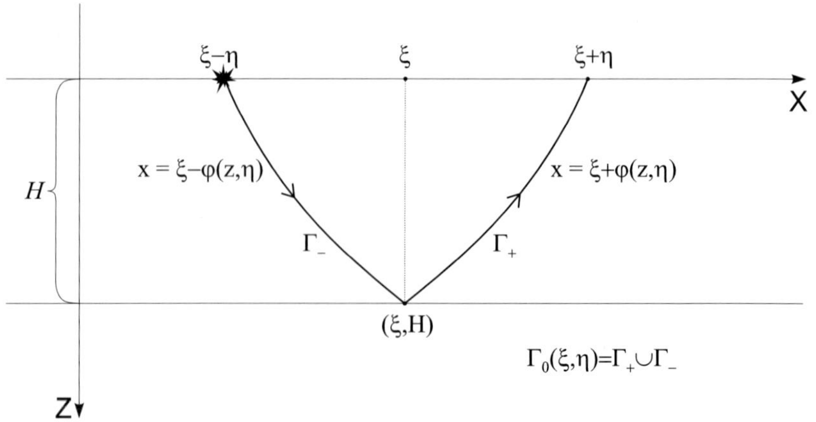

Fig. 4. An illustration to the two-dimensional kinematic problem on reflected rays.

So, when we have the decomposition

$$n = n_0(z) + u(x,z)$$

into known n_0 and unknown $u(x,z)$ parts of slowness, after the linearization, we arrive at the linear integral equation (11),

$$\int\limits_{\Gamma(\xi,\eta)} u ds = f(\xi,\eta),$$

which can, for $\xi \in \mathbb{R}$ and $\eta \geq 0$, be rewritten as

$$\int\limits_0^H \left[u(\xi - \varphi(z,\eta), z) + u(\xi + \varphi(z,\eta), z)\right](1 + {\varphi'_z}^2)^{1/2} dz = f(\xi,\eta), \tag{12}$$

(on using $ds = \sqrt{dx^2 + dz^2} = \sqrt{1 + {\varphi'_z}^2} dz$), where $\Gamma_0 = \Gamma_+ \bigcup \Gamma_-$ is the ray reflected on the boundary $z = H$ at a point (ξ, H), see Figure 4, where $\Gamma_\pm = \{x : x = \xi \pm \varphi(z,\eta)\}$, and where the function φ satisfies

$$\varphi(H,\eta) = 0, \quad \varphi(0,\eta) = \eta, \quad \varphi(z,0) = 0.$$

The reflected ray Γ_0 corresponds to the velocity $V_0(z)$ and hence it is invariant with respect to the lateral shift (since $V_0(z)$ does not depend on x).

Applying now the Fourier transform with respect to X,

$$\hat{f}(\lambda,\eta) = \int\limits_{-\infty}^{\infty} f(\xi,\eta) e^{-i\lambda\xi} d\xi,$$

we find (recall (12) here and assume that u has compact support)

$$\begin{aligned}\hat{f}(\lambda,\eta) &= \int\limits_{-\infty}^{\infty} e^{-i\lambda\xi} \left\{ \int\limits_0^H \left[u(\xi - \varphi, z) + u(\xi + \varphi, z)\right] \sqrt{1 + {\varphi'_z}^2} dz \right\} d\xi \\ &= 2 \int\limits_0^H \hat{u}(\lambda, z) \cos(\lambda\varphi(z,\eta)) \sqrt{1 + {\varphi'_z}^2} dz,\end{aligned}$$

since

$$\begin{aligned}\int\limits_{-\infty}^{\infty} u(\zeta \perp \varphi, z) e^{-i\lambda\xi} d\zeta - \int\limits_{-\infty}^{\infty} u(\zeta \perp \varphi, z) e^{-i\lambda(\xi \pm \varphi)} e^{\mp i\lambda\varphi} d(\xi \perp \varphi) \\ = e^{\mp i\lambda\varphi(z,\eta)} \hat{u}(\lambda, z).\end{aligned}$$

Therefore, we have

$$2\int_0^H \cos(\lambda\varphi(z,\eta))\sqrt{1+\varphi_z'(z,\eta)^2}\hat{u}(\lambda,z)dz = \hat{f}(\lambda,\eta).$$

For every λ we have an integral equation for finding $\hat{u}(\lambda,z)$. In particular, if $V_0 = \text{const}$, then $\varphi = \eta\frac{(H-z)}{H}$, $(\varphi_z')^2 = \left(\frac{\eta}{H}\right)^2$ and so we have

$$\int_0^H \cos\left(\lambda\eta\frac{(H-z)}{H}\right)\hat{u}(\lambda,z)dz = \frac{\hat{f}}{2\left(1+\left(\frac{\eta}{H}\right)^2\right)^{1/2}}.$$

If we let $H - z = p$, and so $z = H - p$, then we find

$$\begin{aligned}\int_H^0 \cos\left(\frac{\lambda\eta p}{H}\right)\hat{u}(\lambda,H-p)(-dp) &= \int_0^H \cos\left(\frac{\lambda\eta p}{H}\right)\hat{u}(\lambda,H-p)dp \\ &= \frac{\hat{f}(\lambda,\eta)}{2\left(1+\left(\frac{\eta}{H}\right)^2\right)^{1/2}}.\end{aligned} \tag{13}$$

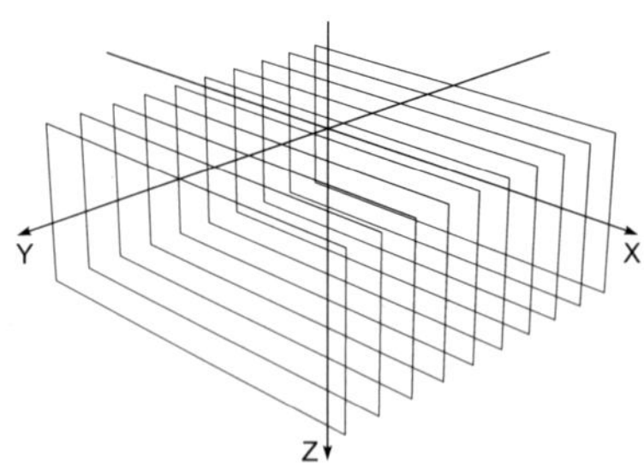

Fig. 5. An illustration for the reconstruction of three-dimensional heterogeneities via a series of two-dimensional problems, slice by slice.

This is the cosine Fourier transform of the function $\psi(p) := \hat{u}(\lambda, H-p)$ for fixed λ, and hence using the fast Fourier algorithms we can easily find the function u. Note that we do not have an exponential instability in this case, provided that we can measure $f(\xi,\eta)$ for all $\xi \in \mathbb{R}$ and all $\eta \geq 0$.

In practice, of course, $\xi \in [0, L]$, $\eta \in [0, N]$ (moreover, ξ, η belong to some grid) and an exponential instability also emerges, but it is not as *bad* as in the first case. If we have the second lateral variable y, which is perpendicular to the plane (x,z) of Figure 4, then we can recover a three-dimensional finite heterogeneity of the type $n(x,y,z) = n_0(z) + u(x,y,z)$, slice by slice in the y dimension, see Figure 5.

Note that this tomography problem (with reflecting boundary) can easily be generalized to the *dipping reflector case*, see Figure 6. This merely requires recalculating the data on the new free boundary which is parallel to the half-space with velocity $V^* > V_0$. The parameters V_0, H, θ, V^* can be found as in the first example.

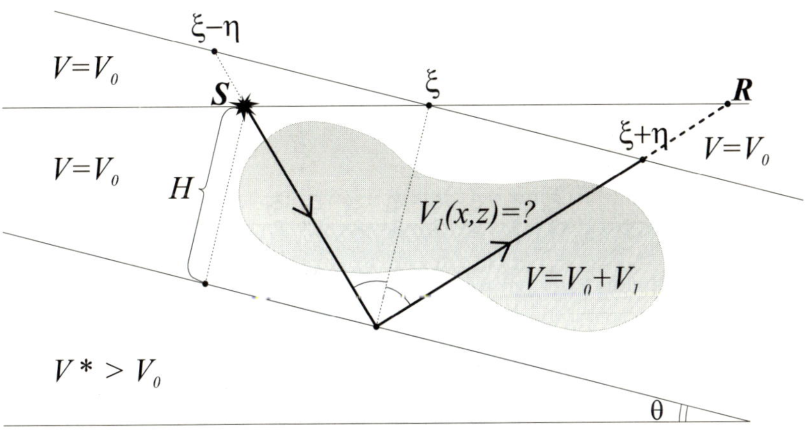

Fig. 6. An example of the dipping reflector case.

In order to find the stable *refracted* 3D seismic problem (in a layer), which reduces to a sequence of 2D Radon problems with weight, we need to use a completely different foliation with *nonlinear* slides. In the case $V_0 = m + bz$, we encounter spherical slices, as we will see in the next section.

In the remainder of this section, we show how we can use our results for recovering also the *attenuation*, by using the so-called *attenuation time*. Comparing the amplitude A of a monochromatic (with the frequency ω) P- or S-wave at a source point x_0 and at a receiver point x_1 we can find its ratio as

$$\frac{A(x_1)}{A(x_0)} = e^{-\omega t^*},$$

and hence we can determine the attenuation time as

$$t^* = \int_{\Gamma(n,x_0,x_1)} \frac{n ds}{Q},$$

where Q is the associated with P-wave or S-wave *quality factor* [14] which describes the attenuation of P- or S-waves, respectively. It is known that for S-waves, the value t^* is greater than for P-waves, which means that S-waves attenuate faster than P-waves. If we put

Example 5. Consider a slowness given by the formula

$$n(x, z) = n_0 + k\big(\sin(3.2\pi x + \sin(3.2\pi z)) + \sin(3.2\pi z + \sin(3.2\pi x))\big),$$

see Figure 9 (a), top. The reconstruction (without noise) is shown in the middle, and the reconstruction from noisy data is shown below. The noise level is 6% of the measured travel time τ which constitutes 56% of the small additive t. The L_2-norm of the small additive of the slowness u is 50% of the L_2-norm of the *background* slowness n_0. Discretization parameters: the interval $[0, L]$ is split into $M = 200$ parts and the vertical interval $[0, H]$ into $N = 200$ parts.

Example 6. Consider a slowness given by the formula

$$\begin{aligned} n(x, z) = n_0 + k\big(&\sin(0.01(\cos(6.4\pi x) + \sin(3.2\pi z))) \\ &\times \sin(0.02(\cos(6.4\pi z) + \sin(3.2\pi x)))\big), \end{aligned}$$

see Figure 9 (c), top, where the reconstruction (without noise) is shown below. Discretization parameters: the interval $[0, L]$ is split into $M = 200$ parts and the vertical interval $[0, H]$ into $N = 200$ parts.

Example 7. Consider a discontinuous slowness given by the function from Figure 9 (b), top. The reconstruction (without noise) is shown in the middle, and the reconstruction from noisy data is shown below. The noise level is 5% of the measured travel time τ which constitutes 30% of the small additive t. The L_2-norm of the small additive of the slowness u is 24% of the L_2-norm of the *background* slowness n_0. Discretization parameters: the interval $[0, L]$ is split into $M = 400$ parts and the vertical interval $[0, H]$ into $N = 400$ parts.

Example 8. Another discontinuous slowness is given by the function from Figure 9 (d), top. We consider this example in order to show the resolution capabilities of the inversion algorithm. The reconstruction (without noise) is shown at the bottom. Discretization parameters: the interval $[0, L]$ is split into $M = 400$ parts and the vertical interval $[0, H]$ into $N = 400$ parts.

3 Local 3D Seismic Tomography on Refracted Rays

Consider a half-space in 3D and let

$$x^0 = (x_1^0, x_2^0) = r(\cos\alpha, \sin\alpha),$$

$$x^1 = (x_1^1, x_2^1) = r(\cos\beta, \sin\beta),$$

$\alpha, \beta \in [0, 2\pi)$, $r \in (0, \rho]$, be two arbitrary points on the circle $|x| = r$ in the (x_1, x_2)-plane.

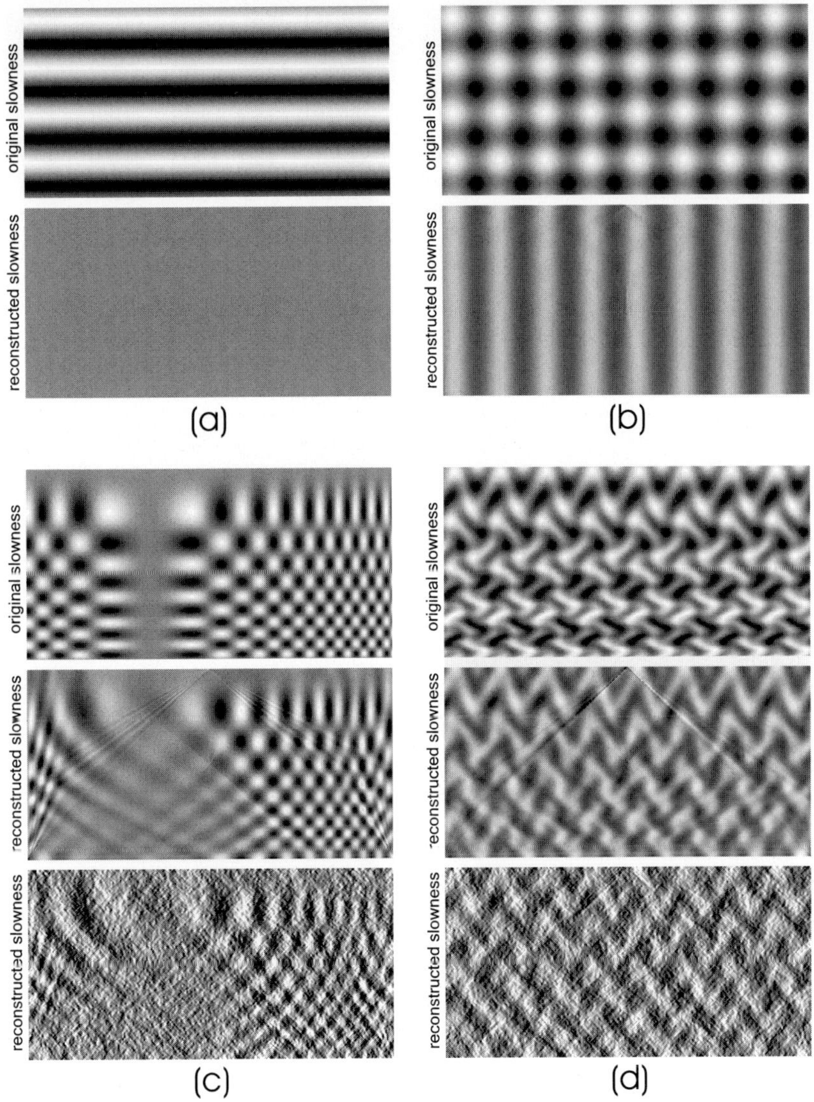

Fig. 8. Illustration for the numerical Examples 1–4. Parts (c) and (d) involve reconstruction from noisy data.

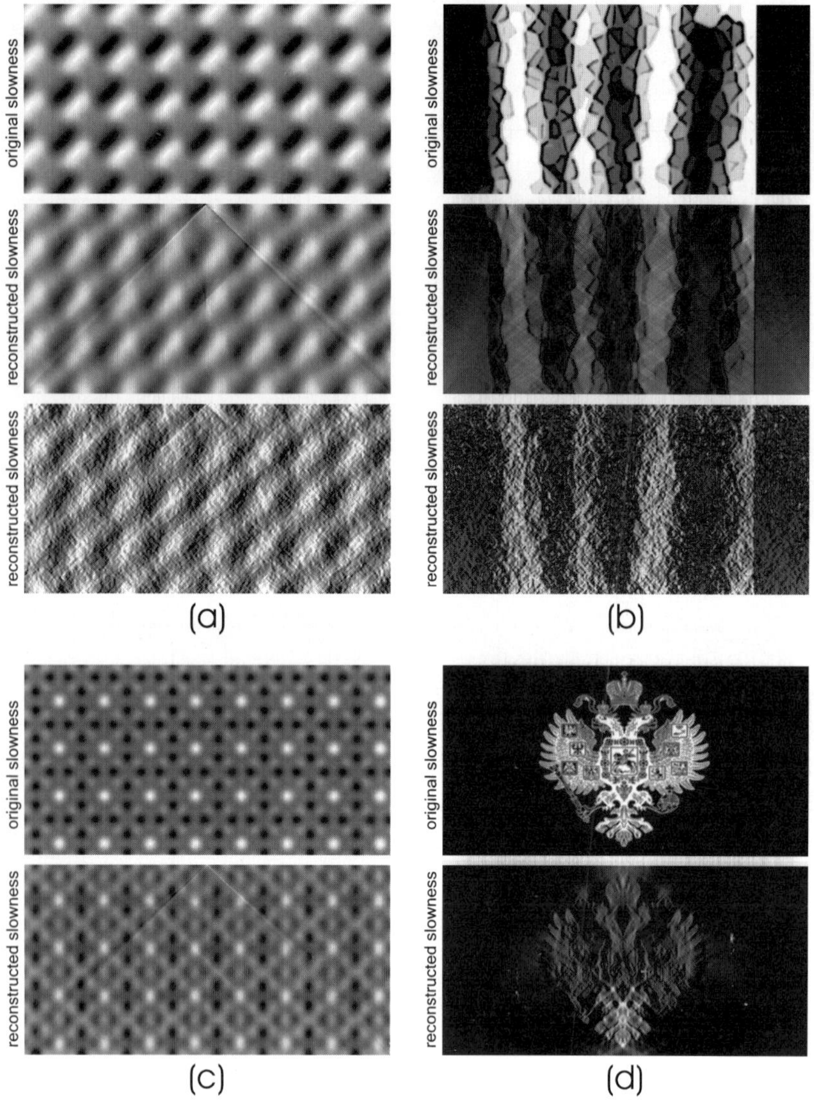

Fig. 9. Illustration for the numerical Examples 5–8. Parts (a) and (b) involve reconstruction from noisy data.

Let $\Gamma = \Gamma(n;\alpha,\beta,r) = \Gamma(x^0,x^1)$ be a ray, corresponding to the slowness $n(x,z)$, that passes through the points x^0 and x^1. We suppose that for any pair of two points, x^0 and x^1, satisfying $|x^0| = |x^1| = r$, $r \in (0,\rho]$, there exists only one ray $\Gamma(x^0,x^1)$ connecting them. Suppose that we know the travel time

$$\tau(\alpha,\beta,r) = \int\limits_{\Gamma(n;\alpha,\beta,r)} n ds := F(n), \tag{14}$$

and would like to find the slowness n. This is a nonlinear integral equation and at the first step we reduce it to a sequence of linear equations using the modified Newton's method. To this end, note that

$$\begin{aligned} F(n+h) - F(n) &= \int\limits_{\Gamma(n+h)} (n+h)ds - \int\limits_{\Gamma(n)} nds \\ &= \int\limits_{\Gamma(n+h)} \big[(n+h) - n\big] ds + \underbrace{\int\limits_{\Gamma(n+h)} nds - \int\limits_{\Gamma(n)} nds}_{o(h) \text{ by Fermat's principle}} \\ &\approx \int\limits_{\Gamma(n)} hds = F'(n)h, \end{aligned}$$

where we use the abbreviation $\Gamma(n;\alpha,\beta,r) \equiv \Gamma(n)$. Recall that the *Fermat principle* states that the travel time along the ray is stationary with respect to the small perturbations in the ray trajectory. Now we rewrite the equation (14), $F(n) = \tau$, as

$$F(n) \equiv \underbrace{F(n_0+u) - F(n_0)}_{\approx F'(n_0)u} + \underbrace{F(n_0)}_{\tau_0} = \tau,$$

so that we have

$$F'(n_0)u = \tau - F(n_0) := g(\alpha,\beta,r) \tag{15}$$

and

$$u = \big[F'(n_0)\big]^{-1}\big(\tau - F(n_0)\big).$$

Note that this is a standard linearization. Now we use the modified Newton iteration method by

$$\begin{aligned} \tau = F(n) &\equiv F(n_k + (n - n_k)) - F(n_k) + F(n_k) \\ &\approx \underbrace{F'(n_k)}_{\approx F'(n_0)}(n - n_k) + F(n_k) \approx F'(n_0)(n - n_k) + F(n_k), \end{aligned}$$

and hence for the next iteration $n_{k+1} := n$ we have

$$n_{k+1} = n_k - \big[F'(n_0)\big]^{-1}\big(F(n_k) - \tau\big).$$

Our next step is to choose $n_0 = n_0(z)$ so that the calculation of $\big[F'(n_0)\big]^{-1}$ would lead to the sequence of usual 2D Radon transform inversions in the discs $|x| < r$, $r \in (0, \rho]$. In the case $n_0 = n_0(z)$ it is easy to see that $\Gamma(n_0)$ is determined by the system of equations

$$p_1 := \langle x, \nu\rangle - h = 0,$$

$$p_2 := z - \varphi(|x|, r) = 0,$$

or $p(x, z) = 0$ in short, where we let $p = (p_1, p_2)$. For the linear velocity $V_0 = m + bz$ we have $n_0 = (m + bz)^{-1}$ and

$$\varphi = \sqrt{r^2 + \left(\frac{m}{b}\right)^2 - |x|^2} - \frac{m}{b}.$$

Since

$$F'(n_0)u = \int\limits_{\Gamma(n_0;\alpha,\beta,r)} uds = \int |\nabla p|\delta(p)u(x, z)dxdz, \tag{16}$$

where $\delta(p) = \delta(p_1)\delta(p_2)$ is a Dirac delta-function concentrated on the curve $p = 0$, where we need to calculate ∇p_j. Here we used the usual formula

$$\int \delta(p(x))u(x)dx = \int\limits_{p=0} \frac{ud\sigma}{|\nabla p|},$$

where $d\sigma$ is a surface element. In general, $p(x) = (p_1(x), p_2(x), \ldots, p_k(x))$, $k \leq n$, $x \in \mathbb{R}^n$, $u \in C_0^\infty(\mathbb{R}^n)$ and

$$|\nabla p| := \sqrt{\det\langle \nabla p_i, \nabla p_j\rangle},$$

$\langle \cdot, \cdot\rangle$ is the scalar product in $\mathbb{R}^n$ and ∇p_j should be linearly independent.

Next we compute $|\nabla p|$ by using the identities

$$\begin{aligned}
\nabla p_1 &= (\partial_x p_1, \partial_z p_1) = (\nu, 0),\\
\nabla p_2 &= (\partial_x p_2, \partial_z p_2) = \left(-\varphi'_{|x|}(|x|, r)\frac{x}{|x|}, 1\right),\\
\langle \nabla p_1, \nabla p_1\rangle &= |\nu|^2 = 1,\\
\langle \nabla p_1, \nabla p_2\rangle &= -\varphi'_{|x|}(|x|, r)\frac{\langle x, \nu\rangle}{|x|} = -\varphi'_{|x|}(|x|, r)\frac{h}{|x|},\\
\langle \nabla p_2, \nabla p_2\rangle &= {\varphi'_{|x|}}^2(|x|, r) + 1,
\end{aligned}$$

and

$$|\nabla p|^2 = \begin{vmatrix} \langle \nabla p_1, \nabla p_1 \rangle & \langle \nabla p_1, \nabla p_2 \rangle \\ \langle \nabla p_2, \nabla p_1 \rangle & \langle \nabla p_2, \nabla p_2 \rangle \end{vmatrix}$$

$$= 1 \times \left(1 + {\varphi'_{|x|}}^2\right) - {\varphi'_{|x|}}^2 \frac{h^2}{|x|^2} = 1 + {\varphi'_{|x|}}^2 \left(1 - \frac{h^2}{|x|^2}\right). \tag{17}$$

We let

$$w^2(x,h) := 1 + {\varphi'_{|x|}}^2 \left(1 - \frac{h^2}{|x|^2}\right), \tag{18}$$

and so from (16) we can conclude

$$\begin{aligned} F'(n_0)u &= \int_{\Gamma(n_0;\alpha,\beta,r)} u d\sigma \\ &= \int |\nabla p| \delta(p) u(x,z) dx dz \\ &= \int |\nabla p| \delta(p_1) \delta(p_2) u(x,z) dx dz \\ &= \int w(x,h) \delta(\langle x, \nu \rangle - h) u(x, \varphi(|x|, r)) dx = g(\alpha, \beta, r), \end{aligned} \tag{19}$$

with $g(\alpha, \beta, r)$ as defined in (15).

In order to obtain an inversion formula for (19), $F'(n_0)u = g$, expressed through the inversion formula of the Radon transform, it is necessary that the weight $w(x,h)$ allows the factorization

$$w(x,h) = w_1(|x|) \times w_2(h^2). \tag{20}$$

Lemma. *The weight $w(x,h)$, given by formula (18), can be represented as a factorization (20), if and only if*

$$n_0(z) = (m + bz)^{-1}.$$

A proof of this result can be found in [5].

We are primarily interested in the case of a *linear sound velocity*, where $n_0(z) = (m + bz)^{-1}$, $b > 0$, and

$$\varphi = \sqrt{r^2 + \left(\frac{m}{b}\right)^2 - |x|^2} - \frac{m}{b},$$

$$\varphi'_{|x|} = -\frac{|x|}{\sqrt{r^2 + \left(\frac{m}{b}\right)^2 - |x|^2}}.$$

By (17), we find

$$
\begin{aligned}
|\nabla p|^2 \equiv w(x,h) &= 1 + {\varphi'_{|x|}}^2 \left(1 - \frac{h^2}{|x|^2}\right) \\
&= 1 + \frac{|x|^2}{r^2 + \left(\frac{m}{b}\right)^2 - |x|^2}\left(1 - \frac{h^2}{|x|^2}\right) \\
&= 1 + \frac{|x|^2}{r^2 + \left(\frac{m}{b}\right)^2 - |x|^2} - \frac{h^2}{r^2 + \left(\frac{m}{b}\right)^2 - |x|^2} \\
&= \frac{r^2 + \left(\frac{m}{b}\right)^2 - h^2}{r^2 + \left(\frac{m}{b}\right)^2 - |x|^2} = w^2(|x|, h).
\end{aligned}
$$

So, using (19), the solution $u(x,z)$ is given by the *inversion formula*

$$
u(x,z) = \sqrt{r^2 + \left(\frac{m}{b}\right)^2 - |x|^2} \times \mathcal{D}^{-1}\left(\frac{g}{\sqrt{r^2 + \left(\frac{m}{b}\right)^2 - h^2}}\right), \tag{21}
$$

where $\mathcal{D}^{-1}$ is the 2D *inverse Radon transform*. We determine the unknown function u from the inversion formula (21) in a disc $x_1^2 + x_2^2 \leq r^2$, thus getting the sought-for refractive index (or slowness) $n_1(x,z) = n_0(z) + u(x,z)$ on the surface of a spherical segment, supported by a circle $x_1^2 + x_2^2 = r^2$.

3.1 Numerical Examples

We assume that the function that represents a sound velocity is strictly increasing (sound velocity increases with depth). This condition ensures that geodesics will bend upward on the surface, where measurements are performed. However, the inversion algorithm works even for media where this requirement is *locally* violated (like in Example 2, see Figures 13 and 14). This function should also be *reasonable* in terms that there are no waveguides and other singularities. The ideal case is when the sound velocity V is close to the linear function $V_0 = m + bz$ that increases proportionally with the depth. In this linear case geodesic lines form arcs of circles, see Figure 10 (a), and so they cover a spherical segment, see Figure 10 (b), provided that sources and receivers lie on the circle in the plane $\{(x,y,z) \in \mathbb{R}^3 : z = 0\}$.

Kinematic data are measured as follows. First one chooses a point on the Earth's surface $\{(x,y,z) \in \mathbb{R}^3 : z = 0\}$ that will be the origin. Then one needs to introduce a finite set of concentric circles with the center at this origin. For each circle we introduce a finite set of points on this circle (preferably equally spaced). In this case, a point can either be a source or a receiver. For a source point, we measure travel times spent by the signal for traveling from the source to the receivers, see Figure 10 (a). We repeat this for all possible source-receiver pairs, see Figure 10 (b). Having kinematic data measured on

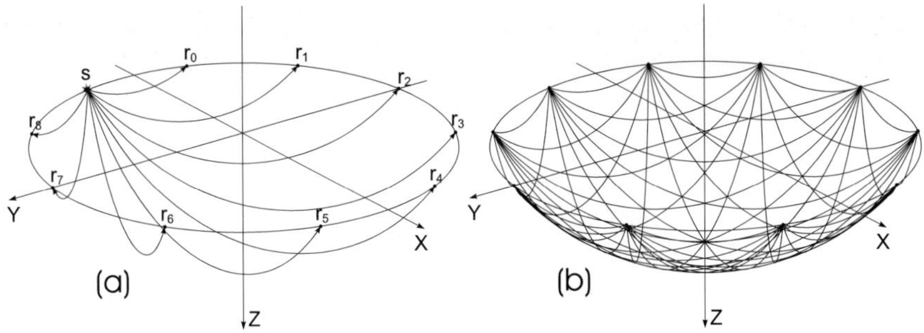

Fig. 10. (a) For the linear sound velocity $V_0 = m + bz$, geodesic lines form arcs of circles. Here $\mathbf{s}$ is a source point and $\mathbf{r_i}$ are the receiver points. (b) Altogether, the geodesic lines cover a spherical segment.

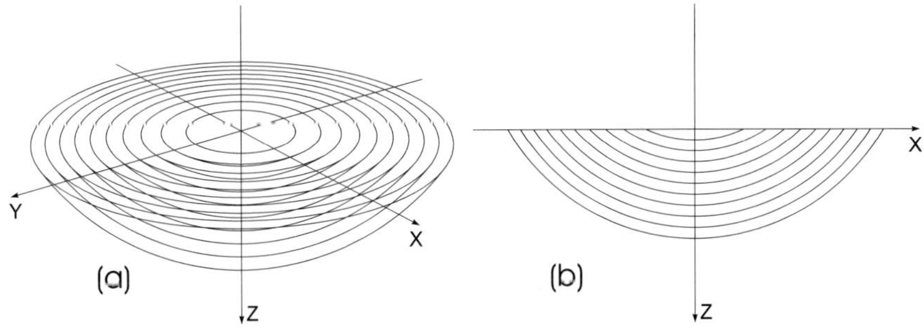

Fig. 11. (a) By measuring kinematic data on a number of concentric circles it is possible to reconstruct the slowness in the whole 3D volume, contained in the most outer spherical segment. (b) Profile view.

a circle, it is possible to reconstruct the slowness on the spherical segment (by using the inversion formula (21)), supported by this circle. Performing kinematic measurements for a number of concentric circles, it is then possible to cover the whole 3D volume, contained in the most outer spherical segment, see Figures 11 (a) and (b).

After travel times are measured we use an inversion formula (21) that reduces our problem to the planar 2D Radon problems in discs. Then we use algorithms from [3] for inverting it.

In the numerical examples presented in this subsection, the parameters for linearization are $m = 0.6$, $b = 1.0$. So, the underlying domain is bounded by the spherical segment with the center at the point $(0, 0, -m/b) = (0, 0, -0.6)$ and the background slowness n_0 is given by $n_0(z) = (m+bz)^{-1}$. Figures 12–17 show the *small additive* $u = n - n_0$ that we seek for and reconstruct. In the following examples, we show reconstructions from the *pure* travel times, not specifically affected by a noise and also from noisy data. Again, even if the noise level is small (compared to the measured travel time τ), it can be comparable in magnitude with the small additive t, that is used for reconstruction. Note also, that even when we do not superimpose any noise, some noise may be due to the numerical solution of the direct kinematic problem.

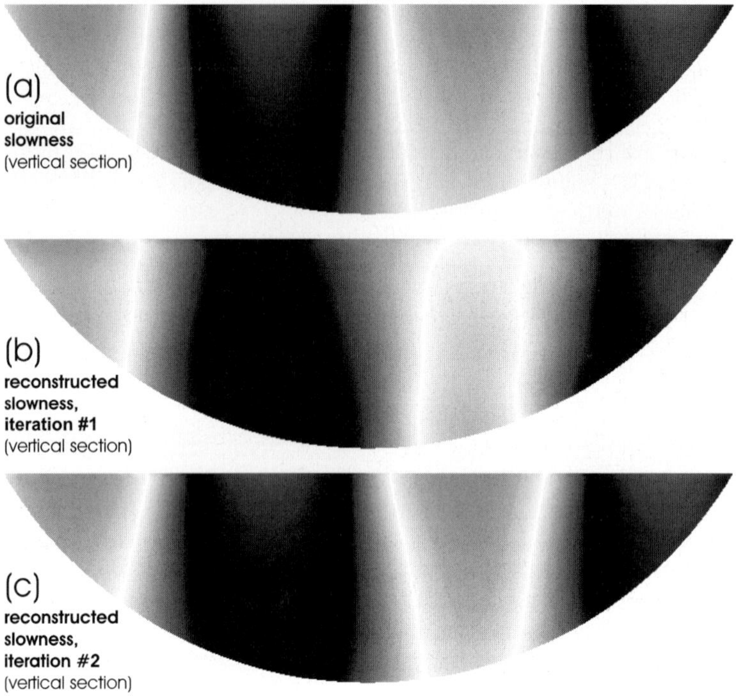

Fig. 12. Original *small additive* of the slowness u, which is the difference between the original unknown slowness n and the given *background* slowness $n_0 = (m+bz)^{-1}$, is shown in a vertical section (a), as well as its reconstructions from the 32 projections on the first (b) and second (c) iterations. For details, see the numerical Example 1. The level of perturbation is 15% of the *background* slowness n_0. 91 spherical layers are used for reconstructing the slowness in a given 3D volume and 32 fan-projections are used for reconstructing the slowness on each layer.

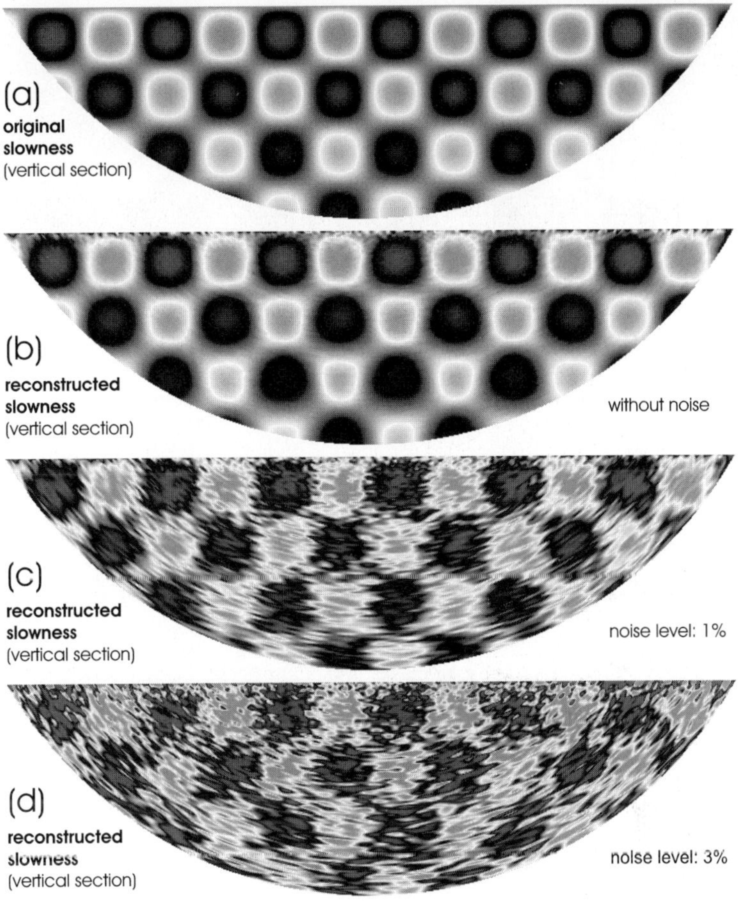

Fig. 13. Original *small additive* of the slowness u, which is the difference between the original unknown slowness n and the *background* slowness $n_0 = (m + bz)^{-1}$, is shown in a vertical section (a), as well as its reconstruction from the 128 projections (b) and from the 128 noisy projections, whose (c) noise level is 1%, (d) noise level is 3%. For details, see the numerical Example 2. The level of perturbation is 5% of the *background* slowness n_0. Such spatial 3D *checkerboard* is usually used for testing the resolution capabilities of the inversion algorithm. 91 spherical layers are used for reconstructing the slowness in a given 3D volume and 128 fan-projections are used for reconstructing the slowness on each layer.

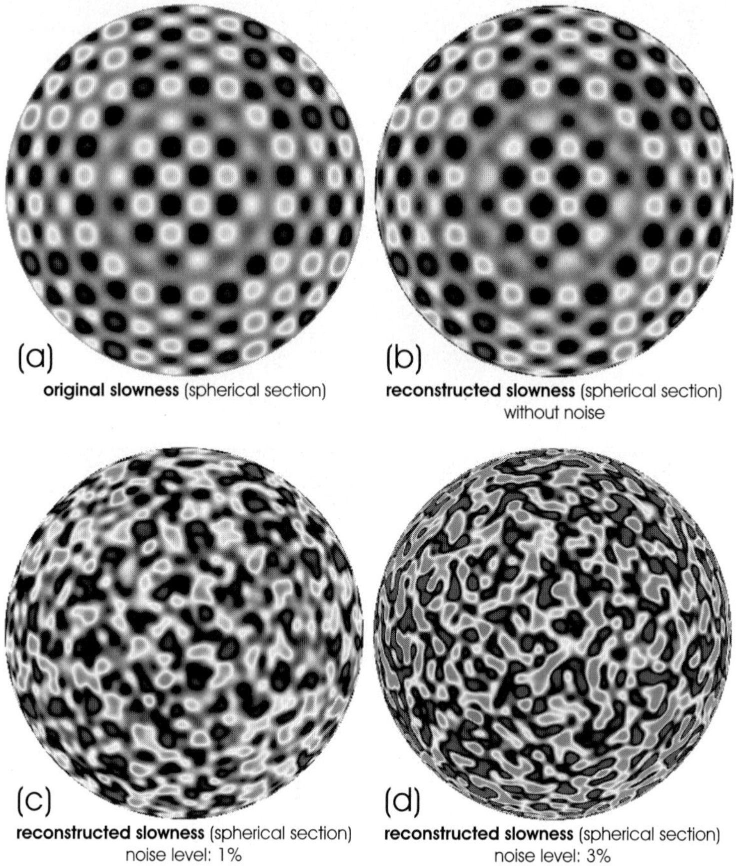

Fig. 14. Original *small additive* of the slowness u, which is the difference between the original unknown slowness n and the *background* slowness $n_0 = (m + bz)^{-1}$, is shown in a spherical section (a), as well as its reconstruction from the 128 projections (b) and from the 128 noisy projections, whose (c) noise level is 1%, (d) noise level is 3%. For details, see the numerical Example 2. The level of perturbation is 5% of the *background* slowness n_0. Such spatial 3D *checkerboard* is usually used for testing the resolution capabilities of the inversion algorithm. 91 spherical layers are used for reconstructing the slowness in a given 3D volume and 128 fan-projections are used for reconstructing the slowness on each layer.

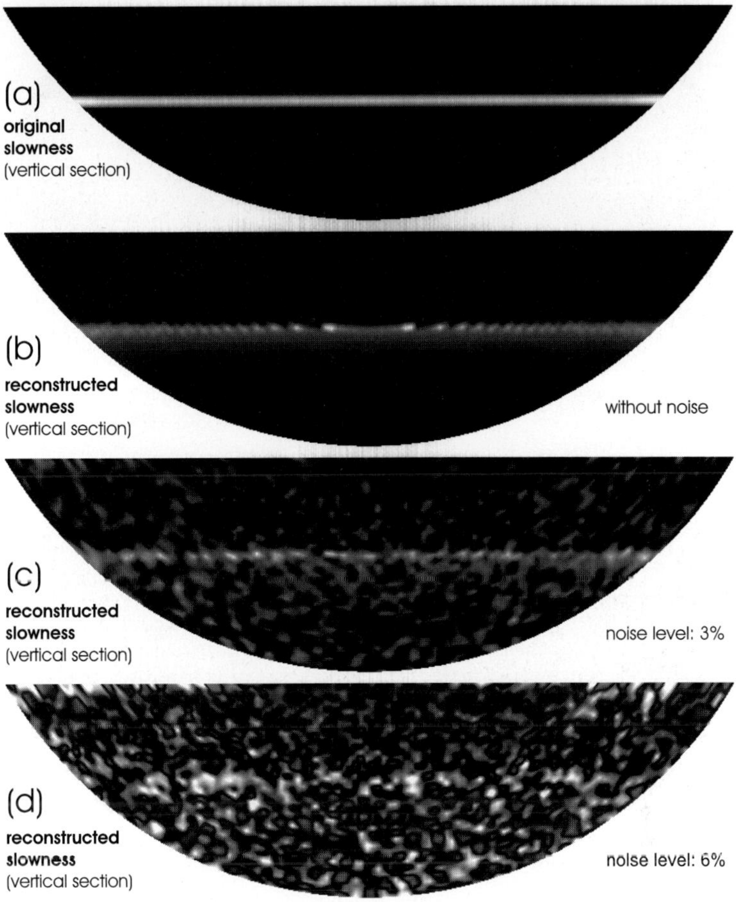

Fig. 15. Original *small additive* of the slowness u, which is the difference between the original unknown slowness n and the *background* slowness $n_0 = (m + bz)^{-1}$, is shown in a vertical section (a), as well as its reconstruction from the 256 projections (b) and from the 256 noisy projections, whose (c) noise level is 3%, (d) noise level is 6%. For details, see the numerical Example 3. The level of perturbation is 24% of the *background* slowness n_0. 37 spherical layers are used for reconstructing the slowness in a given 3D volume and 256 fan-projections are used for reconstructing the slowness on each layer.

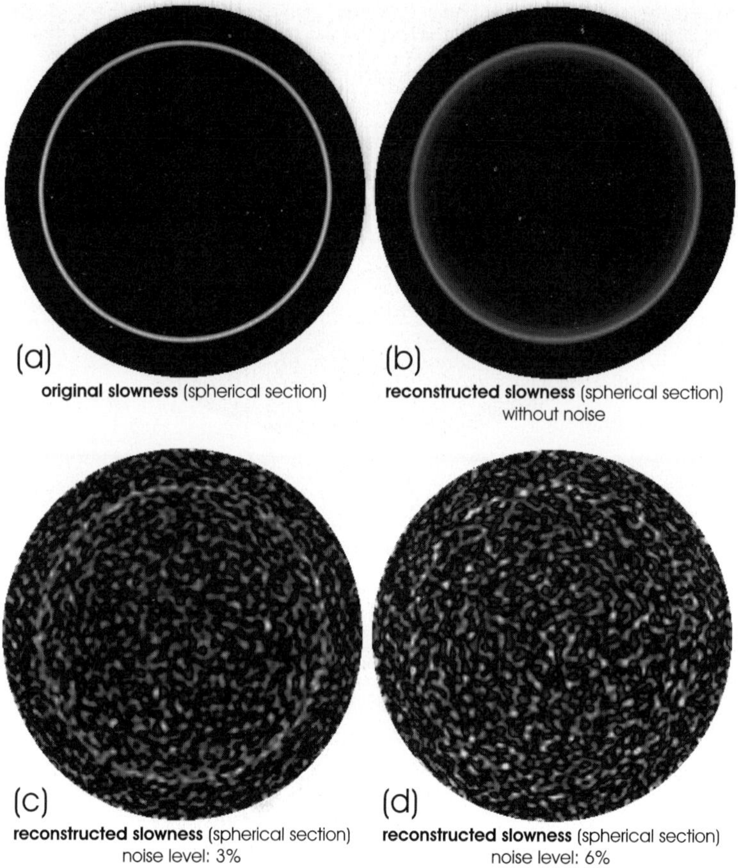

Fig. 16. Original *small additive* of the slowness u, which is the difference between the original unknown slowness n and the *background* slowness $n_0 = (m + bz)^{-1}$, is shown in a spherical section (a), as well as its reconstruction from the 256 projections (b) and from the 256 noisy projections, whose (c) noise level is 3%, (d) noise level is 6%. For details, see the numerical Example 3. The level of perturbation is 24% of the *background* slowness n_0. 37 spherical layers are used for reconstructing the slowness in a given 3D volume and 256 fan-projections are used for reconstructing the slowness on each layer.

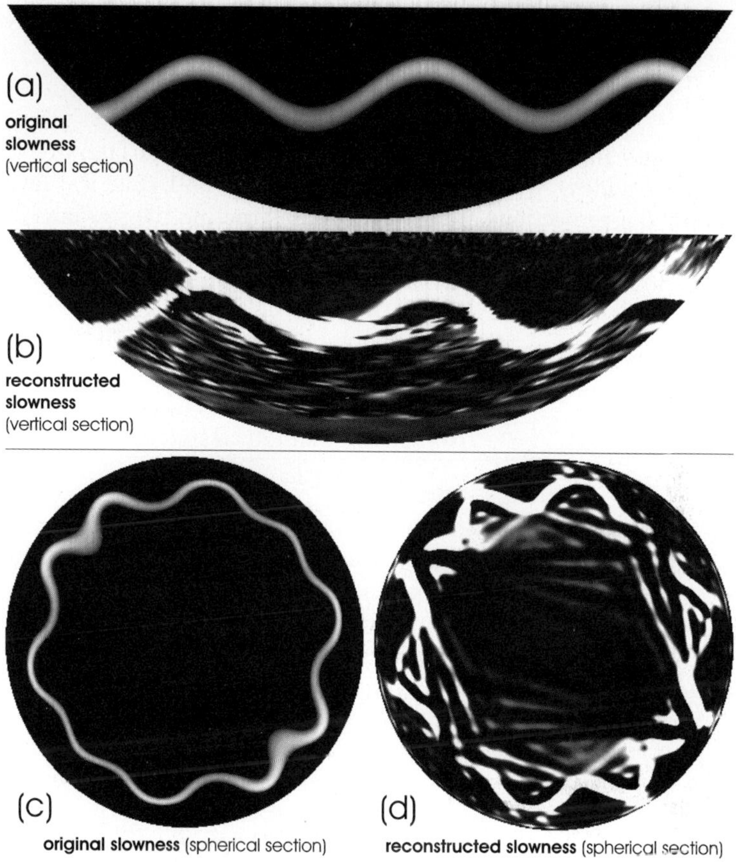

Fig. 17. Original *small additive* u, which is the difference between the original unknown slowness n and the given *background* slowness $n_0 = (m + bz)^{-1}$, is shown in a vertical (a) and a spherical (c) section, as well as its reconstruction from the 128 projections (b) and (d). Note, that some noise comes from the numerical solution of the direct kinematic problem for this nontrivial medium. For details, see the numerical Example 4. The level of perturbation is 5% of the *background* slowness n_0. 91 spherical layers are used for reconstructing the slowness in a given 3D volume and 128 fan-projections are used for reconstructing the slowness on each layer.

Example 1. Consider a slowness $n(x,z)$, $x = (x_1, x_2) \in \mathbb{R}^2$, $z \in \mathbb{R}$ given by the function

$$n(x,z) = \frac{1 + 0.15\sin(5x_1)\sin(5x_2)}{m + bz},$$

that represents the background slowness n_0 with 15% of laterally varying perturbations added, see Figure 12. In this example, 91 spherical layers are used and 32 fan-projections are used for reconstructing the slowness on each layer. The *small additive* $u = n - n_0$ is shown above in (a). The reconstruction on the first iteration is displayed in the middle (b) and the reconstruction on the second iteration is shown below in (c).

Example 2. Consider a slowness $n(x,z)$, $x = (x_1, x_2) \in \mathbb{R}^2$, $z \in \mathbb{R}$ given by the function

$$n(x,z) = \frac{1 + 0.05\sin(20x_1)\sin(20x_2)\sin(20z)}{m + bz},$$

that represents the background slowness n_0 with 5% of perturbations varying in all three dimensions, see Figure 13 for the vertical section and Figure 14 for the spherical section. This example is used to estimate the resolution capabilities of the inversion algorithm. In this example, 91 spherical layers are used and 128 fan-projections are used for reconstructing the slowness on each layer. Parts (a) of the Figures 13 and 14 show the *small additive* $u = n - n_0$, parts (b) show the reconstruction without introducing the noise in the sinogram. Note that some noise comes from the numerical solution of the direct kinematic problem. Parts (c) show a reconstruction from the noisy measurements, the L_2-norm of the noise constituted 1% of the L_2-norm of the measured travel time $\tau = \tau_0 + t$. Note that for the inversion algorithm we use only the small additive t, $|t| \ll \tau$. Parts (d) show a reconstruction from noisy data, where the noise level is 3% of the measured travel times.

Example 3. Consider a slowness $n(x,z)$, $x = (x_1, x_2) \in \mathbb{R}^2$, $z \in \mathbb{R}$ given by the function

$$n(x,z) = \frac{1}{m + bz - F_1(z)},$$

where

$$F_1(z) = \begin{cases} 0.2\exp\left(\frac{0.02^2}{(z-0.25)^2 - 0.02^2}\right)\exp(1), & \text{if } |z - 0.25| \le 0.02, \\ 0, & \text{otherwise.} \end{cases}$$

This example (see Figure 15 for the vertical section and Figure 16 for the spherical section) simulates a horizontal fault surface (the amplitude of this *discontinuity* constitutes approximately 24% of the sound velocity at the same depth) over the medium with linearly increasing sound velocity. In this example, 37 spherical layers are used and 256 fan-projections are used for reconstructing the slowness on each layer. Parts (a) of the Figures 15 and 16 show the *small additive* $u = n - n_0$, parts (b) display the reconstruction without introducing the noise in the sinogram. Note that some noise comes from the numerical solution of the direct kinematic problem. Parts (c) show a reconstruction from the noisy measurements, the L_2-norm of the noise constituted 3% of the L_2-norm of the measured travel time $\tau = \tau_0 + t$. Note that for the inversion algorithm we use only the small additive t, $|t| \ll \tau$. Parts (d) display a reconstruction from noisy data, where the noise level is 6% of the measured travel times.

Example 4. Consider a slowness $n(x,z)$, $x = (x_1, x_2) \in \mathbb{R}^2$, $z \in \mathbb{R}$ given by the function

$$n(x,z) = \frac{1}{m + bz - F_2(z)},$$

where

$$F_2(z) = \begin{cases} \frac{e}{25} \exp\left(\frac{0.04^2}{(F_3(x_1,x_2,z)-0.23)^2-0.04^2}\right) & \text{if } |F_3(x_1,x_2,z) - 0.23| \leq \frac{1}{25}, \\ 0, & \text{otherwise,} \end{cases}$$

and

$$F_3(x_1, x_2, z) = 0.08 \sin(10x_1) \sin(10x_2) + z.$$

This example (see Figure 17 for the vertical and spherical sections) simulates a *bumpy* fault surface (the amplitude of this *discontinuity* constitutes approximately 5% of the sound velocity at corresponding depths) with complicated shape over the medium with linearly increasing sound velocity. In this example, 91 spherical segments are used and 128 fan-projections are used for reconstructing the slowness on each layer. Parts (a) and (c) of Figure 17 show the *small additive* $u = n - n_0$, parts (b) and (d) show the reconstruction without introducing the noise in the sinogram.

Acknowledgement

The work was partly supported by the European Union through the research and training network NetAGES, contract no. IST-1999-29034.

References

1. G.E. Backus (1964) Geographical interpretation of measurements of average phase velocities of surface waves over great circular and great semi-circular paths. Bull. Seismol. Soc. Am. **54**, 571–610.
2. N.N. Bernshtein and M.L. Gerver (1980) Conditions for distinguishability of metrics by hodographs: methods and algorithms for seismic data interpretation. Computational Seismology, vol. 13, Moscow, Nauka, 50–73.
3. A.A. Boukhgueim (2003) *Numerical Algorithms for Attenuated Tomography in Medicine and Industry.* Ph.D. dissertation, University of Vienna.
4. A.L. Bukhgeim (1983) On one algorithm of solving the inverse kinematic problem of seismology. Numerical Methods in Seismic Investigations, Nauka, Novosibirsk (in Russian), 152–155.
5. A.L. Bukhgeim, S.M. Zerkal', and V.V. Pikalov (1983) On one algorithm for solution of a 3D inverse kinematic problem of seismology. Methods for Solution of Inverse Problems. Novosibirsk, 38–47.
6. A.M. Dziewonski (1999) Earth's mantle in three dimensions. *Seismic Modelling of Earth Structure*, E.G. Boschi, Ekström and A. Morelli (eds.), Editrice Compositori, Bologna, 507–572.
7. A.M. Dziewonski and D.L. Anderson (1981) Preliminary reference Earth model. Phys. Earth. Planet. Inter. **25**, 297–356.
8. A.M. Dziewonski and D.L. Anderson (1984) Seismic tomography of the Earth's interior. American Scientist **72**, 483–494.
9. A.M Dziewonski, T.A. Chou, and J.H. Woodhouse (1981) Determination of earthquake source parameters from waveform data for studies of global and regional seismicity. J. Geophys. Res. **86**, 2825–2852.
10. A.M. Dziewonski, B.H. Hager, and R.J. O'Connel (1977) Large-scale heterogeneities in the lower mantle. J. Geophys. Res. **82**, 239–255.
11. A.M. Dziewonski and J.H. Woodhouse (1987) Global images of the Earth's interior. Science **236**, 38–47.
12. P. Funk (1916) Über eine geometrische Anwendung der Abelschen Integralgleichung. Math. Ann. **77**, 129–135.
13. G. Herglotz (1905) Über die Elastizität der Erde bei Berücksichtigung ihrer variablen Dichte. Z. für Math. Phys. **52**(3), 275–299.
14. H. Kanamori (1970) Velocity and Q of mantle waves. Phys. Earth. Planet. Int. **2**, 259–275.
15. M.M. Lavrentiev, V.G. Romanov, and V.G. Vasiliev (1970) *Multidimensional Inverse Problems for Differential Equations.* Lecture Notes in Mathematics **167**. Springer-Verlag.
16. R.G. Mukhometov (1977) A problem of reconstructing 2D Riemannian metric and integral geometry. Dokl. An. SSSR. vol. 296, no. 2, 279–283.
17. J. Radon (1917) Über die Bestimmung von Funktionen durch ihre Integralwerte längs gewisser Mannigfaltigkeiten. Ber. Verh. Sachs. Akad. Wiss. Leipzig. Math. Nat. Kl. **69**, 262–277.
18. B.A. Romanowicz (1991) Seismic tomography of the Earth's mantle. Ann. Rev. Earth Planet. Sci. **19**, 77–99.
19. B.A. Romanowicz (1995) A global tomographic model of shear attenuation in the upper mantle. J. Geophys. Res. **100**, 12375–12394.

20. V.A. Sharafutdinov (1993) *Integral Geometry of Tensor Fields.* Novosibirsk, Nauka.
21. V.A. Udias (1999) *Principles of Seismology.* Cambridge University Press.
22. E. Wiechert and K. Zoeppritz (1907) Über Erdbebenwellen. Nachr. Königl. Gesellschaft Wiss. Göttingen **4**, 415–549.
23. J.H. Woodhouse and A.M. Dziewonski (1984) Mapping the upper mantle: three-dimensional modeling of Earth structure by inversion of seismic waveforms. J. Geophys. Res. **89**, 5953–5986.
24. R.L. Woodward and G. Masters (1991) Upper mantle structure from long-period differential travel times and free oscillation data. Geophys. J. Int. **109**, 275–293.

Part III

Reservoir Modelling and Simulation

From 3D Seismic Facies to Reservoir Simulation: An Example From the Grane Field

Alexis Carrillat and Brice Vallès

Schlumberger Stavanger Research, P.O. Box 8013, N-4068 Stavanger, Norway

Summary. A new *seismic to simulation* workflow is proposed, where the aim is the reduction of the overall turn-around time, from seismic data acquisition to reservoir model building and simulation. To this end, new automated procedures are established: firstly, for discriminating seismic data into three-dimensional seismic facies, and secondly, for building a voxel-based reservoir model.

This chapter is divided into three parts. In the first part, automated three-dimensional seismic facies mapping is discussed, where both the stratigraphic and the structural framework of the seismic data are reflected. The resulting seismic facies are then identified with lithologies by calibration against well data.

In the second part, automated voxel grid extraction for reservoirs is explained. The required input is the voxel size together with the top and bottom horizons delimiting the reservoir extents. The calibrated three-dimensional seismic facies are then used to associate each voxel with porosity and permeability values. This last automated step results in a voxel-based reservoir model.

Finally, in the third part, an application of the new workflow is presented. To this end, a case study for the Grane field is used. The selected simulation scenario models a three-phase reservoir life.

1 Three-Dimensional Seismic Facies Model

1.1 Introduction

In this chapter, we propose a new workflow for reservoir model building and updating. First, we shall review the evolution and current state in quantitative interpretation for reservoir geophysics. We mainly look at the role of automated interpretation in model building. The new method of automated interpretation presented here, has the power to help geoscientists perform interpretation tasks faster and more quantitatively. This quantitative information is translated efficiently to the reservoir model. A full workflow from seismic to simulation is demonstrated on the Grane field using this new quantitative interpretation paradigm.

The main idea in this new method is to accelerate the seismic to simulation process by establishing a new streamlined workflow that brings the seismic closer to the reservoir model. One of the aims is to involve the geoscientist at a higher level, allow him/her to spend more time on the analysis, and less on tedious and repetitive interpretative tasks. In this new approach, the interpreter gets involved in powerful post stack image processing methods that can condense large volumes of data to relatively simple patterns, or enhance and reveal subtle key features. These patterns and features can be easily visualized and analysed in their true three-dimensional nature, and then be selected and calibrated with well data and finally be forwarded to the reservoir model.

With increasing seismic data volumes, real-time reservoir data and well data being collected, today's geoscientists involved with hydrocarbon exploration and reservoir characterization face a need for improved and faster interpretation in the model building workflows. The interpreter is expected to comprehend an increasing number of seismic volumes, processed and migrated with different parameters, acoustic impedance data, AVO (amplitude versus offset) data plus a plethora of attributes for reservoir description. In addition, multi-sets of time-lapse seismic data come into play for dynamic reservoir modelling. At the other end, reservoir engineers are expected to run various scenarios and update the model with soft and hard data as they are acquired.

1.2 From Attributes to Geology

Starting from post stack seismic volumes in time domain and synthesizing all relevant information is challenging. It is quite common to overlook some aspects of the seismic data because of time constraints in operations. In the first part of this chapter, we present a procedure for combining this information into a three-dimensional (3D) geological model using 3D seismic facies classification [7, 8]. This method enables 3D multi-attribute seismic data analysis, and uses an interpreter-guided neural network to produce a 3D geological model of the reservoir. Optimally, the workflow would impose that the seismic volumes, from which the attributes are generated, are converted to the depth domain prior to the 3D classification. Most methods of 3D seismic classification are primarily based on the analysis of waveforms bounded between two mapped horizons, with the results displayed in 2D cross section or map view [1, 10]. A major disadvantage of this approach is that it *collapses* the 3D data onto a single 2D surface for the purposes of display and interpretation. Thus, three-dimensional geological features of different levels/age identified from within the data volume are superimposed over each other on a single 2D map/section. Therefore it is difficult to assess their true 3D form, cross cutting relationships and connectivity, information that is critical to the understanding of reservoir volume and heterogeneity (Figure 1).

In our approach, the 3D nature of the seismic data is preserved. The analysis is made using a neural network algorithm producing a 3D classification output. The value of this approach to 3D seismic texture mapping has already been demonstrated in the analysis of gas chimneys [25, 34]. Distinct advantages of the approach are that: (1) it is independent from and requires no previous horizon interpretation, (2) multiple attributes are simultaneously

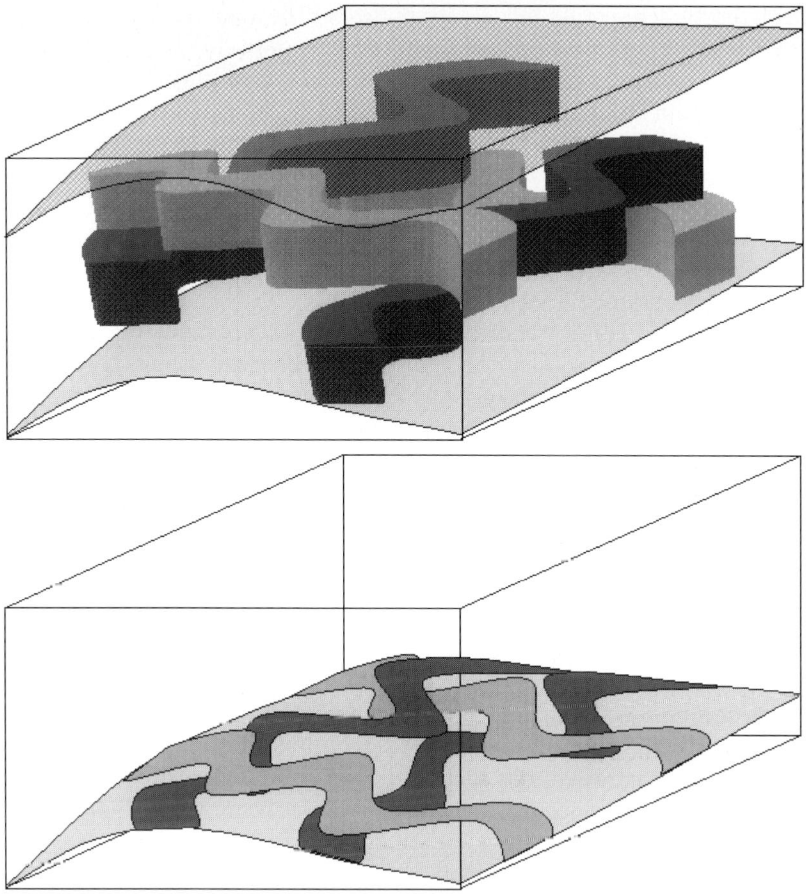

Fig. 1. Grid-based classification between two stratal surfaces or horizons produces an inaccurate representation of the geological features such as channels distributed in the volume by collapsing the information onto a grid. Connectivity of drainable pay zone may be overestimated and heterogeneity misevaluated.

analysed, (3) the technique is independent from the dip of the reflectors, (4) the automation offers fast turn-around and reproducible results, (5) it provides a method of uncertainty evaluation, (6) the interpreter can qualitatively and quantitatively evaluate and edit training data and/or import/remove input data cubes in order to fine-focus results, (7) the results are displayed as a volume in 3D space so that the true shapes and relationships of the geobodies (defined later) can be analysed. The latter is considered to be of fundamental importance since it can provide new geological information making for the rapid assessment of geobody volume and connectivity and better prediction/understanding of the reservoir. (8) The classification cube can be translated to a reservoir model. By doing so, the reservoir model conserves the geometrical primitives given by 3D seismic data. (9) It provides a mean to evaluate more quickly and quantify risk and uncertainty for well planning and production scenarios.

Among these seismic attributes, those derived from the complex seismic trace analysis, developed in [41], have been extensively used in the industry for reservoir characterisation and monitoring. These attributes are derivatives of the basic seismic measurements, meaning that the use of complex trace analysis is simply a way to present and analyse a limited amount of information contained in the seismic trace.

The standard Hilbert transform attributes [41] are given by the complex trace analysis, and produce the well-known reflection strength (amplitude of the envelope), instantaneous phase, instantaneous frequency, or apparent polarity. In this approach, the information obtained is time, amplitude, frequency or attenuation, and is used as an input into industry's standard grid-based classification. Additional information can be derived directly from the reflection amplitude or from summation of amplitude value within intervals. This is the case for composite amplitude, average absolute amplitude, root-mean-square amplitude, number of zero crossings, and of minima or maxima.

More recently, numerous case studies have demonstrated the value of wave shape classification using neural networks. With this method, the seismic trace is decomposed into components of amplitude and frequency information such as in VRS (Volume Reflection Spectrum) attributes [39].

Surface derived attributes such as dip, azimuth, curvature, edge and illumination may be included in the classification scheme or used on their own to address and reveal structural information.

AVO attributes have also widely proven their discrimination power. Combined with classification they help identifying the contribution of offset traces for lithology, pressure and fluid mapping in definition of static reservoir models [2], as well as in 4D scenarios and dynamic reservoir models [21]. Static reservoir models refer to the rock properties of the model (porosity and permeability distribution), while dynamic reservoir models refer to the fluid saturation and pressure, and their relative changes over time. Combined attribute analysis and rock physical models guarantee a successful strategy to calibrate and understand attribute responses. It allows differentiating pressure, saturation and fluid movements from noise. For reservoir monitoring from seismic data, attribute analysis and rock physical models provide a mean to address repeatability issues when analysing repeated surveys in 4D seismic.

As mentioned before, there are a number of limitations due to grid-based calculation of attributes that are carried into the classification, and thus to the final interpretation and analysis. By nature, reservoirs are not limited to layer-cake geology and even when compensated for lateral variation in thickness, grid-based attributes are affected by the seismic response at tuning locations, pinch-outs, truncations, etc. Numerous case studies, in particular 4D analyses, have proven that in such cases, structural features can be dominating and may perturb detection of the subtle variations in the signal.

Recent research and development efforts have been focused on improving seismic attribute extraction invariant to or steered by the local orientation (dip and azimuth) of the reflectors (cf. the chapter [35] of Randen and Sønneland). This new approach provides a way to honour the stratigraphic orientation of reflections and has determined a new way to handle reservoir characterisation and monitoring by the use of volume attributes.

Combining the volume attributes into a simpler model is performed using 3D classification of seismic data [40], which produces a synthesised model reducing the large data set to a simpler pattern. Using this condensed data, geologists and interpreters work at a higher level on interpreting the classification results and use this output as a guide and as a living interpretation model. It allows analysis and editing of geobodies based on quantitative properties, such as volume, position, orientation, connectivity and more.

These seismic facies geobodies, whether they are sedimentary, structural, diagenetic, or fluid related constitute an attempt to identify, isolate and extract geological/rock physical objects or properties in one single coherent modelling entity. The system level allows interactive visualization and analysis of these geobodies and their more efficient transfer to the reservoir model.

1.3 Seismic Facies Analysis

The procedure for seismic sequence stratigraphic and facies interpretation is well established, and is an evolution of the basic principles outlined by the seminal papers of Payton [31]. In order to clarify our use of terminology, we first refer to original definitions of seismic stratigraphy or seismic sequence analysis. Seismic sequence analysis is the seismic identification and interpretation of depositional sequences. As defined in [26, 27, 28], a depositional sequence is *"a stratigraphic unit composed of a relatively conformable succession of genetically related strata and bounded at its top and base by unconformities or their correlative conformities"*, see Figure 2. From the same authors, seismic facies analysis corresponds to the description and geological interpretation of mappable three-dimensional seismic unit composed of groups of reflections whose parameters and signal characteristics are relatively homogeneous and differ from those of surrounding facies units [26, 27, 28]. This particular description of seismic facies refers to the definition of texture or pattern from signal processing.

In our approach, we consider that reflections of seismic waves within sedimentary rock bodies produce an image of their external shape and internal configuration or texture. The study of these external shapes and internal textures is also referred to as seismic facies analysis. For further details on this, we refer to the chapter [37] of Schlaf and Randen. For instance a meandering channel produces an external shape that can be filled with different seismic

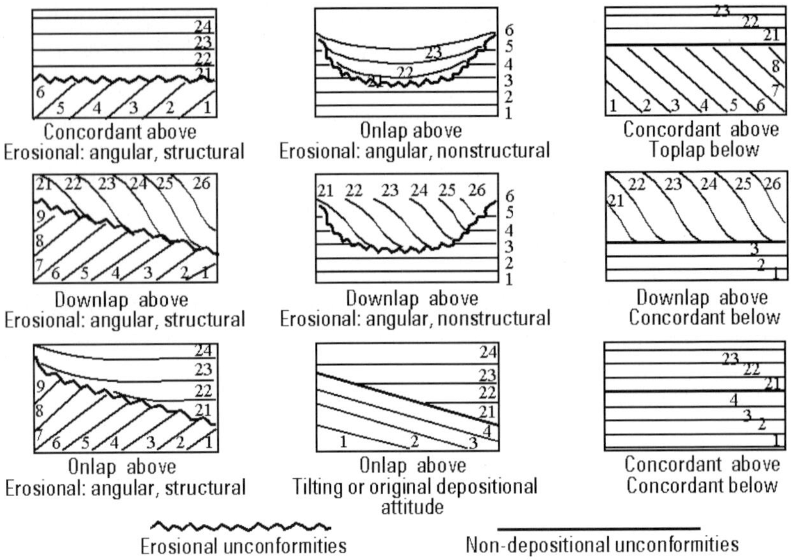

Fig. 2. Seismic reflection configurations that define conformities and unconformities (due to Brown and Fischer [6]).

textures or facies (onlap, divergent, mounded, prograded, complex, chaotic and transparent fills), the same remark applies to carbonate buildups (reefs and mounds) which present an external shape and various internal seismic patterns.

1.4 Seismic Texture Attributes

Automated 3D seismic facies mapping must honour the stratigraphic and structural framework of the seismic data. Similarly to the interpreter's eye that follows the local orientation of reflectors and stratigraphy across a seismic section, the texture attributes have to be dip-steered. For details, see the chapter [35] of Randen and Sønneland. The *dip-steering* constraint is achieved by using 3D attributes that follow and capture the seismic patterns. These attributes are referred to as 3D texture attributes, since they are able to describe the reflector-geometry in a small 3D neighbourhood (local orientation-guided multitrace attribute). The combination of the 3D texture attributes, containing orientation and/or continuity information, identifies the stratigraphic patterns in the seismic data.

3D seismic texture attributes presented in this chapter are subdivided into two groups. The first includes kinematic texture attributes that capture the reflector orientation or the reflector continuity information. The second defines dynamic texture attributes that capture features in the seismic signal such as spectral representations or amplitude behaviour. More information about texture attributes can be found in the chapter [35] of Randen and Sønneland.

The traditional approach of seismic attribute computation is to extract attributes along vertical traces, irrespective of any dipping nature of the reflections. This industry standard clearly implies a risk of introducing artefacts, when the stratigraphic pattern is not layer cake and flat. As a more consistent alternative, seismic texture attributes compensate for the dip and azimuth or make the attribute extraction invariant to local dip and azimuth. In addition, they are genuine 3D with no trace bias as opposed to coherency and semblance attributes. Moreover, these attributes are amplitude-invariant.

A geometrical tensor is used in 3D seismic texture attributes for dip and azimuth estimation. This local dip and azimuth estimation (local orientation estimation) approach is based on three steps (see [35, Section 2]):

- Gradient vector estimation.
- Local gradient covariance matrix estimation.
- Principal component analysis. The principal eigenvector represents the normal to the local reflection dip and azimuth.

Kinematic Texture Attributes. The geometrical tensor is used as an input to other processes, in particular kinematic texture attributes and some discontinuity and fault attributes.

The flatness attribute is a measure in three dimensions of the degree to which local seismic reflections are flat. The flatness is orientation-invariant, i.e., planar reflections produce the same attribute response, no matter whether they are dipping or horizontal. This attribute is genuine 3D and amplitude-invariant.

The Gabor filter bank produces a set of cubes, which show the frequency characteristics of pseudo-traces orthogonal to the local stratification. The orthogonal pseudo-traces are derived using the local dip and azimuth estimate.

The edge enhancement attribute is a measure of the *edginess* of the seismic reflections calculated parallel to the local stratification. The edge enhancement attribute enhances spatial discontinuities by measuring changes in the signal amplitude. The edge enhancement attribute uses the local dip estimate of the reflection layers. The local dip estimate represents a plane, and by projecting the vector with derivatives, onto this plane, changes that are nearly perpendicular to the reflector will produce vectors with small magnitude, whereas changes in the direction of the reflector will produce vectors with larger magnitudes. The edge enhancement attribute is genuine 3D, orientation-invariant, but amplitude-dependent. The latter may be relevant when the amplitude correlates with the signal quality.

The variance attribute is also used, even if it does not share all characteristics of 3D kinematic texture attributes. This attribute uses the local variance as a measure of signal unconformity or discontinuity. The variance is computed for each sample in small horizontal sub-slices. If the slice is within an unbroken reflection layer, the amplitude variance will be small, whereas amplitude changes due to discontinuous horizon will result in a large variance.

Dynamic Texture Attributes. Dynamic texture attributes are generated from the original seismic cube. The volume reflection spectrum (VRS) attributes realize spectral analysis of the reflectivity response for each seismic trace [39]. Each trace is characterized in terms of its eigenvalue (spectral attribute) and the associated eigenvector (orthogonal, polynomial), to approximate the reflection amplitude along the trace in a least squares sense. For texture mapping a set of discrete spectral VRS coefficients is combined into a composite spectral representation.

1.5 Seismic Facies Classification Workflow

The complete classification workflow involves several steps (see Figure 3): (1) Generation of the seismic texture attribute cubes. (2) Running a classification in an unsupervised mode (A) and inspection of the results. (3) Running a supervised classification with user-defined selection of training points (B), and examination of these points in N-dimensional space to ensure adequate clustering and minimal overlap between the training points. (4) Evaluation of the supervised classification output, using (i) checking the result in 3D and comparing the results against the seismic/attribute cubes used and, (ii) the uncertainty analysis of the classification with probability cubes. (5) The removal/incorporation of training points/seismic attribute cubes prior to making a revised classification. (6) Final tuning of the neural network parameters. (7) Visualization and interpretation of the data and the analysis of facies associations and structures in 3D space. (8) Finally, the calibration of the results against well data allows establishing a deterministic link between seismic facies and lithology.

Steps 5-7 are necessarily iterative and it should be expected that several modifications to both training data picking and/or selection of input data (attribute cubes) are required before a satisfactory result is achieved. In order to speed up the process, the workflow can be first run on a representative subvolume. After the training data have been picked, checked and parameters optimised, the neural network classification can then be run on the full volume of interest (Figure 3).

3D Neural Network Classification. Neural networks have often been used to analyse data, and to recognize patterns within data [9]. Automated 3D mapping of seismic textures through 3D classification [40] allows combining geometric and signal attributes information into a 3D seismic facies cube. In the procedure, the three-dimensional classification is done unsupervised as a first step to identify the natural clustering of the data. This initial step allows the interpreter to guide the picking of training data and choice of specific

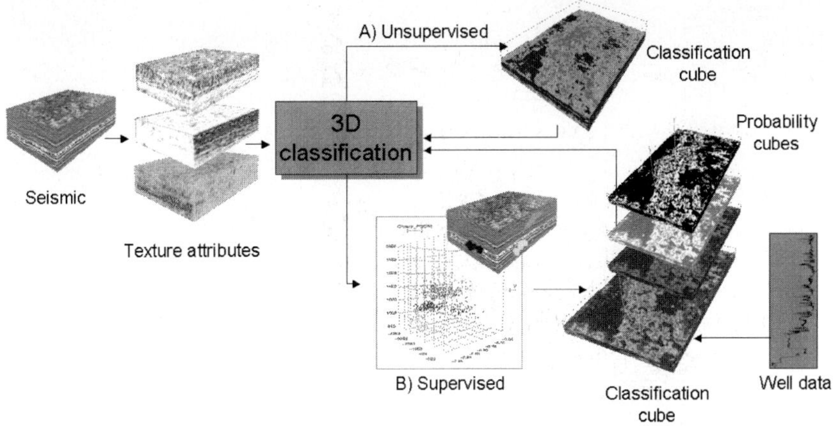

Fig. 3. [Reproduced in colour in Plate 16 on page 433.] Workflow for seismic facies mapping using texture attributes and 3D classification based on neural network algorithm.

features for running supervised classification as a second step (Figure 3). The supervised classification process allows the interpreter to select training data on different seismic facies (calibration samples) within the volume of interest. Once the interpreter is satisfied with the choice of the training points, these are used to train the artificial neural network. In the supervised classification, both the inputs and the outputs are provided. The results of the classification are then output together with probability cubes or confidence measures for each of the different classes identified, expressed as the distance from the cluster centre or by the means of validation data (Figure 3).

Training Data Picking. Training data are picked on the original seismic or on attribute cubes. These training data define calibration samples for seismic facies (Figure 4) that are picked in the 3D space of the seismic data. Statistically, a large amount of training data (several hundreds to thousands of samples) makes the classification more stable than a small number. This is due to the fact that training data are locally correlated. A large number of training data will span a larger portion of the class distribution and hence be more representative.

The training data can be inspected in attribute-space [40], an N-dimensional space, where N is the number of attributes used in the training. Separated clusters in attribute-space imply good discrimination between the seismic stratigraphic patterns that need to be mapped (Figure 5). Potential outliers in the clusters can be readily identified and quality controlled since there is a one-to-one correspondence between the training points in attribute-space and the training points in the seismic cube. Mispicked training data can therefore

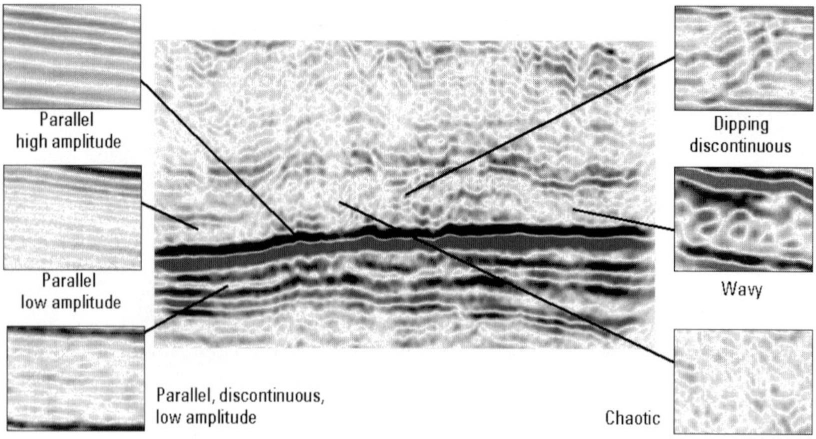

Fig. 4. [Reproduced in colour in Plate 17 on page 433.] Picking of training data done by digitising portions of seismic patterns on seismic cross-sections. In this example six seismic facies or textures were defined, and several calibration samples used for each facies. Training data are picked on different sections within the zone of interest.

easily be deleted. In some cases, principal component analysis allows a better selection of the attributes set by avoiding correlated attributes.

Once training data have been picked and checked, the neural network can be trained and classification be run on a subvolume for evaluation. After parameter optimization, the classification is run on the full volume of interest.

Facies Analysis and Calibration with Well Data. The interpretation phase involves analysis of patterns, facies associations and calibration with well data. The 3D mapping of seismic facies defines a preliminary geological/structural model based on seismic patterns that needs to be calibrated. As different lithologies, sedimentary bodies or structural features may produce the same seismic pattern, the calibration of the patterns is a significant step and involves all available knowledge from the reservoir under study. Geophysical information (AVO data, amplitude attributes, acoustic impedance), well data and geological knowledge such as depositional, structural, diagenetic, or reservoir fluid characteristics, pressure and saturation distribution, will be incorporated and treated in that step (Figure 6).

Well information is used deterministically to assign a lithology to every seismic facies of the model. This process can be iterative or hierarchical and allows, first locally and then more completely, linking seismic facies with the corresponding lithologies or rock properties, as knowledge is gained. In the end, seismic facies and geobodies based on texture attributes honour stratigraphic and structural information and carry lithological and/or fluid

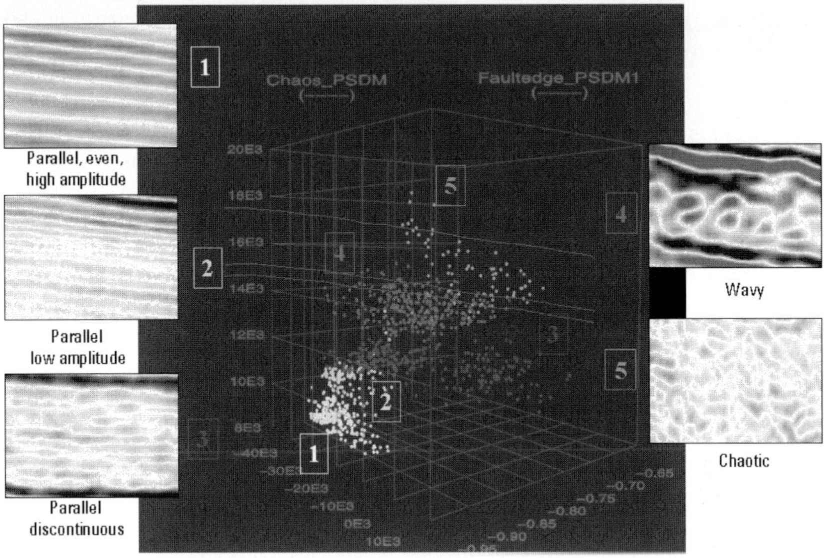

Fig. 5. [Reproduced in colour in Plate 18 on page 434.] Training data distribution in attribute space. Each class defines a cluster in the attribute space, where training data can be checked and edited.

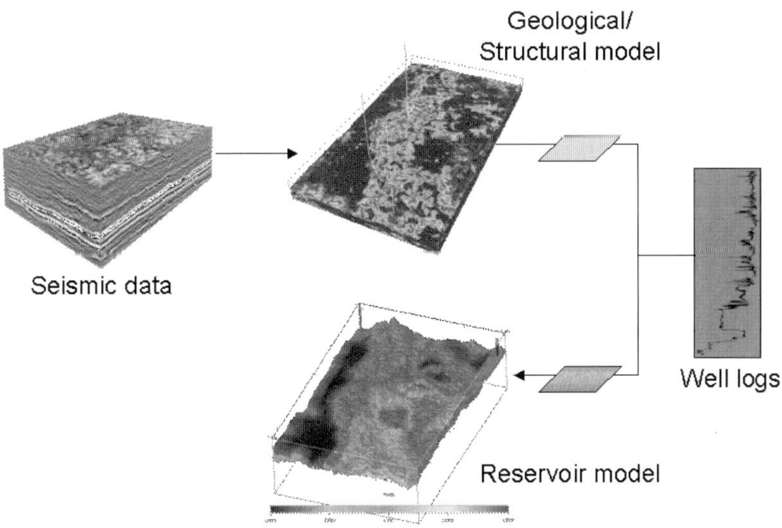

Fig. 6. [Reproduced in colour in Plate 19 on page 434.] Classification of seismic facies based on texture attributes defines a geological/structural model. Using iterative and hierarchical classification capability, seismic facies can be calibrated and assigned to lithology and fluids using well data or another set of attributes such as amplitude-based, or AVO data.

property information. This detailed structural framework information is used during the property population of reservoir models.

2 Building a Voxel Model

Once a three-dimensional seismic facies model has been generated and calibrated with well data, the next step consists in translating the facies model more efficiently to the simulation domain. The challenge faced is to minimize the manual work and to preserve as much as possible the detailed representation carried by the three-dimensional seismic facies model in order to improve the seismic to simulation workflow.

It is shown in the following subsections that combining the voxel model approach, introduced hereafter, together with three-dimensional seismic facies results fulfil all of the above specifications.

2.1 Definitions

As defined in [20], *"a voxel is the cubic unit of volume centered at the integral grid point. As a unit of volume, the voxel is the three-dimensional counterpart of the two-dimensional pixel, which represents a unit of area"*, see Figure 7. Volumes with parallelepiped rectangle shape and not only cubes are being considered.

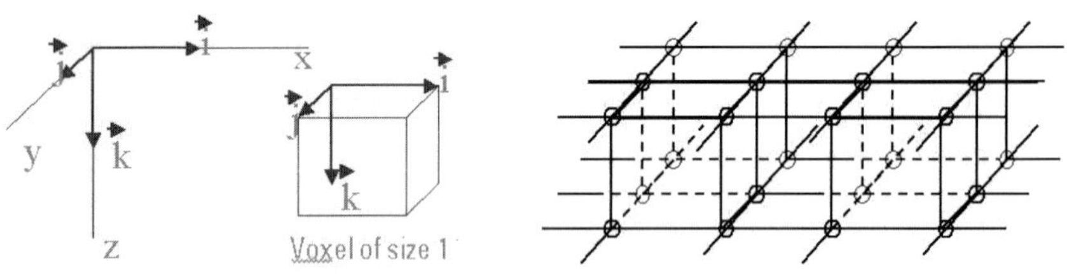

Fig. 7. Representation of voxels. Left: an elementary voxel. Right: a simple voxel grid composed of six voxels and divided into three rows (x direction) and two columns (y direction).

Consequently, a voxel grid is defined as a three-dimensional numerical mesh composed of voxels. Figure 7 (right) illustrates a simple voxel grid.

A voxel model then defines a *voxelized* voxel grid, i.e., a voxel grid associated with numeric values representing some *"measurable properties or independent variables of the real phenomenon or object residing in the unit volume represented by the voxels"* [20]. These properties or variables can be of different types: sampled data, computed data, simulation results etc.

2.2 Workflow

Nowadays the reservoir model building workflow is mainly divided into four steps: (1) definition of the reservoir boundaries (and especially the top and bottom horizons), (2) building a geologically true structural model (layering, faults etc.), (3) population of properties (porosity, permeability etc.) by kriging or co-kriging for instance and then (4) upscaling the model in order to obtain a reservoir model that can be handled by today's computer capabilities. Many steps in this workflow require manual work, in particular the task of building the structural model.

The method described here proposes a new approach to automate parts of the work and so to reduce the manual work. This leads to a speed-up of the whole *seismic to simulation* process and so this allows faster and easier model updating.

The voxel-based workflow can be divided into three main steps:

(1) automated grid and fault extraction, classification of seismic facies;
(2) definition of the reservoir model geometry and its automated extraction;
(3) property voxelization which associates physical property values to each numerical cell.

Figure 8 summarizes and compares the main steps of the two workflows.

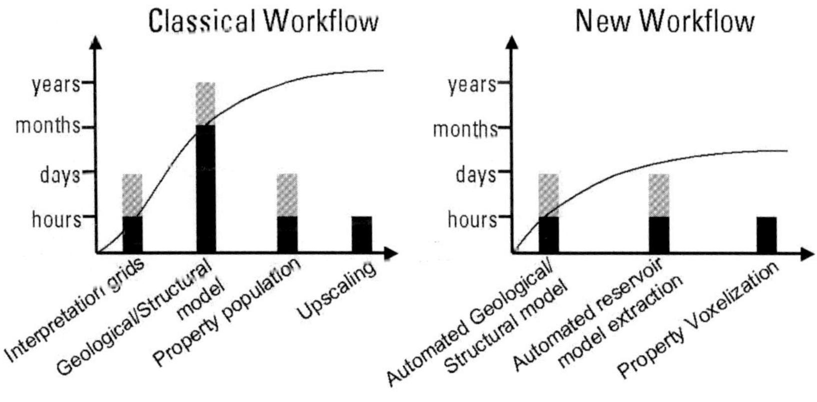

Fig. 8. Workflows. Left: the classical workflow. Right: the voxel model workflow. The abscissa summarizes the main steps in a chronological order. The ordinate gives an estimate of the time that each step roughly requires. For each step, black means the time usually needed and grey the time required in some particular cases.

(1) Automated Grid and Fault Extraction, Seismic Facies Classification. The first requirement for building a reservoir model is to determine its extents and characteristics.

Horizons are usually seen as interfaces between subsurface layers of rocks of different properties. Thus reservoirs are discriminated in the depth extent from seismic cubes by their top and bottom horizons corresponding to their upper and lower limits, respectively. Horizons are usually manually or semi-automatically interpreted by analysing seismic data. However, new tools have recently emerged that allow an automated extraction of horizons from seismic. These new techniques are based on classification of a set of seismic waveform attributes along extrema (minima or maxima) in a seismic post stack cube. For further details on this, we refer to [4, 5].

When analysing seismic data, geoscientists are often able to identify horizons, other than the top and bottom ones, inside the reservoir. Combining them with available well-log data helps to refine the reservoir description. That information is then used later when building the reservoir model.

Faults can be of different types: sealing or permeable. These properties can change through the life of the reservoir. When sealing and intersecting each other to form a closed geometry, faults divide the reservoir into several isolated compartments. When permeable, fluids can migrate through these faults. During the reservoir production life, fault properties can change due to, for instance, pressure changes on either side of the fault-zone, thereby leading to different reservoir flow behaviour. This in turn would affect production scenarios. Therefore, fault systems and fault properties are of great importance for reservoir simulation studies. As for horizons, mapping of faults from seismic cubes can be a time-consuming and difficult task. New technologies have also been established to automate the extraction of faults from seismic, thereby allowing geoscientists to focus more on the analysis of fault systems and less on the mapping and editing. For further details on automated fault extraction, we refer to the chapter [32] of Pedersen, Skov, Randen, and Sønneland.

The first part of the present chapter describes how three-dimensional seismic facies can help characterizing the reservoir. Other techniques are available, as described in the chapters [29, 36]. These facies, also called geobodies, represent some typical geological features, which in turn can be linked to lithologies, thereby giving a more detailed and better structural description of the reservoir. The approach adopted here for linking three-dimensional seismic facies to lithologies is explained later in the property voxelization step.

(2) Definition of the Reservoir Model Geometry and Automated Extraction. The next step is to define the volume of interest. It can be the whole reservoir or just part of it. Vertically, the reservoir is delimited by the top and bottom horizons defined in the previous step. In the lateral extents the reservoir can be delimited either by these two horizons, when they intersect each other, or by faults.

If one is willing to study only a limited zone of the reservoir then the corresponding extents have to be explicitly provided. Altogether these boundaries are closing a volume that will be extracted from the previously defined set of seismic cubes.

It is noteworthy that for reservoir simulation purposes, the focus is set on the reservoir zone and usually includes neither part of the overburden nor part of the neighbouring areas in order to devote as many numerical grid cells to the reservoir simulation study as possible.

Today the numerical simulation grids used are structured. Then, their total number of cells will correspond to the product of the chosen discretizations in the inline, crossline and depth directions. Next generation simulators may use *unstructured* grids. Numerical methods on unstructured grids are described in the chapter [19] of Käser and Iske. These techniques would allow to reduce the total number of grid cells needed for the simulations and consequently for reducing the required CPU time.

Reservoir simulators usually handle two different types of grid geometry: Block-Centred (BC) and Corner-Point (CP), as illustrated in Figure 9. These geometries have either tetragons in two dimensions or hexahedrons in three dimensions as grid cell shapes.

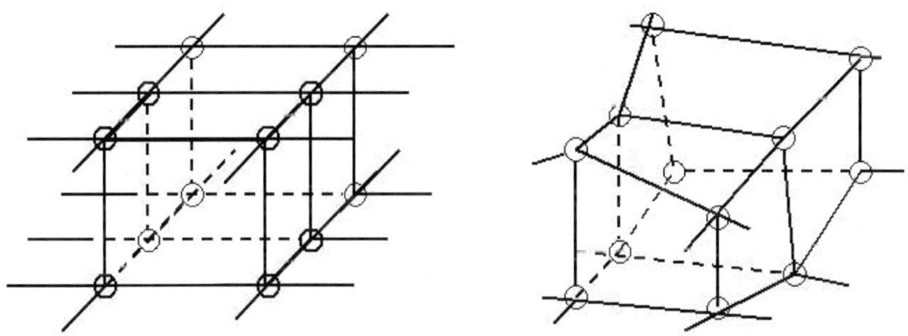

Fig. 9. Comparison of Block Centred (BC) and Corner Point (CP) grid geometries. BC grids (left) are simply formed by cells with parallelepiped shapes only. Cells forming CP grids (right) can have less restricted shape thereby allowing CP grids to reproduce most kind of geological structure.

BC geometries need only four parameters for each grid cell to be defined, namely: the cell size in the X, Y and Z direction (dX, dY, dZ) and the depth at which its top face is located. For the latter parameter, when all cells are identical, it is sufficient to only provide the depths at which the top face of the cells constituting the first layer of the model is situated. These values are considered as reference depths. For the other layers, the depths are given by summing the cell size in the Z direction (dZ) to the depth of the above layer.

Because of their simple description, BC grid cells have their upper and lower faces flat and horizontal while the other faces are all flat and vertical. It is then notable that BC grid cells and voxels are alike in shape. Hence voxel grids can be seen as BC grid geometries.

CP geometries require more parameters to be built. First, the co-ordinate lines and the corner depths notions must be introduced. The co-ordinate lines are the edges of each column of cells; they are straight lines but do not need to be vertical. These lines are defined by providing the X, Y and Z coordinates of their two end points, located above and below the horizons representing the reservoir. The cells are then defined by setting their corner depths along the co-ordinate lines. Hence, each grid cell is defined by four co-ordinate lines and eight corner depths. Such description allows CP grid cells for having less restricted shapes compared to BC grid cells. It is then possible to model more accurately complex geological structures compared to BC geometries. More details about CP geometries can be found in the chapter [12] of Farmer, or in the paper [33] by Ponting, or in the book [43].

The second main difference between BC and CP geometries is the cell connection across fault planes. Figure 10 illustrates this concept for both BC (left) and CP (right) grid geometries in a two-dimensional case.

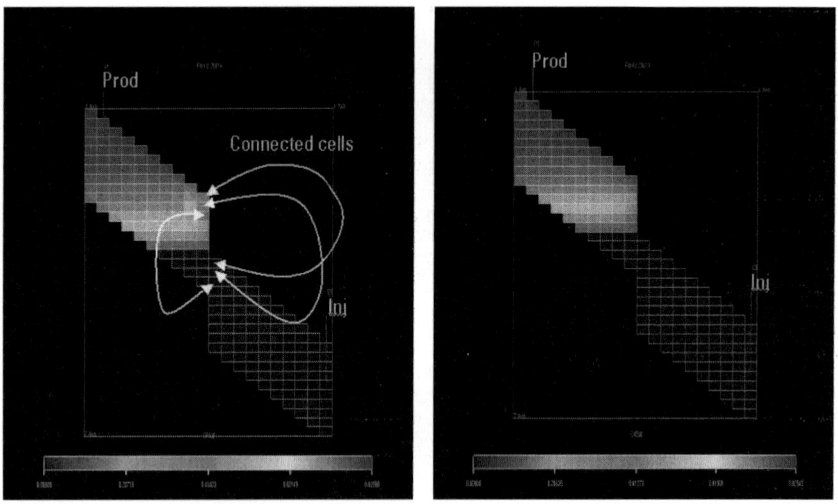

Fig. 10. [Reproduced in colour in Plate 20 on page 435.] Cell connection across fault plane. Left: Block Centred (BC) grid. Right: Corner Point (CP) grid. Colour scale given by oil saturation values. Fault plane is not taken into account with the BC grid. Cells are connected according to their grid indices and not their physical location in space.

Because of its simple description, BC grid geometry uses only grid indices to calculate transmissibility factors from one cell to the other. It is assumed that neighbouring cells sharing the same grid indices are connected thereby allowing fluids to flow from one cell to the other. This is the reason why the cells across the fault plane are connected as shown in Figure 10 (left). On the contrary, CP grid geometry contains enough information to calculate the overlap between neighbouring cells meaning that only cells physically sharing an interface are effectively connected. Connections across fault planes are then correctly modelled as seen in Figure 10 (right).

In order to combine the advantages of both grid geometries, namely the ease of mesh building (BC) and modelling correctly cell connections across fault planes (CP), the present method first creates a numerical mesh in a BC description, before it exports it as a CP grid.

The voxel grid extraction is automated as follows: the inputs are the voxel size (dX, dY, dZ), the top and bottom horizons. The number of discretization, in the inline and crossline directions, is given by the ratio of the reservoir extents in each direction with dX and dY, respectively. In the depth direction, the biggest vertical distance between the two horizons is calculated first. The number of discretization in the depth direction will then be the ratio of that length with dZ.

The top and bottom reservoir horizons are usually not parallel to each other. The voxel grid will extend typically beyond the reservoir or volume of interest. Hence, it is necessary to exclude all voxels that are not lying in between the two horizons from numerical simulations. This is done by marking them as non-active cells. In practice, these voxels are associated with zero porosity value.

The voxel size is chosen such that it matches, as closely as possible, the size of the smallest three-dimensional seismic facies, while keeping the total number of voxels small enough to allow time-effective numerical simulations. The output is then a voxel grid in physical coordinates (X, Y, Z) exported as a CP grid geometry where the co-ordinate lines are all vertical and the corner depths of the voxels are regularly sampled along these co-ordinate lines. Figure 11 illustrates the automated geometry extraction.

When the reservoir is made of several isolated compartments, it is necessary to implement the faults delimiting them into the reservoir model. As mentioned in the previous step, faults are either extracted automatically or manually interpreted from seismic data and represented as rather smooth surfaces.

When overlaid on the voxel grid, these surfaces are intersecting the voxels as it can be seen in Figure 12 (top). One approach to solve this problem would be to establish a set of rules defining which voxels faces make the best representation of the fault surface. This approach may although produce a stair-cased representation of the fault surface in the voxel model. In this respect, the CP geometry offers a better representation of faults.

To some extent, the voxel model offers some advantages when compared to standard CP geometry models, where each fault necessitates accurate structural modelling and individual closure of the volume compartments. This structural modelling tends to be cumbersome when complex structures, such

impedance, or volume of clay, etc. This can be done by means of empirical models, to be found in the literature.

In this case, the seismic facies are used to compensate for the loss in detailed geological representation inherent to the building of the voxel grid (Figure 12). As mentioned in the first part of this chapter, three-dimensional seismic facies and lithologies can be linked by means of well log data such as gamma-ray, density, resistivity. The voxelization process then assigns, to the voxel grid, porosity and permeability values from the property cubes constrained by the seismic facies The latter alternative is used in this chapter.

At this stage, the voxel model does not provide all answers for accurate modelling of complex structures, neither does the corner-point grid. However, it offers at least in its nature the possibility to capture discontinuities and heterogeneities evidenced in the 3D seismic facies classification with no manual interaction. This advantage is significant when compared to classification maps for instance. Even though these maps can be relatively easily integrated to reservoir models with modern software, they carry a wrong representation of the spatial heterogeneity within formations or layers by collapsing 3D information onto a 2D map.

Finally, the voxel model is flexible and can be tuned to provide the best representation of the heterogeneity of the reservoir. As mentioned in the automated geometry extraction step, this is achieved by comparing the size of the voxel with the size of 3D seismic facies. Yet when more than one type of 3D seismic facies is contained within one single voxel, and when the overall number of voxels can not be increased, an averaging of the lithological properties of the 3D seismic facies should be performed during the voxelization step. Related to the size of the voxel, a potential limitation of the voxel models would be faced when modelling isolated thin beds or sloping thin layers. In the current state, this aspect is better handled by corner point grids. More thorough research is needed to solve these difficulties.

3 Test Case: Grane Field

The general workflow of this new paradigm is now illustrated by a case study of the Grane field in the Norwegian sector of the North Sea. The seismic facies are related to lithologies by means of well log data and are then used to voxelize the grid with rock physical properties. Finally the dynamic part is covered with streamline simulation.

3.1 Quantitative Interpretation and 3D Seismic Facies Model

The Grane Field is located at the eastern margin of the Viking Graben on the western flank of the Utsira high. The Grane Field consists of Palaeocene turbidite sandstones (Heimdal Formation), which originate from the East Shetland Platform at the West of the field. The deposition and geometry of the turbidite sand lobes were mostly controlled by the original basin morphology, which forced the sand to deposit along a main N-S axis parallel to the Utsira High. The Heimdal Formation (Fm.) consists of high-density turbidite sandstones that were mostly deposited into the deepest part of the basin. In the Grane Field, the sand lobes of the Heimdal Fm. have undergone syn- and post-depositional deformation generating a complex geometry.

The main sand body appears to be penetrated from below (Lower Lista Fm.) by shale streaks potentially associated with faults, slumping or diapirism. It is characterised by very irregular top reservoir geometry due to sand injectites extruded from the main sand body into the overlying shales. The objective is to build a facies model, using 3D seismic facies classification, that captures the complex geometry and lithology variability, and then to translate efficiently this information to a reservoir voxel model. Due to heavy oil and limited contrast between sands and shales in the Heimdal Fm., seabed multicomponent seismic data were acquired in 2001 over the Grane field. These seismic data are used for the quantitative interpretation and seismic facies modelling. By the time of analysis, the Grane field was in appraisal phase. The production started in September 2003.

Seismic multicomponent (4C) data are acquired from the seabed with ocean-bottom cables containing receivers with three orthogonal geophones, usually one vertical (PZ) and two horizontal, and a hydrophone. This device allows recording and sorting of P-wave and S-wave. The P-wave is the elastic body wave in which particle motion is in the direction of propagation. P-waves are referred to as compressional wave and acoustic mode (PP) in our case study. The S-wave is a body wave in which particle motion is perpendicular to the direction of propagation. S-waves are generated by the incidence of P waves on interfaces at other than normal incidence, whereupon they are referred to as converted waves [38] and converted mode (PS) in our case study.

The general inversion scheme for reservoir characterisation and delineation from seismic multicomponent data involves the transformation of the converted shear waves (PS data) to PP time domain. This operation is performed in order to have the multi-component data with the same time reference for allowing direct comparison and analysis of both PP and PS data. The transformation of PS data to PP time requires a detailed analysis of the overburden, and interpretation of correlative reflection events on both data sets. The next step in the general inversion scheme involves 3D seismic facies analysis using seismic texture attributes as described earlier.

The multi-component data have been classified into seven types of seismic facies. These facies include flat parallel continuous high amplitude, flat parallel continuous low amplitude, discontinuous low amplitude, discontinuous high amplitude, dipping continuous, transparent facies and chaotic patterns. Seismic facies analysis and direct comparison of the acoustic mode (PP) and the converted mode (PS) show that in parts of the Grane Field the converted wave mode (PS) provides an improved image of the reservoir sands. Top and base reflectors of reservoir sands show new features neither observable in streamer data nor in the PP data. In particular, potential fluid effects are identified in the PP data and could mask the depositional or structural features.

In this case study, well data are used to calibrate seismic facies with the main reservoir facies (oil saturated sand, water saturated sand, sand injections, shales and shales streaks). The results of the 3D seismic facies classification are shown in Figure 13 along a horizon-based slice just above the bottom of the reservoir sands.

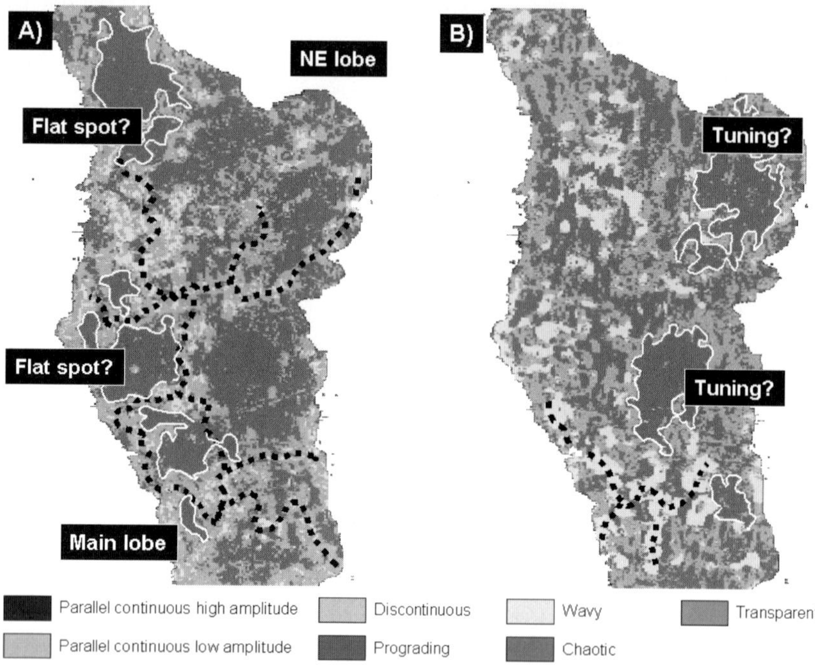

Fig. 13. [Reproduced in colour in Plate 23 on page 437.] Map views of seismic facies 3D classification 8 ms above the base reservoir from (A) PP data (acoustic mode), and (B) PS data (converted shear wave mode) viewed in PP time domain. The dotted lines represent linear deformation features. The white outlines highlight a potential flat spot on PP data (A) and a tuning effect or diagenesis effect on PS data (B).

From the PP data (Figure 13 A), the following features can be identified:

- a widespread high amplitude flat feature is extending on the western part of the main sand lobe and sparsely visible in the eastern part of the field.
- distinct linear patterns characterized by discontinuous, wavy and prograding facies are distributed within zones of flat parallel continuous facies.

From the PS data (Figure 13 B), the following features can be identified:

- flat high amplitude response is visible less commonly in the western part of the main sand lobe compared to the PP data. In fact, in the western part, the PS data show abundant distribution of discontinuous and wavy facies.
- a more distinct flat high amplitude response is visible in the eastern part of the main lobe and in the northeastern lobe than in the PP data seismic facies classification.

In addition, seismic facies classification of PS data sliced above the top of the reservoir sands shows an organised distribution of the dipping or prograding facies along the western margin of the main lobe as well as over the thickest part of the reservoir sands. The dipping seismic facies captures local change in reflector geometry at and above the top reservoir sands. Most of these disturbances at the top of the reservoir are characterised by wing shaped reflections visible mostly on the western margin of the main lobe. Locally some V-shaped reflections are visible over the thickest zones of the sand lobe (Figure 14).

New Interpretation Model. The widespread flat high amplitude feature visible on the PP data only in the western part of the main lobe coincides with the detected oil-water contact (OWC) from well data (Figure 15). On the other hand, the flat high amplitude anomaly coming brighter in the PS data is located higher up than the regional OWC and corresponds with the zone where the reservoir sands thin out towards east and can be either interpreted as tuning effect, or due to a change in rock physical properties. The main elongated element characterised by discontinuous and prograding facies visible mostly on PP data is interpreted as a shale prone deformational/depositional element. The crescent features are interpreted as major slump scars, related to the Utsira High uplifting or sediment instability. These crescents are marked discontinuities in the base reservoir reflector and probably filled by shales.

Facies Model and Calibration with Well Data. While the comparison of PP versus PS seismic image seems to indicate a potential fluid contact in the western part of the sand lobe fitting the observed OWC at well location, additional information, such as VP/VS ratio (the ratio of velocities of P-wave to S-wave, and gives equivalent information as measurements of Poisson's ratio), density logs, and porosity logs have confirmed the complexity of the

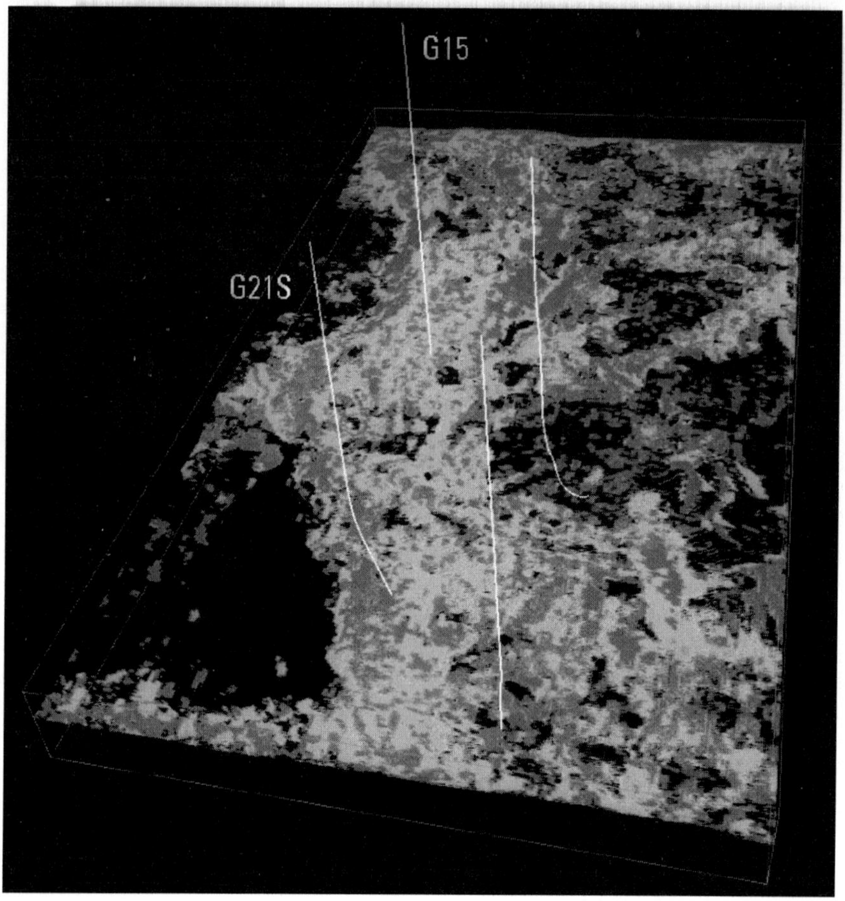

Fig. 14. [Reproduced in colour in Plate 24 on page 438.] Distribution of potential sand injections mapped by the red seismic facies from PS data above the top of the reservoir (some facies are set to transparent on the display). The injected sands are mostly visible along the margins of the main sand lobe and locally above it. The four wells displayed show evidence of injected sands above the main reservoir body.

model. Locally, fluid, diagenesis and difference in the sorting of the sand, or a combination of effects are still kept as potential causes for the flat features revealed by the classification.

The most discontinuous seismic facies forming the elongated lineaments at the base of the reservoir on the PP data (Figure 16) are correlated with shale prone zones logged along horizontal wells. These shale prone zones are most likely related to internal deformation and shale injections or shale streaks intruded from below the reservoir sands, probably from the Lower Lista Fm.

Cross plotting of seismic facies located along horizontal well sections against log data allowed to diagnose a predominant lithology and sand/shale

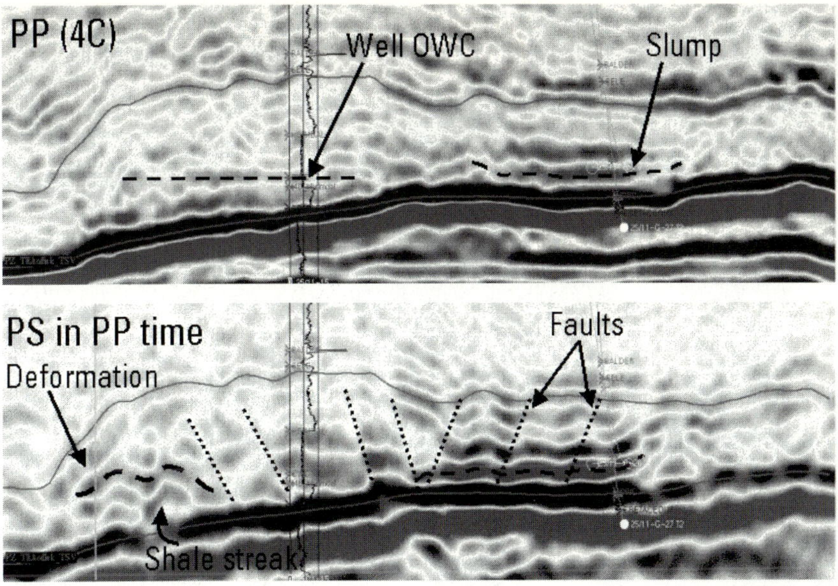

Fig. 15. [Reproduced in colour in Plate 25 on page 439.] Seismic cross-sections of multicomponent data: PP data (upper section) and PS data in PP time (lower section).

ratio. Each seismic facies shows a distinct histogram distribution of gamma ray values (cf. the Appendix, part 1.).

Each of the seismic facies is interpreted individually and is calibrated against well data to be related to a lithology or rock type. Depending on the position within the reservoir, one seismic facies may be associated with different lithologies or rock types. This is due to the non-unique relationship of the seismic facies with geology. For instance, discontinuous and dipping seismic facies within the reservoir level do not represent the same lithologies at top and bottom the reservoir. They are most likely shale prone zones, at the base and within the main reservoir body, identified as discordant reflections patterns and stratigraphic anomalies. Discordant reflections and stratigraphic disturbances, and in particular V-shaped ones, represent potential sand injectites above the top of the reservoir. This observation allows us to define a detailed zonation within the reservoir stratigraphy and laterally into a southern and northern area of the sand lobe.

As a result, the facies model has captured the complex geometry of the reservoir, including the sand injectites extruded from the main body at the top into the overburden. Being able to map these sand injections and translate them into the reservoir model is important because they affect the top reservoir geometry, and might change the connectivity and flow properties in a production scenario [18]. Similarly, the heterogeneous distribution of the

Table 1. Porosity for the Grane field. From well data information three-dimensional seismic facies have been linked to lithologies, each of them having typical porosity values. Some differences have been noticed between the south part and the north part of the reservoir. Three dimensional seismic facies: (1) Flat continuous high amplitude, (2) Flat continuous low amplitude, (3) Discontinuous low amplitude, (4) Discontinuous high amplitude, (5) Transparent, (6) Dipping continuous reflector, (7) Discontinuous-chaotic. NA: no available values.

Location	Seismic facies	Lithology	Porosity (%)	
			range	1st mode
South	1	sand	29-36	34
	2	water filled sand	26-33	30
	3	shale mixed with sand	1-15	8
	4	sand	22-38	34
	5	shale mixed with sand	1-15	8
	6	shale	1-5	3
	7	shale	1-5	3
North	1	sand	29-39	35
	2	water filled sand	32-38	35
	3	shale mixed with sand	1-15	8
	4	sand	29-39	35
	5	shale mixed with sand	1-15	8
	6	shale	1-5	3
	7	shale	1-5	3

Table 2. Permeability for the Grane field. From well data information three-dimensional seismic facies have been linked to lithologies, each of them having typical permeability values. Some differences have been noticed between the south part and the north part of the reservoir and between its upper and under part. The facies and their corresponding lithologies are defined in Table 1.

Location	Seismic facies	Permeability (mD) / upperpart		Permeability (mD) / underpart	
		range	1st mode	range	1st mode
South	1	1000-14500	8500	1000-14500	2500
	2	1000-12000	7000	1000-12000	2000
	3	NA	NA	NA	NA
	4	1000-14000	7000	1000-14000	2000
	5	NA	NA	NA	NA
	6	0-1	0.5	0-1	0.5
	7	0-1	0.5	0-1	0.5
North	1	1500-13000	8000	1500-13000	5000
	2	1500-13000	8000	1500-13000	5000
	3	NA	NA	NA	NA
	4	1500-13000	8000	1500-13000	5000
	5	NA	NA	NA	NA
	6	0-1	0.5	0-1	0.5
	7	0-1	0.5	0-1	0.5

The acoustic impedance I can be expressed as

$$I = 1000 V_P \rho_b . \tag{2}$$

The bulk density can be written as a function of porosity φ, matrix density ρ_M and fluid density ρ_f:

$$\rho_b = (1 - \varphi)\, \rho_M + \varphi \rho_f \tag{3}$$

Because of a deliberate choice for simplification and because the reservoir is rather thin, the use of an empirical relation for sandstone and standard values for the matrix density and fluid density have been made. These latter values are 2.65 g/cm^3 for the matrix density (corresponding to pure sand) and 1.00 g/cm^3 for the fluid density (corresponding to water).

The acoustic impedance seismic cube is obtained by inversion of seismic data using the Best Feasible Approximation (BFA) algorithm [42]. Combining (1) and (2) with (3), the porosity can be expressed as a function of impedance:

$$\varphi = -0.606 \left(1.4947 \left(\frac{I}{1000} \right)^{0.207} - 2.65 \right) . \tag{4}$$

Then for each voxel, the results given by (4) are compared with the expected range of porosity given by the lithologies. When a mismatch is observed (the porosity value obtained with (4) lays outside the expected range), the porosity value will be constrained by drawing a new value within the expected range. The porosity ranges given in Table 1 correspond to the minima and maxima values of porosity distribution for each seismic facies. The first mode of these ranges correspond to the center of the porosity distribution cluster. The probability distribution function (pdf) used depends on the relative position of the first mode within the porosity range. These pdf are either a Gaussian function (first mode at the center of the range), or a Gamma function.

Permeability has been calculated by using one relationship, established by Pape [30], where permeability k is function of porosity only. For shaly sandstone the relation used is given by

$$k = 1.013 \times 10^{-3} \left(6.2\varphi + 1493\varphi^2 + 58\,(10\varphi)^{10} \right) . \tag{5}$$

Similarly to porosity, permeability values have been constrained according to the expected range of permeability values obtained from lithologies. Permeability ranges and first modes corresponding to each seismic facies for the different geographical locations are given in Table 2.

It is noteworthy that the use of three-dimensional seismic facies information during the property voxelization step is not unique. It would have been possible to use a rock physical transform (such as Han's relations for shaley sandstones) giving porosity as a function of P-wave velocity and volume of clay (VC) [24]; the calibration of seismic facies with well data enabling the discrimination of the facies into different lithologies for which ranges of VC can be estimated (cf. the Appendix, part 1.).

Advantages and Limitations of the Voxel Model. In the Grane case study, well data indicated a regional OWC (oil water contact) with equal pressure in all wells as initial conditions. From these observations it is inferred that there seems to be no or only few barriers and no compartmentalization of the reservoir sand lobe.

The voxel grid built for the present study is composed of 449,256 voxels with 98 in the inline direction, 139 in the crossline direction and 33 in the depth direction. The reservoir itself is represented by 101,014 active voxels. The size of a single voxel is $(50 \times 50 \times 4)$ meters. This size is chosen according to the size of the smallest seismic facies, which is typically 2 to 3 times bigger than the seismic scale of $(25 \times 12.5 \times 4)$ meters.

Due to the observations and because of the small voxel size chosen, the inclusion of faults in the reservoir model does not appear as a priority for this case study. As mentioned earlier and as seen in Figure 12, the detailed heterogeneities of the reservoir captured with the tree-dimensional seismic facies are brought forward to the reservoir voxel model thereby implicitly modelling the discontinuities.

Nonetheless, further thorough research is needed in order to investigate how accurately these discontinuities are effectively captured and modelled in the voxel model.

3.3 Simulation Results and Comments

The effect of three-dimensional seismic facies on the voxel model were evaluated by using two case studies. On the one hand, a voxel model based on acoustic impedance only, and on the other hand a voxel model based on both acoustic impedance and three-dimensional seismic facies were built.

Figure 18 shows porosity at the top of the reservoir obtained for the two voxel models. The second voxel model (Figure 18 (b)) does not only show a less scattered representation of the porosity but also that high porosity values reproduce better the sand lobes.

Figure 19 compares the horizon map of the top reservoir with the porosity obtained for the second voxel model, at the same location. Here, a fairly good correspondence between the position of the sand lobes and the high porosity zones is observed.

The reservoir started producing in September 2003; no production history was then available to compare with simulation results. For the purpose of this study, it has been decided to mimic a typical three-phase reservoir life. First, gas injection at the top of the reservoir pushes the oil towards the producing wells. Second, after some days of production injection of water near the bottom of the reservoir is done. The oil is then squeezed between gas

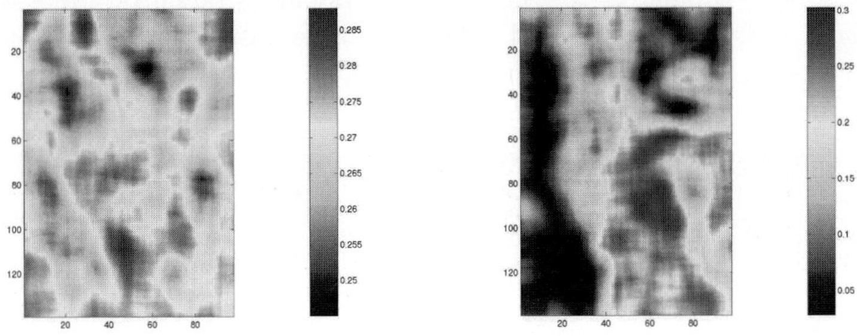

(a) (b)

Fig. 18. [Reproduced in colour in Plate 27 on page 441.] Comparison of porosity at the top of the reservoir for the two voxel models. (a) Voxel model based on acoustic impedance only. (b) Voxel model based on acoustic impedance and three-dimensional seismic facies results.

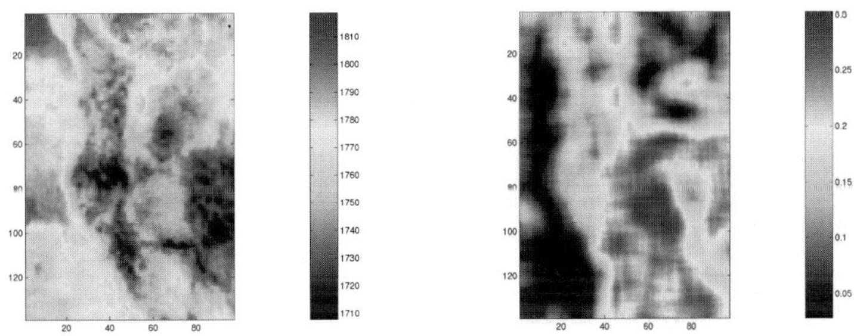

(a) (b)

Fig. 19. [Reproduced in colour in Plate 28 on page 441.] (a) Illustration of the top reservoir horizon (in ms). (b) Porosity distribution along the top of the reservoir for the voxel model based on acoustic impedance and three-dimensional seismic facies results.

and water injections and swept towards the producing wells. A total of seven wells were used: one gas injector, one water injector and five oil-producing wells. After building the reservoir model, production wells have to be implemented together with injection wells. This step is a time-consuming task. However, as it comes to this implementation, the classical workflow and the voxel-based workflow are identical. This is the reason why the corresponding implementation details are omitted in this chapter.

In order to illustrate the simulation step of the new workflow, a streamline-based method simulator is used. Its advantages are twofold: the streamlines indicate how the fluid is moving inside the reservoir and the required

CPU time is smaller than conventional simulators based on finite difference method. Batycky [3] observed that streamline-based method are one to two order of magnitude faster than finite difference methods for simple cases.

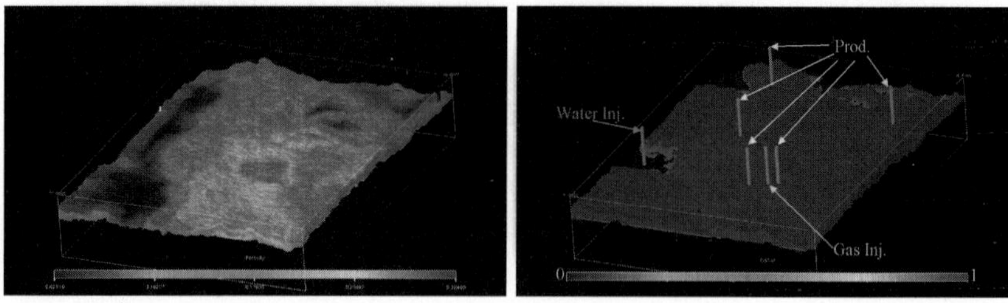

(a) (b)

Fig. 20. [Reproduced in colour in Plate 29 on page 442.] Voxel model based on acoustic impedance and three-dimensional seismic facies results. (a) Porosity distribution; (b) initial oil saturation distribution together with the locations of the wells.

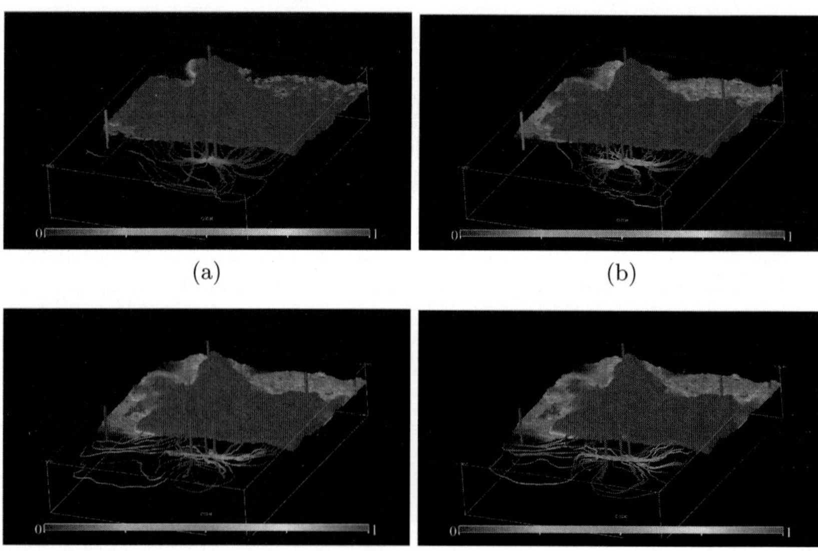

(a) (b)

(c) (d)

Fig. 21. [Reproduced in colour in Plate 30 on page 442.] Voxel model based on acoustic impedance and three-dimensional seismic facies results. Simulation results for the oil saturation at day (a) 300, (b) 1800, (c) 3300, and (d) 4800 of the production.

The computer used for simulation is a Sunblade 1000 with two CPUs at 750 MHz and 4 gigabytes memory. About 30 CPU hours were needed to run a three phases streamline simulation for 4800 production days. Figure 20 (a) shows a three-dimensional view of the porosity distribution obtained for the voxel model based on acoustic impedance and three-dimensional seismic facies results. Figure 20 (b) shows the initial oil saturation together with the location of the seven wells used for the simulation.

Simulation results at four different time steps are shown in Figure 21. This simulation illustrates the main steps of a heavy oil reservoir life where the production strategy is to inject gas first and then squeeze the oil in between gas injection at the top and water injection at the bottom. Figure 21 (top row) shows results of the first stage when the reservoir undergoes gas injection only; at the beginning of the production (a) and before water injection starts (b). The two pictures in the bottom row of Figure 21 show simulation results when the reservoir undergoes both water and gas injections. No clear differences can be made between these two results. This illustrates the stage where new production strategy should be made in order to enhance the oil recovery. It could be either increasing wells production and/or injection rates, either planning new production and/or injection wells.

4 Conclusion and Final Remarks

A new methodology for building reservoir models from seismic data based on three-dimensional seismic facies has been demonstrated on the Grane field. This new approach enables a workflow from seismic domain to simulation domain, by bridging the two close together through the voxelization process.

The workflow specifications were to build a model directly from the seismic data as fast and as easy as possible, such that updating at real time can be done. This requires developing an automated procedure in order to reduce the manual (time consuming) work. It has been shown that:

- A three-dimensional seismic facies model could be built from supervised classification of seismic textures that captures the detailed structural framework of the reservoir and its complex architecture.
- The seismic facies have been calibrated against well data thereby providing a link between facies and lithologies or fluids.
- The reservoir grid geometry has been pre-defined in a given number of voxels according to seismic facies scale to compensate for the lack of structural details intrinsic of the voxel grid.
- A property voxelization step allowed to populate the model with rock properties based on (1) empirical models linking porosity and permeability to acoustic impedance, or (2) empirical models and additional constraints given by the calibration of the seismic facies with well data.

- Finally, after defining the static model, a simulation scenario is run with a set of producers, gas and water injectors.

The next step would be to go further: from seismic to prediction. This implies that a 4D history matching is successfully done. The high quality of the history matching is a prerequisite to assess the reliability of the reservoir model and therefore its forecasting capability.

This method from seismic to simulation exploiting three-dimensional seismic facies and voxelization process opens new perspectives on reservoir model building, multiple scenarios realizations and faster or right time model updating.

This new voxel-based workflow accelerates the seismic to simulation process. It is aimed at building and/or updating a reservoir voxel model as soon as new data are acquired and give good enough simulation results in real time so that drilling engineers or field engineers can take immediate decisions. This prevents risks, it reduces collateral costs, and it improves benefits. Moreover, this voxel-based workflow could easily be adapted to the new simulator generation, working with unstructured mesh, requiring only slight modifications.

Acknowledgement

The work was partly supported by the European Union through the research and training network NetAGES, contract no. IST-1999-29034. We wish to thank Anna-Lena Skjerdingstad, Jan Petter Fjellanger, and the Grane license owners Norsk Hydro, ConocoPhillips, ExxonMobil and Petoro AS for collaboration and permission to publish data. We also wish to thank Lars Sønneland and Michael Nickel for fruitful discussions. We finally want to acknowledge Christopher Farmer for advice and suggestions, as well as Trygve Randen and Vincent Kretz for proofreading of this chapter.

References

1. S.K. Addy (1998) Neural network classification method helps seismic interval interpretation. Oil & Gas Journal **96**(37), 47–59.
2. F. Aminzadeh, J. Barhen, C.W. Glover, and N.B. Toomarian (2000) Reservoir parameter estimation using a hybrid neural network. Computers & Geosciences **26**(8), 869–875.
3. R.P. Batycky, M.J. Blunt, and M.R. Thiele (1997) A 3d field-scale streamline-based reservoir simulator. SPE Reservoir Engng., Nov. 1997, 246–254.
4. H.G. Borgos, T. Skov, T. Randen, and L. Sønneland (2003) Automated geometry extraction from 3D seismic data. In Expanded Abstr., 73rd Ann. Int. Mtg., Soc. Expl. Geophys., Dallas, USA, 26-31 Oct., 2003, 1541–1544.
5. H.G. Borgos, T. Skov, and L. Sønneland (2004) Automated structural interpretation through classification of seismic horizons. This volume.

6. L.F. Brown and W.L. Fisher (1980) Seismic stratigraphic interpretation and petroleum exploration: continuing education course note series. Chapter 16. American Association of Petroleum Geologists.
7. A. Carrillat, T. Randen, L. Sønneland, and G. Elvebakk (2002) Automated mapping of carbonate mounds using 3D seismic texture attributes. In Expanded Abstr., 72nd Ann. Int. Mtg., Soc. Expl. Geophys., Salt Lake City, USA, 6-11 Oct., 2002, 552–555.
8. A. Carrillat, T. Randen, L. Sønneland, and G. Elvebakk (2002) Stratigraphic mapping of carbonate mounds using 3D texture attributes. In Extended Abstr., 64th Conference & Exhibition, Eur. Ass. Geosc. & Eng., Florence, Italy, 27-30 May, 2002.
9. J. Dayhoff (1990) *Neural Network Architectures: An Introduction.* Van Nostrand Reinhold, New York.
10. J.M. Van de Sande (1996) Prediction of reservoir parameters from 3D seismic data for the Zechstein 2 carbonate play in the northeast Netherlands. *AAPG Studies in Geology No. 42 and SEG Geophysical Developments Series No. 5*, P. Weimer and L. Davis (eds.), Tulsa, 1996, 197–204.
11. P.M. Doyen (1988) Porosity from seismic data: a geostatistical approach. Geophysics **53**(10), 1263–1275.
12. C.L. Farmer (2004) Geological modelling and reservoir simulation. This volume.
13. Q.J. Fisher and R.J. Knipe (1998) Fault sealing processes in siliciclastic sediments. *Faulting, Fault Sealing and Fluid Flow in Hydrocarbon Reservoirs*, G. Jones, Q.J. Fisher, and R.J. Knipe (eds.), vol. 147 of Geological Society Special Publications, 117–134.
14. K.A. Foxford, J.J. Walsh, J. Watterson, I.R. Garden, S.C. Guscott, and S.D. Burley (1998) Structure and content of the Moab fault zone, Utah, USA, and its implications for fault seal prediction. *Faulting, Fault Sealing and Fluid Flow in Hydrocarbon Reservoirs*, G. Jones, Q.J. Fisher, and R.J. Knipe (eds.), Geological Society Special Publications, vol. 147, 87–103.
15. G.H.F. Gardner, L.W. Gardner, and A.R. Gregory (1974) Formation velocity and density - the diagnostic basics for stratigraphic traps. Geophysics **39**, 770 780.
16. P. Goovaerts (1997) *Geostatistics for Natural Resources Exploration.* Oxford University Press, New York.
17. B.K. Hegstad and H. Omre (2001) Uncertainty in production forecasts based on well observations, seismic data, and production history. SPE 74699, 409–423.
18. M. Huus, D. Duranti, C.G. Guargena, P. Prat, K. Holm, N. Steinsland, B.T. Cronin, A. Hurst, and J. Cartwright (2003) Sandstone intrusions: detection and significance for exploration and production. First Break **21**, 33–42.
19. M. Käser and A. Iske (2004) Reservoir flow simulation by adaptive ADER schemes. This volume.
20. A. Kaufman, D. Cohen, and R. Yagel (1993) Volume graphics. Computer **26**(7), 51–64.
21. M. Landrø, H. Hafslund Veire, K. Duffaut, and N. Najjar (2003) Discrimination between pressure and fluid saturation changes from marine multicomponent time-lapse seismic data. Geophysics **68**(5), 1592–1599.

22. G. Mangerud, T. Dreyer, L. Søyseth, O. Martinsen, and A. Ryseth (1999) High-resolution biostratigraphy and sequence development of the Palaeocene succession, Grane field, Norway. *Development and Evolution of the Wessex Basin*, J.R. Underhill (ed.), vol. 133 of Geological Society, London, 167–184.
23. T. Manzocchi, J.J. Walsh, P. Nell, and G. Yielding (1999) Fault transmissibility multipliers for flow simulation models. Petroleum Geoscience **5**, 53–63.
24. G. Mavko, T. Mukerji, and J.Dvorkin (1998) *The Rock Physics Handbook*. Cambridge University Press.
25. P. Meldahl, R. Heggland, B. Bril, and P. de Groot (1999) The chimney cube, an example of semi-automated detection of seismic objects by directive attributes and neural networks: part 1: methodology. In Expanded Abstr., 69th Ann. Int. Mtg., Soc. Expl. Geophys., Houston, USA, 1999.
26. R.M. Jr. Mitchum and P.R. Vail (1977) Seismic stratigraphic interpretation procedure. *Seismic Stratigraphy - Applications to Hydrocarbon Exploration*, C. E. Payton (ed.), vol. 26, American Association of Petroleum Geologists Memoir, 135–143.
27. R.M. Jr. Mitchum, P.R. Vail, and J.B. Sangree (1977) Stratigraphic interpretation of seismic reflection patterns in depositional sequences. *Seismic Stratigraphy - Applications to Hydrocarbon Exploration*, C. E. Payton (ed.), vol. 26, American Association of Petroleum Geologists Memoir, 117–133.
28. R.M. Jr. Mitchum, P.R. Vail, and S. Thompson (1977) The depositional sequence as a basic unit for stratigraphic analysis. *Seismic Stratigraphy - Applications to Hydrocarbon Exploration*, C. E. Payton (ed.), vol. 26, American Association of Petroleum Geologists Memoir, 53–62.
29. E. Monsen, T. Randen, L. Sønneland, and J.E. Odegard (2004) Geological model building: a hierarchical segmentation approach. This volume.
30. H. Pape, C. Clauser, and J. Iffland (1999) Permeability prediction based on fractal pore-space geometry. Geophysics **64**(5), 1447–1460.
31. C.E. Payton, editor (1977) *Seismic Stratigraphy - Applications to Hydrocarbon Exploration*. American Association of Petroleum Geologists Memoir, vol. 26.
32. S.I. Pedersen, T. Skov, T. Randen, L. Sønneland (2004) Automatic fault extraction using artificial ants. This volume.
33. D.K. Ponting (1992) Corner point geometry in reservoir simulation. *The Mathematics of Oil Recovery*, P.R. King (ed.), Clarendon Press, Oxford, 45–65.
34. T. Randen, E. Monsen, A. Abrahamsen, J.O. Hansen, J. Schlaf, and L. Sønneland (2000) Three-dimensional texture attributes for seismic data analysis. In Expanded Abstr., 70th Ann. Int. Mtg., Soc. Expl. Geophys., 2000.
35. T. Randen and L. Sønneland (2004) Atlas of 3D seismic attributes. This volume.
36. S.K. Richardsen and T. Randen (2004) Mapping 3D geo-bodies based on level set and marching methods. This volume.
37. J. Schlaf and T. Randen (2004) Introduction to seismic texture. This volume.
38. R.E. Sheriff (2002) *Encyclopedic Dictionary of Applied Geophysics*. Fourth edition, *Geophysical References*, vol. 13, Society of Exploration Geophysicists, Tulsa, USA.
39. L. Sønneland (1994) Volume reflection spectral analysis. Technical report, Schlumberger Geco-Prakla.

40. L. Sønneland, P.O. Tennebø, T. Gehrmann, and O. Yrke (1994) 3D model-based Bayesian classification. In Expanded Abstr., 64th Ann. Int. Mtg., Soc. Expl. Geophys., 510–511.
41. M.T. Taner, F. Koehler, and R.E. Sheriff (1979) Complex seismic trace analysis. Geophysics **44**(6), 1041–1063.
42. P.O. Tennebø, H.H. Veire, L. Sønneland, C. Signer, and B. Reymond (1998) Inversion of 4D seismic with best feasible approximation. Society of Exploration Geophysicists, expanded abstract, New Orleans, USA, 52–53.
43. J.F. Thompson, B.K. Soni, and N.P. Weatherill (1999) *Handbook of Grid Generation*. CRC Press, Boca Raton.
44. G. Yielding, G. Freeman, and B. Needham (1997) Quantitative fault seal prediction. American Association of Petroleum Geologists Bulletin **81**, 897–917.

Appendix

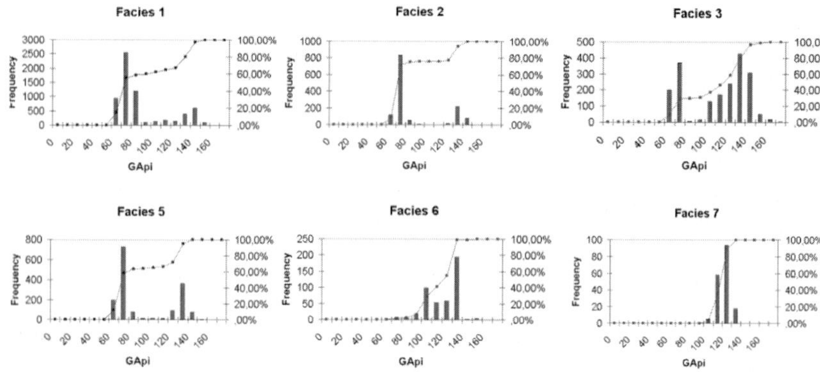

1. Gamma Ray histograms and cumulative percentage for well G26 and the respective seismic facies located along the horizontal section of the well.

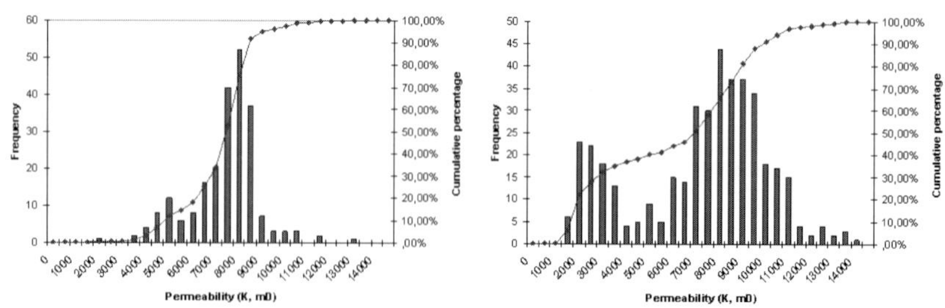

2. Permeability histograms and cumulative percentage for well G15 (left) and G21S (right).

Reservoir Flow Simulation by Adaptive ADER Schemes

Martin Käser and Armin Iske

Munich University of Technology, Department of Mathematics, Germany

Summary. In this chapter, an extension of ADER schemes is presented in order to solve both linear and nonlinear scalar conservation laws on unstructured triangulations. The proposed scheme is conservative and belongs to the class of finite volume schemes. It combines high order reconstruction techniques with a high order flux evaluation method to update cell average values through fluxes across cell interfaces. The ADER approach results in an explicit, one-step scheme based on the solution of generalized Riemann problems across cell interfaces. Moreover, the triangulation is adaptively modified during the simulation to effectively combine high order accuracy with locally refined meshes and therefore reduce the computational costs. The required adaption rules for the refinement and coarsening of the triangular mesh rely on a customized error indicator. Numerical experiments confirm the expected orders of accuracy and show the good performance of the proposed scheme for linear and nonlinear problems. Finally, the adaptive ADER schemes are applied to a test case from oil industry, which plays an important role in the modelling of fluid flow in petroleum reservoirs.

1 General Overview

Modern approaches of constructing conservative, very high order numerical methods for hyperbolic conservation laws are typically based on the Finite Volume approach combined with *essentially non oscillatory* (ENO) or *weighted essentially non-oscillatory* (WENO) reconstruction techniques. We mention, that the Discontinuous Galerkin Finite Element method [8] is a very attractive alternative and has found many applications within the last few years. Harten, Engquist, Osher, and Chakravarthy [15] introduced a one-dimensional cell average version of the original ENO Finite Volume scheme. Later, Harten and Chakravarthy [14], Abgrall [1], and Sonar [34] extended the finite volume formulation of ENO schemes to unstructured triangular meshes. The central idea of ENO schemes is to select the *smoothest* stencil out of several possible ones and then reconstruct the solution from cell averages with high order accuracy, e.g. by using high order polynomials. By constructing such nonlinear schemes, as required by Godunov's theorem [12], the growth of spurious oscillations can successfully be avoided. The more advanced WENO schemes were first suggested by Liu, Osher, and Chan [25] and Jiang and Shu [19]. Very recently, Friedrich [11] constructed WENO schemes on unstructured meshes based on the dual mesh of a triangulation. In the WENO approach, a set of different possible reconstruction polynomials is used in order to construct a specific weighted sum of polynomials. In contrast to ENO schemes the main advantages of WENO schemes are the better convergence to steady-state solutions and the increased spacial accuracy, especially in smooth regions of the solution.

High order accuracy in time is typically achieved through multi-stage Runge-Kutta methods. However, to retain the monotonicity properties of the resulting schemes the chosen time discretisation requires to be *total variation diminishing* (TVD) as observed by Shu [32] and Shu and Osher [33]. Ruuth and Spiteri [29] recently showed, that such TVD Runge-Kutta methods have barriers to the order of time accuracy and consequently to the order of the entire scheme. Thus, in most practical implementations a third order TVD Runge-Kutta method is used even for very high order WENO schemes, i.e., the spatial order of accuracy is much larger than three.

A new approach introduced by Toro, Millington, and Nejad in [38] and further developed by Titarev in [36, 39] is the so-called *ADER* approach, which is an explicit one-step finite volume scheme of **A**rbitrary high order using high order **DER**ivatives of piecewise polynomial reconstructions. In [36, 38, 39] a very high order version of the classical Godunov scheme [12] is constructed, which leads to an arbitrary high order of accuracy in both space and time. In fact, ADER schemes can be interpreted as high order generalizations of the classical Godunov scheme. In the last few years, the use of ADER schemes has gained considerable popularity in the field of gas and aerodynamics, e.g. [30, 31], especially for linear advection problems, and currently constitutes a very active research area [35], also for nonlinear problems, systems of hyperbolic equations, and advection-reaction equations [41].

2 Introduction

In this chapter, high order WENO schemes on unstructured triangulations are combined with the ADER approach introduced in [38] to solve scalar, *linear* and *nonlinear* conservation laws of the form

$$\frac{\partial u}{\partial t} + \nabla f(u) = 0\,, \tag{1}$$

where for some domain $\Omega \subset \mathbb{R}^2$, and a compact time interval $I = [0, T]$, $T > 0$, the function $u : I \times \Omega \to \mathbb{R}$ is unknown and $f(u) = (f_1(u), f_2(u))^T$ denotes the *flux tensor*. Furthermore, *adaptive mesh refinement* is included to balance computational cost and approximation quality. This is important, in particular, in the vicinity of discontinuities, that typically occur in solutions of hyperbolic problems and can be accurately resolved by locally refined meshes. Therefore, our aim is to combine the ADER approach with adaptive mesh refinement in order to obtain highly accurate results at reasonable computational costs.

In general, a high order extension of the classical scheme of Godunov [12] consists of the three basic operations:

- polynomial reconstruction of the solution from cell average values,
- evaluation of fluxes across interfaces between adjacent cells,
- update (evolution) of cell average values in each cell.

This chapter is arranged by following these main steps. An introduction to the reconstruction of high order polynomials from cell average values on unstructured triangulations is given in Section 3. The corresponding WENO reconstruction is addressed and followed by a detailed description of the stencil selection algorithm. The finite volume formulation of the governing equation (1) on unstructured triangulations is outlined in Section 4 together with the presentation of the high order flux evaluation technique using the ADER approach. The remaining update of the cell average values is discussed in Section 5. Finally, the accuracy of the proposed scheme is evaluated in Section 6 and its good performance is confirmed by numerical experiments concerning reservoir flow simulation in Section 8.

3 Reconstruction of High Order Polynomials

The reconstruction of high order polynomials on unstructured triangulations is much more difficult than the reconstruction on one-dimensional intervals or multi-dimensional Cartesian grids. In fact, polynomial reconstruction from scattered data requires the solution of multi-dimensional interpolation problems, which typically tend to be ill-conditioned. This problem becomes even more critical with increasing order of the reconstruction.

To keep the notation short we use multi-indices, i.e., $\alpha = (\alpha_1, \alpha_2)$ with $\alpha_i \in \{0, 1, 2, ...\}$, $i = 1, 2$, and $x = (x_1, x_2) \in \mathbb{R}^2$. Moreover, we let $|\alpha| = \alpha_1 + \alpha_2$ and $x^\alpha = x^{(\alpha_1,\alpha_2)} = x_1^{\alpha_1} x_2^{\alpha_2}$. In the following, let $\mathcal{P}_n$ denote the set of bivariate polynomials of degree at most n. Note that the set $\mathcal{P}_n$ is a vector space of dimension $N(n) = \frac{1}{2}(n+1)(n+2)$, and, for any $b \in \mathbb{R}^2$, the functions $\{(\cdot - b)^\alpha\}_{|\alpha| \leq n}$ constitute a basis of $\mathcal{P}_n$. Therefore, any polynomial $p \in \mathcal{P}_n$ can uniquely be expressed by a monomial expansion of the form

$$p(x) = \sum_{|\alpha| \leq n} a_\alpha (x - b)^\alpha \,, \tag{2}$$

with coefficients $a_\alpha \in \mathbb{R}$, $|\alpha| \leq n$. We remark, that the monomial expansion is not suitable for practical computations, but often useful for theoretical purposes.

We assume that the computational domain $\Omega \subset \mathbb{R}^2$ is discretized by a conforming triangulation $\mathcal{T}$ (cf. [26, Subsection 3.3.1]), given by a set $\mathcal{T} = \{T_\ell\}_\ell$ of triangles $T_\ell \subset \Omega$, $\ell = 1, ..., \#\mathcal{T}$. In the finite volume framework each (i) triangle, (ii) cell, or (iii) control volume T_ℓ carries a *cell average value*

$$\bar{u}_\ell(t) = \frac{1}{|T_\ell|} \int_{T_\ell} u(t,x)\,dx\,, \tag{3}$$

where $|T_\ell|$ is the area of triangle T_ℓ and $u(t,x)$ is the solution of (1).

Now, in the reconstruction we consider solving the following problem.

Given the polynomial degree n and cell average values $\bar{u}_{\ell_k}$, $k = 1, ..., N$, $N = \dim \mathcal{P}_n$, of the function u on each control volume T_{ℓ_k}, find a polynomial $p \in \mathcal{P}_n$, that satisfies the interpolation conditions

$$\begin{aligned} \bar{p}_{\ell_1} &= \bar{u}_{\ell_1}\,, \\ \bar{p}_{\ell_2} &= \bar{u}_{\ell_2}\,, \\ &\vdots \\ \bar{p}_{\ell_N} &= \bar{u}_{\ell_N}\,. \end{aligned}$$

This problem has a unique solution, iff the Vandermonde matrix

$$M = \left(\overline{(\,\cdot\, - b)^\alpha}_{\ell_k} \right)_{1 \le k \le N,\, |\alpha| \le n} \tag{4}$$

is non-singular. In this case, we call the set $\mathcal{S} = \{T_{\ell_k}\}_{1 \le k \le N}$ of triangles a $\mathcal{P}_n$-unisolvent or *admissible* stencil. Note that the matrix M may be ill-conditioned, due to the geometry of the chosen stencil. This is further supported by Abgrall, who shows in [1], that the condition number of the above system matrix is $\mathcal{O}(h^{-n})$, where $h = \sqrt{|T_\ell|}$ is a measure for the local mesh width and n the degree of the polynomial space. So for large degree n and for small mesh width h, the corresponding linear equation system is ill-conditioned.

As a remedy to this stability problem, Friedrich [11] considered introducing a scaling factor $s = (\sqrt{|T_\ell|})^{-1}$ in order to obtain a condition number independent from h. Therefore, the standard monomial representation (2) is changed to

$$p(x) = \sum_{|\alpha| \le n} s^{|\alpha|} \tilde{a}_\alpha (x - b)^\alpha\,, \tag{5}$$

where $\tilde{a}_\alpha$ are the scaled coefficients of p. Now by using the scaled polynomial expansion (5), the condition number of the corresponding coefficient matrix does no longer depend on the mesh width h. However, it is still unclear, if this expansion is sufficient to provide a robust reconstruction procedure for strongly distorted, unisotropic meshes. For this reason, Abgrall suggests in [1] to use a polynomial expansion based on *barycentric coordinates.*

In order to briefly explain this standard stabilization technique, let the set $\mathcal{S}_n = \{T_1, T_2, ..., T_N\}$, $n \ge 3$, denote an admissible stencil for $\mathcal{P}_n$, $n \ge 1$. Then, there is a substencil $\mathcal{S}_1 \subset \mathcal{S}_n$ containing three triangles from $\mathcal{S}_n$, say $\{T_1, T_2, T_3\}$, such that $\mathcal{S}_1$ constitutes an admissible stencil for $\mathcal{P}_1$. In this case, there are three unique linear polynomials $\Lambda_1, \Lambda_2, \Lambda_3 \in \mathcal{P}_1$ satisfying

$$\bar{\Lambda}_{i_j} = \delta_i^j \,, \qquad 1 \le i, j \le 3, \qquad \text{with} \quad \sum_{i=1}^{3} \Lambda_i(x) \equiv 1$$

These polynomials $\Lambda_1, \Lambda_2, \Lambda_3$ are said to be the *barycentric coordinates* of the sgtencil $\mathcal{S}_1$. Now any polynomial $p \in \mathcal{P}_n$ can uniquely be expressed as a linear combination of the form

$$p(x) = \sum_{|\alpha| \le n} b_\alpha \Lambda^\alpha(x), \qquad \text{where } \Lambda^\alpha = \Lambda_1^{\alpha_1} \Lambda_2^{\alpha_2}. \tag{6}$$

Due to the scale-invariance of the barycentric coordinates $\Lambda_1, \Lambda_2, \Lambda_3$, the condition number of the matrix, arising from the expansion (6), is independent of the local mesh width h, see [1]. We remark that this observation is very crucial when working with adaptive mesh refinement, where very small cells may appear. Indeed, the representation (6) is, due to its robustness, particularly suited for adaptive mesh refinement, even for strongly distorted meshes.

An alternative way to overcome stability problems is suggested in [3, 28]. This approach works with an overdetermined system. Instead of using exactly N neighbouring cells, this works with a slightly larger stencil to enhance the robustness of the reconstruction. In our computations, we typically use 4 cells for linear, 8 cells for quadratic, and 13 cells for cubic reconstruction. To obtain a *conservative* scheme, one needs to satisfy the condition

$$\bar{p}_\ell = \bar{u}_\ell \tag{7}$$

on any cell T_ℓ, where the reconstruction polynomial p is computed.

In this case, a linear least-squares problem with the linear equality constraints (7) needs to be solved (cf. [22, Chapter 21]). Finally, we remark that the entries of the system matrix M in (4) can be computed by using quadrature rules for triangles, that are exact for the desired polynomial degree n. A detailed list of quadrature rules for triangles is given in [9].

3.1 WENO Reconstruction

During the last decade, WENO methods have extensively been used for one-dimensional problems, and they have also gained popularity for problems on multi-dimensional Cartesian grids. The general idea of ENO and WENO schemes is to chose several stencils S_i, $i = 1, ..., k$, where k denotes the number of stencils, and to compute the corresponding reconstruction polynomials p_i. The ENO approach selects *one* least oscillatory polynomial. In contrast, WENO methods work with a weighted sum

$$p(x) = \sum_{i-1}^{k} \omega_i p_i(x) \,,$$

where the ω_i are positive, data-dependent, and normalized weights, such that $\sum_{i=1}^{k} \omega_i = 1$. Originally introduced in [19, 25], the WENO approach was extended to unstructured meshes in [11, 16]. In order to compute the weights ω_i, we have to clarify how the oscillation of the corresponding polynomial p_i is measured. Numerical tests in [11, 16] have demonstrated, that a suitable *oscillation indicator* for a polynomial p on a triangular cell T is given by

$$\mathcal{I}_i = \sum_{1 \leq |\alpha| \leq n} \int_T |T|^{|\alpha|-1} \left(D^\alpha p_i(x) \right)^2 dx \,, \tag{8}$$

where $D^\alpha = \frac{\partial^{|\alpha|}}{\partial x_1^{\alpha_1} \partial x_2^{\alpha_2}}$ is the α-th partial derivative operator with respect to x_1 and x_2. The factor $|T|^{|\alpha|-1}$ eliminates effects due to the local mesh width. Then, the weights ω_i can be calculated through

$$\omega_i = \frac{\tilde{\omega}_i}{\sum_{i=1}^{k} \tilde{\omega}_i} \qquad \text{with} \qquad \tilde{\omega}_i = (\epsilon + \mathcal{I}_i)^{-r} \,. \tag{9}$$

Here ϵ is a small positive number to avoid division by zero. Usually, numerical results are not very sensitive to the choice of ϵ. In general, however, larger values of ϵ are better suited for smooth problems but may generate small oscillations near shocks, whereas smaller ϵ are better suited for discontinuous problems. In the literature, typically values of $\epsilon \in [10^{-6}, 10^{-2}]$ are chosen. In our computations, we let $\epsilon = 10^{-5}$. The positive integer r in (9) is a measure of the sensitivity of the weights with respect to the oscillation indicator of (8). As r tends to infinity, the WENO scheme behaves like a classical ENO scheme. On the other hand, if r tends to zero, the oscillation indicator has almost no effect on the weights, which means that the scheme becomes an oscillatory, or even unstable scheme. In most applications we find $r \in [2, 8]$, and in our implementation we chose $r = 4$, which turns out to be large enough to essentially avoid oscillations near discontinuities, but small enough to improve upon the classical ENO scheme.

3.2 Stencil Selection

So far, we have assumed to have admissible stencils. However, as shown by Sonar in [26, Subsection 3.6.1], selecting a small number of admissible stencils on unstructured triangulations is not a trivial task as there is a large number of possible stencils to choose from. Furthermore, for polynomials of degree greater than 1 it is unknown, if there is a geometrical property on unstructured meshes indicating if a chosen stencil is admissible or not.

However, some essential aspects for selecting a stencil have to be taken into account:

- the stencils should be local,
- the number of stencils should be small to keep computational cost small,
- in smooth regions it is necessary that the stencil is well-centered with respect to the cell T_ℓ to obtain a good approximation quality,
- in non-smooth regions one-sided stencils have to be selected to avoid interpolation across discontinuities leading to oscillations.

Considering the construction of stencils on unstructured triangulations, it is convenient to define the neighbourhood of adjacent triangles as introduced in [34].

Definition 8. *Let $\mathcal{T}$ be a conforming triangulation. Then for any triangle $T_\ell \in \mathcal{T}$ the set*

$$\mathcal{K}_N^0(T_\ell) = \{T \in \mathcal{T} : T \cap T_\ell \text{ is edge of } T_\ell \text{ and } T \neq T_\ell\}$$

is called the `von Neumann neighbourhood`[1] *of T_ℓ and all triangles $T \in \mathcal{K}_N^0(T_\ell)$ are* `level-0 von Neumann neighbours` *of T_ℓ.*

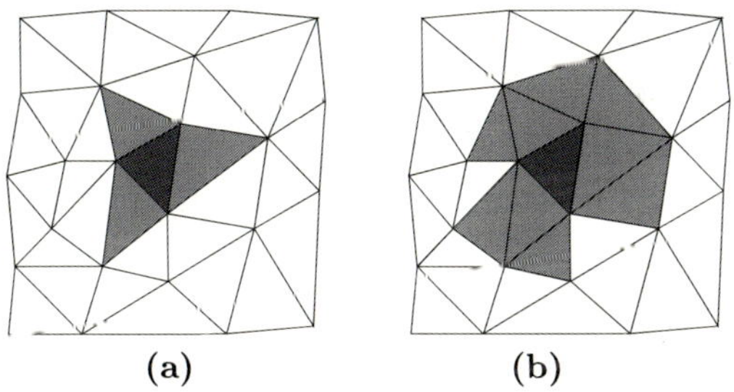

Fig. 1. A triangle T (dark shaded) with von Neumann neighbours (light shaded) of level-0 **(a)** and extended by von Neumann neighbours of level-1 **(b)**.

An extended von Neumann neighbourhood of level-1 can be constructed by merging von Neumann neighbourhoods of the original level-0 von Neumann neighbours, i.e.,

$$\mathcal{K}_N^1(T_\ell) = \left(\bigcup_{T \in \mathcal{K}_N^0(T_\ell)} \mathcal{K}_N^0(T) \right) \setminus T_\ell \,,$$

as shown in Figures 1 **(a)** and **(b)**. This way, we can extend the von Neumann neighbourhoods level by level, until a desired number of cells, i.e., a

[1]This is a standard term used in the theory of cellular automata.

4 Finite Volume Formulation

Consider a two-dimensional, scalar conservation law of the form (1) with solution $u(t,x)$. Within the finite volume framework, each discrete value of the function u is viewed as a cell average $\bar{u}_\ell$ over a cell T_ℓ . The advantage of the finite volume approach is, that any kind of mesh can be used, i.e., the shape of the control volume can be chosen arbitrarily. Here, we work with a conforming triangulation $\mathcal{T}$ with cells $T_\ell \in \mathcal{T}$, $\ell = 1, ..., \#\mathcal{T}$, for which the integral form of the conservation law (1) has the form

$$\frac{d}{dt}\int_T u(t,x)\,dx + \int_{\partial T} \mathbf{F}(t,s)\cdot\mathbf{n}(s)\,ds = 0\,, \tag{10}$$

with the outer normal vector $\mathbf{n}(s)$ and the flux

$$\mathbf{F}(t,s) = \begin{pmatrix} f_1(u(t,x(s))) \\ f_2(u(t,x(s))) \end{pmatrix},$$

where the boundary ∂T of the triangle T is parameterized by arclength s (cf. [24, Chapter 23]). Integrating (10) over the time interval $[t^n, t^{n+1}]$, where $\tau = t^{n+1} - t^n$ is the time step length, and using the definition of cell averages in (3), we derive a finite volume scheme of the form

$$\bar{u}_\ell^{n+1} = \bar{u}_\ell^n - \frac{\tau}{|T_\ell|}\sum_{j=1}^{3}\hat{F}_{\ell,j}^n\,, \tag{11}$$

where the *numerical flux* $\hat{F}_{\ell,j}^n$ across each cell boundary $\partial T_{\ell,j}$, $j = 1,2,3$, of the cell T_ℓ during the time interval $[t^n, t^{n+1}]$ is the time-averaged physical flux given by

$$\hat{F}_{\ell,j}^n = \frac{1}{\tau}\int_{t^n}^{t^{n+1}}\left(\int_{\partial T_{\ell,j}} \mathbf{F}(t^n,s)\cdot\mathbf{n}_{\ell,j}\,ds\right)dt\,.$$

The time integral and the integral along the j-th edge of triangle T_ℓ can be computed exactly by using a suitable Gaussian quadrature rule. Therefore, the numerical flux can be computed through the weighted sum

$$\hat{F}_{\ell,j}^n = \sum_{k=1}^{N_t}\alpha_k|\partial T_{\ell,j}|\sum_{h=1}^{N_x}\beta_h\mathbf{F}(u(t_{G_k}, x_{G_h}))\cdot\mathbf{n}_{\ell,j}\,, \tag{12}$$

where α_k and β_h are the weights of the Gaussian quadrature rule and t_{G_k} and x_{G_h} are the corresponding integration points with respect to time and space. N_t and N_x are the numbers of integration points. The situation for a third order approximation using two Gaussian integration points in time and space is illustrated in Figure 5. It is clear, that in order to evaluate the

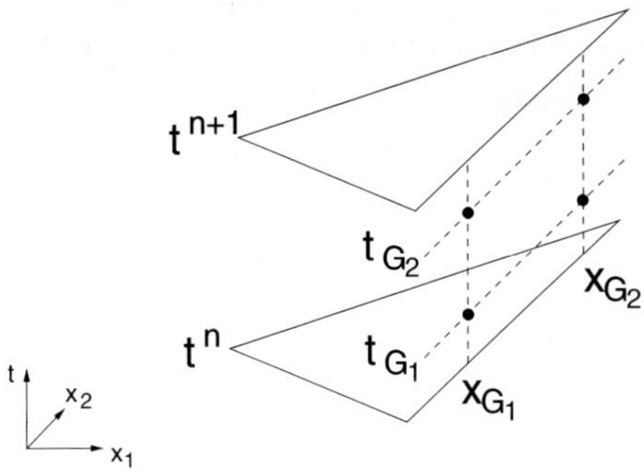

Fig. 5. Example of a third order accurate flux evaluation across a triangle edge from time t^n to t^{n+1} using two Gaussian integration points in space and time.

flux function $\mathbf{F}$ at a particular Gaussian integration point we have to find the function value $u(t_{G_k}, x_{G_h})$, the so-called *state*[2] of the solution, at these points. This is accomplished through the ADER approach, which is extended to unstructured meshes and discussed in detail in the following subsection.

4.1 Flux Evaluation via ADER

Originally, Toro, Millington, and Nejad introduced a method in [38] termed ADER to construct arbitrary high order finite volume schemes for scalar, *linear* conservation laws utilizing high order derivatives. Just very recently, these schemes were extended to scalar, *nonlinear* conservation laws in one dimension by Titarev and Toro in [36, 39, 41] and were applied to problems on multi-dimensional, Cartesian grids by Schwartzkopff, Munz and Toro in [30, 31]. A first attempt to extend ADER schemes for linear conservation laws from structured grids to unstructured triangulations was taken in a preliminary, unpublished note by Munz and Schneider [27]

The main ingredients of the proposed ADER scheme are:

- the WENO technique to reconstruct high order polynomials without creating spurious oscillations;
- the solution of generalized Riemann problems based on
  - the time Taylor expansion of the solution,
  - the Lax-Wendroff procedure replaces time by space derivatives,
  - the solution of conventional, derivative Riemann problems;
- the high order flux evaluation across cell interfaces and updating of cell averages.

In the following, we discuss in detail the construction of ADER schemes for *linear* and *nonlinear* problems. According to the WENO reconstruction procedure in Section 3, the solution $u(t, x)$ at the discrete time $t = t^n$ is

[2] *State* is the common term for the solution of a Riemann problem.

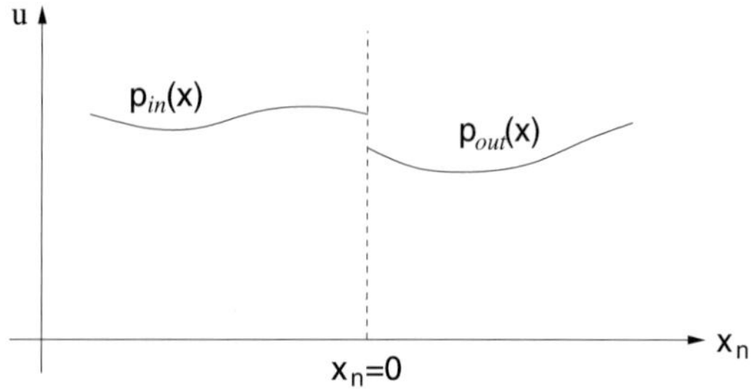

Fig. 6. The generalized Riemann problem along the outward pointing unit vector with reconstructed polynomials $p_{\mathrm{in}}(x)$ and $p_{\mathrm{out}}(x)$ approximating the solution u.

represented by polynomials p_ℓ, $\ell = 1, ..., \#\mathcal{T}$, on each triangular cell $T_\ell \in \mathcal{T}$. In general, these polynomials are different on each cell, leading to a piecewise polynomial approximation of $u(t, x)$ with discontinuities across the cell interfaces. The situation of having two constant functions separated by a discontinuity is usually called a conventional *Riemann problem* (RP). The solution of a RP is a fundamental tool in the development of finite volume methods and is discussed in many textbooks, e.g. [24, 37]. In general, a RP is defined by the governing hyperbolic equation together with a particular initial condition (IC). As mentioned above this IC is usually given by two constant functions and the solution of the RP can be computed through various techniques [24, 37]. However, as our functions p_ℓ, which are separated by the cell interfaces, are not necessarily constant, the situation is more difficult and is called the *generalized Riemann problem* (GRP). Depending on the order m of the designed ADER scheme we will call these schemes ADERm schemes. Therefore, the ADER1 scheme is the classical Godunov scheme [12] of first order (m=1). We remark, that the degree of the reconstruction polynomials for an ADERm scheme is $m-1$. Denoting the corresponding GRP more precisely, we have to solve a GRP$_{m-1}$ at the cells interfaces when using an ADERm scheme[3].

In order to apply the ideas in [36, 39], we reduce the multi-dimensional GRP at the Gaussian integration points x_{G_h} of a cell interface (see Figure 5) to a one-dimensional GRP oriented perpendicular to the interface, i.e., along the outer normal **n** as displayed in Figure 6. The GRP is described by the governing PDE and the IC of $u(t, x)$ at the local time $t = 0$ by

$$\text{PDE:} \quad \frac{\partial u}{\partial t} + \nabla f(u) = 0\,, \tag{13}$$

$$\text{IC:} \quad u(0, x) = \begin{cases} p_{\mathrm{in}}(x)\,, & \text{for} \quad x_n < 0\,, \\ p_{\mathrm{out}}(x)\,, & \text{for} \quad x_n > 0\,. \end{cases}$$

[3] GRP$_0$ denotes the conventional Riemann Problem (RP) with two constant functions as initial condition.

where x_n is a local coordinate oriented along the outer normal $\mathbf{n}$ with origin at the Gaussian integration point x_{G_h}. The two polynomials belonging to the actual cell and the adjacent cell are denoted as p_{in} and p_{out}, respectively.

Recalling equation (12) and Figure 5, we are looking for the solution of the one-dimensional GRP at a Gaussian integration point t_{G_k} in time, i.e., for $u(t,\cdot)$ at an intermediate time $t \in [t^n, t^{n+1}]$. Now, one of the central ideas of the ADER approach is to approximate this solution at m-th order accuracy at the cell interface $x_n = 0$ via the Taylor series expansion in time around the initial time $t = 0$ given through

$$u(t,0) \approx u(0,0) + \sum_{k=1}^{m-1} \frac{t^k}{k!} \frac{\partial^k}{\partial t^k} u(0,0)\,. \tag{14}$$

The time derivatives in (14) can be replaced by space derivatives by applying the *Lax-Wendroff procedure*[4] in order to make the maximum use of the information given by the governing PDE in (13).

4.2 The Lax-Wendroff Procedure

Originally, this technique of substituting time by space derivatives using the governing PDE (13) itself was considered in [23] and goes back to the famous theorem of Cauchy and Kowalewski more than hundred years ago. We remark that for nonlinear problems this procedure can become quite tedious in contrast to the linear case, as the number of terms required to express the time derivatives grows rapidly with the order of the derivative. However, as shown in [39], these terms are necessary in order to guarantee the desired order of accuracy of the designed ADERm scheme. To be more precise, let the characteristic speeds with respect to the space dimensions x_1 and x_2 be given by

$$\lambda_1(u) = \frac{\partial f_1(u)}{\partial u} \quad \text{and} \quad \lambda_2(u) = \frac{\partial f_2(u)}{\partial u}\,. \tag{15}$$

Using the notation $\lambda_i'(u) = \frac{\partial \lambda_i(u)}{\partial u}$, $\lambda_i''(u) = \frac{\partial^2 \lambda_i(u)}{\partial u^2}$ etc., $i = 1, 2$, and $u_t = \frac{\partial u}{\partial t}$, $u_{tt} = \frac{\partial^2 u}{\partial t^2}$, $u_{x_1} = \frac{\partial u}{\partial x_1}$, etc., we can write the governing PDE (13) in two dimensions as

$$u_t + \lambda_1(u) u_{x_1} + \lambda_2(u) u_{x_2} = 0\,. \tag{16}$$

Therefore, to replace the first order time derivative in (14) by space derivatives using the Lax-Wendroff procedure, we simply have to solve (16) for u_t leading to

$$u_t = -\lambda_1(u) u_{x_1} - \lambda_2(u) u_{x_2}\,. \tag{17}$$

The higher order time derivatives of u can now be computed successively by partial differentiation of (17) with respect to t. For example, an ADER4 scheme requires time derivatives up to order three in (14) given by

[4] Sometimes also referred to as the *Cauchy-Kowalewski procedure*.

$$u_{tt} = -\lambda_1'(u)u_t u_{x_1} - \lambda_1(u)u_{tx_1} - \lambda_2'(u)u_t u_{x_2} - \lambda_2(u)u_{tx_2}\,,$$

$$\begin{aligned} u_{ttt} = &-\lambda_1''(u)u_t^2 u_{x_1} - \lambda_1'(u)\Big(u_{tt}u_{x_1} + 2u_t u_{tx_1}\Big) - \lambda_1(u)u_{ttx_1} \\ &-\lambda_2''(u)u_t^2 u_{x_2} - \lambda_2'(u)\Big(u_{tt}u_{x_2} + 2u_t u_{tx_2}\Big) - \lambda_2(u)u_{ttx_2}\,. \end{aligned}$$

Note, that the above expressions for the time derivatives include mixed derivatives with respect to time and space. These can also be expressed by space derivatives by successively differentiating (17) with respect to x_1 and x_2. For example, for the ADER4 scheme we get

$$u_{tx_1} = -\lambda_1'(u)u_{x_1}^2 - \lambda_1(u)u_{x_1x_1} - \lambda_2'(u)u_{x_1}u_{x_2} - \lambda_2(u)u_{x_1x_2}\,,$$

$$u_{tx_2} = -\lambda_1'(u)u_{x_1}u_{x_2} - \lambda_1(u)u_{x_1x_2} - \lambda_2'(u)u_{x_2}^2 - \lambda_2(u)u_{x_2x_2}\,,$$

$$\begin{aligned} u_{tx_1x_1} = &-\lambda_1''(u)u_{x_1}^3 - 3\lambda_1'(u)u_{x_1}u_{x_1x_1} - \lambda_1(u)u_{x_1x_1x_1} \\ &-\lambda_2''(u)u_{x_1}^2 u_{x_2} - \lambda_2'(u)\Big(u_{x_1x_1}u_{x_2} + 2u_{x_1}u_{x_1x_2}\Big) - \lambda_2(u)u_{x_1x_1x_2}\,, \end{aligned}$$

$$\begin{aligned} u_{tx_1x_2} = &-\lambda_1''(u)u_{x_1}^2 u_{x_2} - \lambda_1'(u)\Big(u_{x_1x_1}u_{x_2} + 2u_{x_1}u_{x_1x_2}\Big) - \lambda_1(u)u_{x_1x_1x_2} \\ &-\lambda_2''(u)u_{x_1}u_{x_2}^2 - \lambda_2'(u)\Big(u_{x_1}u_{x_2x_2} + 2u_{x_2}u_{x_1x_2}\Big) - \lambda_2(u)u_{x_1x_2x_2}\,, \end{aligned}$$

$$\begin{aligned} u_{tx_2x_2} = &-\lambda_1''(u)u_{x_1}u_{x_2}^2 - \lambda_1'(u)\Big(u_{x_1}u_{x_2x_2} + 2u_{x_2}u_{x_1x_2}\Big) - \lambda_1(u)u_{x_1x_2x_2} \\ &-\lambda_2''(u)u_{x_2}^3 - 3\lambda_2'(u)u_{x_2}u_{x_2x_2} - \lambda_2(u)u_{x_2x_2x_2}\,, \end{aligned}$$

$$\begin{aligned} u_{ttx_1} = &-\lambda_1''(u)u_t u_{x_1}^2 - \lambda_1'(u)\Big(u_t u_{x_1x_1} + 2u_{x_1}u_{tx_1}\Big) - \lambda_1(u)u_{tx_1x_1} \\ &-\lambda_2''(u)u_t u_{x_1}u_{x_2} - \lambda_2'(u)\Big(u_{tx_1}u_{x_2} + u_{x_1}u_{tx_2} + u_t u_{x_1x_2}\Big) - \lambda_2(u)u_{tx_1x_2}\,, \end{aligned}$$

$$\begin{aligned} u_{ttx_2} = &-\lambda_1''(u)u_t u_{x_1}u_{x_2} - \lambda_1'(u)\Big(u_{tx_1}u_{x_2} + u_{x_1}u_{tx_2} + u_t u_{x_1x_2}\Big) - \lambda_1(u)u_{tx_1x_2} \\ &-\lambda_2''(u)u_t u_{x_2}^2 - \lambda_2'(u)\Big(u_t u_{x_2x_2} + 2u_{x_2}u_{tx_2}\Big) - \lambda_2(u)u_{tx_2x_2}\,. \end{aligned}$$

The problem remaining is to determine the space derivatives at the quadrature points at the cell interface, i.e., to solve the one-dimensional GRP illustrated in Figure 6. In [39], Toro and Titarev suggest to use the *boundary extrapolated values*

$$u_\ell = \lim_{x \to x_{G_h}^-} p_{\text{in}}(x)\,, \tag{18}$$

$$u_r = \lim_{x \to x_{G_h}^+} p_{\text{out}}(x)\,, \tag{19}$$

which represent the values obtained by evaluating the polynomials reconstructed inside and outside the actual cell interface at a Gaussian quadrature point x_{G_h}. We remark, that according to the one-dimensional representation in Figure 6 the inside and outside of a cell are referred to as *left* and *right*, respectively. Therefore, we obtain a conventional Riemann problem GRP_0 with constant functions u_ℓ and u_r of the form

$$\text{PDE:} \quad \frac{\partial u}{\partial t} + \nabla f(u) = 0\,, \tag{20}$$

$$\text{IC:} \quad u(0,x) = \begin{cases} u_\ell & \text{for} \quad x_n < 0, \\ \\ u_r & \text{for} \quad x_n > 0\,. \end{cases} \tag{21}$$

The solution of the above GRP_0 (20),(21) is described in many text books, e.g. [24, 37], and its solution u^* is usually called the *Godunov state.* With this state u^* the characteristic speeds in (15) can be evaluated and used to linearize the governing equation (16). As shown in [38], the linearized equation (16) also holds for all space derivatives $q^\alpha = D^\alpha u$, $|\alpha| \le m - 1$, where $D^\alpha = \frac{\partial^{|\alpha|}}{\partial x_1^{\alpha_1} \partial x_2^{\alpha_2}}$ is the α-th partial derivative operator. Similar to (18) and (19), boundary extrapolated values for the derivatives can be defined by

$$\begin{aligned} q_\ell^\alpha &= \lim_{x \to x_{G_h}^-} D^\alpha p_{\mathrm{in}}(x) \\ q_r^\alpha &= \lim_{x \to x_{G_h}^+} D^\alpha p_{\mathrm{out}}(x) \end{aligned}$$

and we can formulate a series of *linear* conventional Riemann problems of the form

$$\text{PDE:} = \frac{\partial}{\partial t} q^\alpha + \lambda_1(u^*) \frac{\partial}{\partial x_1} q^\alpha + \lambda_2(u^*) \frac{\partial}{\partial x_2} q^\alpha = 0\,,$$

$$\text{IC:} = q^\alpha(0,x) = \begin{cases} q_\ell^\alpha & \text{for} \quad x_n < 0, \\ \\ q_r^\alpha & \text{for} \quad x_n > 0\,. \end{cases}$$

Since each of these linear Riemann problems has a solution, all terms on the right hand side of (14) are well-defined. Therefore, the state $u(t_{G_k}, x_{G_h})$ can be computed through the expansion (14) for any local integration point (t_{G_k}, x_{G_h}) in space and time as indicated in Figure 5. Note, that one of the fundamental ideas of using an ADERm scheme is to solve an GRP_{m-1} by solving one RP, which is linear or nonlinear depending on the governing equation, and a series of *linear* RPs. The number of these linear RPs is $m-1$ for one-dimensional problems and $\frac{1}{2}m(m+1)-1$ for two-dimensional problems. We remark, that the leading term in the computation of the ADER state in (14) is the classical Godunov state u^* itself. The remaining terms in (14) are the correction terms to enhance the approximation quality. Therefore, ADER schemes can be interpreted as high order generalizations of the classical Godunov scheme in [12]. In fact, ADER schemes enable us to evaluate the flux in (12) with arbitrary high order accuracy by using an arbitrary high order accurate state u given by (14) at the cell interface. Finally, this flux is the information we need in order to update the cell average values as summarized in the following section.

5 Update of Cell Average Values

Recalling the numerical scheme in (11) we see, that given the fluxes across all cell boundaries, we can update the cell average values $\bar{u}_\ell^{n+1}$ for time $t = t^{n+1}$ via a one-step explicit scheme. In contrast to multi-stage TVD Runge-Kutta schemes, typically used in combination with high order WENO techniques as presented in [16], we do not need the intermediate stages. Therefore, we can reduce the required computational costs quite significantly, as we have to go through the reconstruction procedure only once. We remark that in multi-stage TVD Runge-Kutta schemes one reconstruction step is necessary for each intermediate stage. Furthermore, Ruuth and Spiteri have just shown recently in [29], that TVD Runge-Kutta of arbitrary high order cannot be constructed in a straight forward manner. In contrast, ADER schemes can be extended to arbitrary high order by simply adding higher order terms in (14) and therefore the order of accuracy is basically limited by the available computing power and machine precision.

It is well-known, that explicit time discretization schemes, such as the proposed ADER scheme, have to satisfy rather severe restrictions on the time step τ due to the Courant-Friedrich-Levy (CFL) condition[5]. Loosely speaking, disturbances from one cell boundary must not reach another boundary within one time step. Let ρ_ℓ be the radius of the inscribed circle of a triangular cell T_ℓ serving as a measure of its diameter (see Figure 7), and let

$$\lambda_\ell^{(\max)} = \max_{1 \le j \le 3N_x} |\lambda_{1,j}(u) \cdot n_{1,j} + \lambda_{2,j}(u) \cdot n_{2,j}|$$

be the maximum normal characteristic speed appearing at the $3N_x$ Gaussian integration points along the cell interfaces. As shown in previous work [36, 38, 39] ADER schemes are stable up to a CFL-number of 1 for structured, Cartesian grids. Therefore, we restrict the time step size τ in our computations by a similar CFL-condition

$$\tau \le \min_{1 \le l \le \#\mathcal{T}} \frac{\rho_\ell}{\lambda_\ell^{(\max)}},$$

for unstructured triangulations. This is similar to the idea in [26, Subsection 3.4.1].

[5] CFL conditions on TVD Runge-Kutta schemes can even be more severe [29].

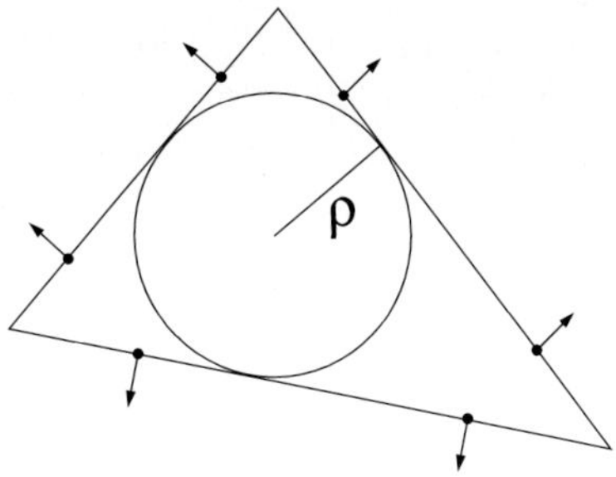

Fig. 7. The radius ρ of the inscribed circle of a triangular cell together with its $N_x = 2$ integration points per edge, where the maximum characteristic speed normal to each edge is evaluated.

6 Performance of ADER schemes

In the last few years, ADER schemes were developed and analysed mainly for one-dimensional linear and nonlinear problems [36, 38, 39] and applied to multi-dimensional problems on fixed rectangular Cartesian meshes, e.g. in [30, 31]. Here we investigate the performance of the proposed ADER schemes for linear and nonlinear problems on unstructured triangulations by determining their convergence properties numerically. Furthermore, we consider their efficiency with respect to computational cost depending on the order of the scheme.

6.1 Experimental Orders of Convergence

In this section, the experimental order of convergence of the proposed ADER schemes on two-dimensional linear and nonlinear advection problems are determined numerically in order to compare them with the theoretically expected orders.

Linear Advection. For the linear problem we solve the two-dimensional equation

$$u_t + u_{x_1} + u_{x_2} = 0\,,$$

a linear example of the general equation (1) with the initial condition

$$u_0(x) = u(0,x) = \sin\Big(2\pi(x_1 + x_2)\Big)$$

on the computational domain $\Omega = [-0.5, 0.5] \times [-0.5, 0.5]$. The computations are carried out for the time interval $I = [0, 1]$. We remark, that we use periodic boundary conditions, such that the reference solution $\tilde{u}(1, x)$ at the end of the simulation time $t = 1$, is identical to the initial condition, i.e., $u_0(x) \equiv u(1, x)$.

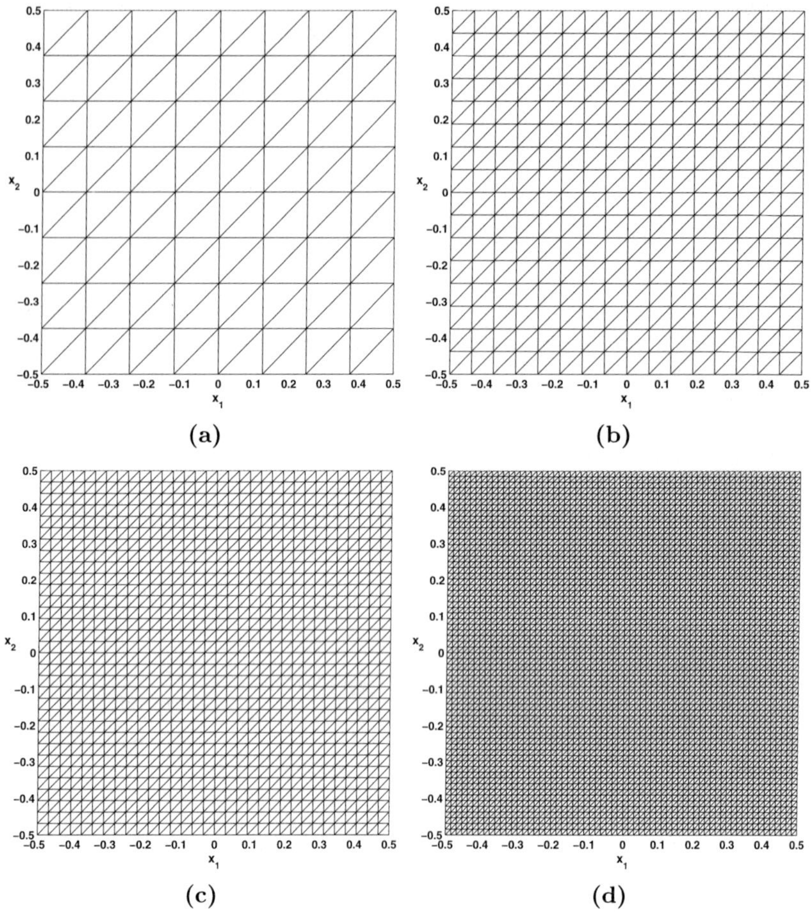

Fig. 8. The sequence of the four regular meshes with their mesh widths **(a)** $\mathbf{A}_0$ ($h = 0.125$), **(b)** $\mathbf{A}_1$ ($h = 0.0625$), **(c)** $\mathbf{A}_2$ ($h = 0.03125$), **(d)** $\mathbf{A}_3$ ($h = 0.015625$).

In order to study the influence of the mesh irregularity, i.e., the distortion of the mesh on the accuracy of the numerical results, we compute the solution on sequences of three different triangular meshes. Mesh **A** is a regular mesh obtained by adding the diagonal line in each square (see Figure 8), mesh **B** is an irregular mesh (see Figure 9) obtained by slightly distorting mesh **A**, and mesh **C** is a strongly distorted irregular mesh (see Figure 10). All sequences of the three meshes consist of five successive refinement levels and are constructed by uniformly refining the coarsest mesh, namely by subdividing each triangular cells into four similar smaller ones. The refinement level of a particular mesh is indicated by subscripts, e.g. $\mathbf{A}_0$ denotes the original mesh **A**, whereas $\mathbf{C}_3$ indicates the third refinement of mesh **C**. The first four refinement levels are displayed in Figures 8, 9, and 10. Note,

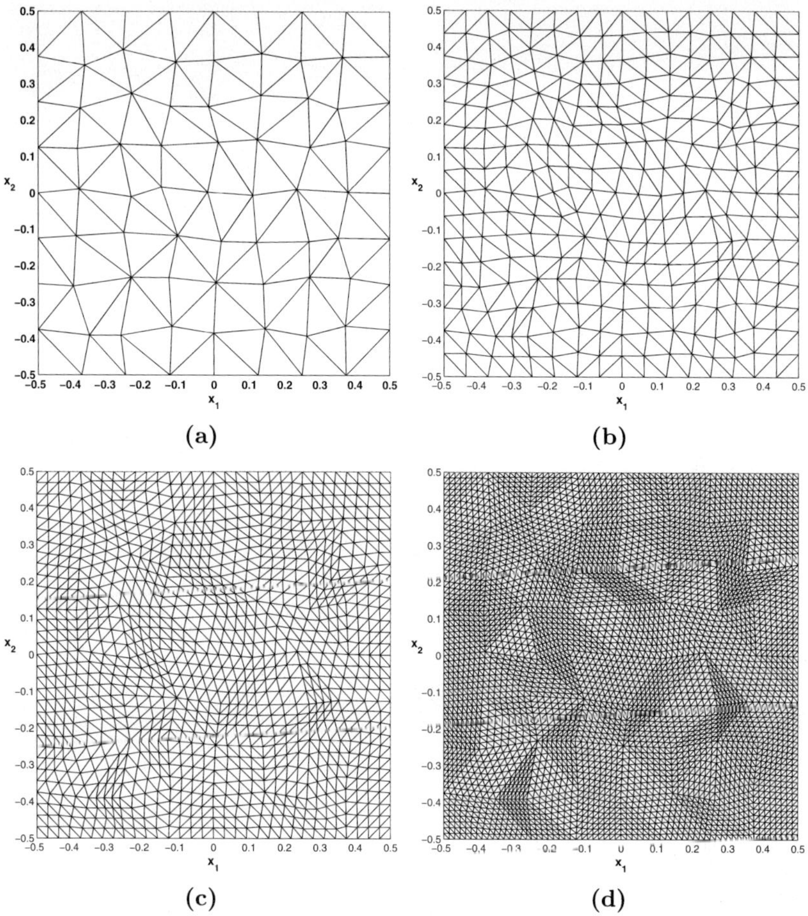

Fig. 9. The sequence of the four slighly irregular meshes with their mesh widths **(a)** $\mathbf{B}_0$ ($h = 0.125$), **(b)** $\mathbf{B}_1$ ($h = 0.0625$), **(c)** $\mathbf{B}_2$ ($h = 0.03125$), **(d)** $\mathbf{B}_3$ ($h = 0.015625$).

that only for the regular mesh $\mathbf{A}$ the mesh spacing h is representative for the entire mesh, whereas h can only be a rough indicator of the mesh width for meshes $\mathbf{B}$ and $\mathbf{C}$. However, all meshes consist of the same number of cells in the corresponding refinement level in order to keep the computational cost independent of the the mesh irregularity.

All computations are carried out for ADER2, ADER3, and ADER4 schemes, where we use nine stencils as constructed by the WENO reconstruction procedure of Subsection 3.2 , i.e., three *centered* stencils, three stencils in the forward sectors $\mathcal{F}_j$ (see Figure 2), and three in the backward sectors $\mathcal{B}_j$ (see Figure 3) as discussed in Subsection 3.2. The stencils consist of 4, 8, or 13 cells for the ADER2, ADER3, or ADER4 schemes, respectively. The time step τ is set to $\tau = 0.025$ for the computations on meshes $\mathbf{A}_0$ and $\mathbf{B}_0$

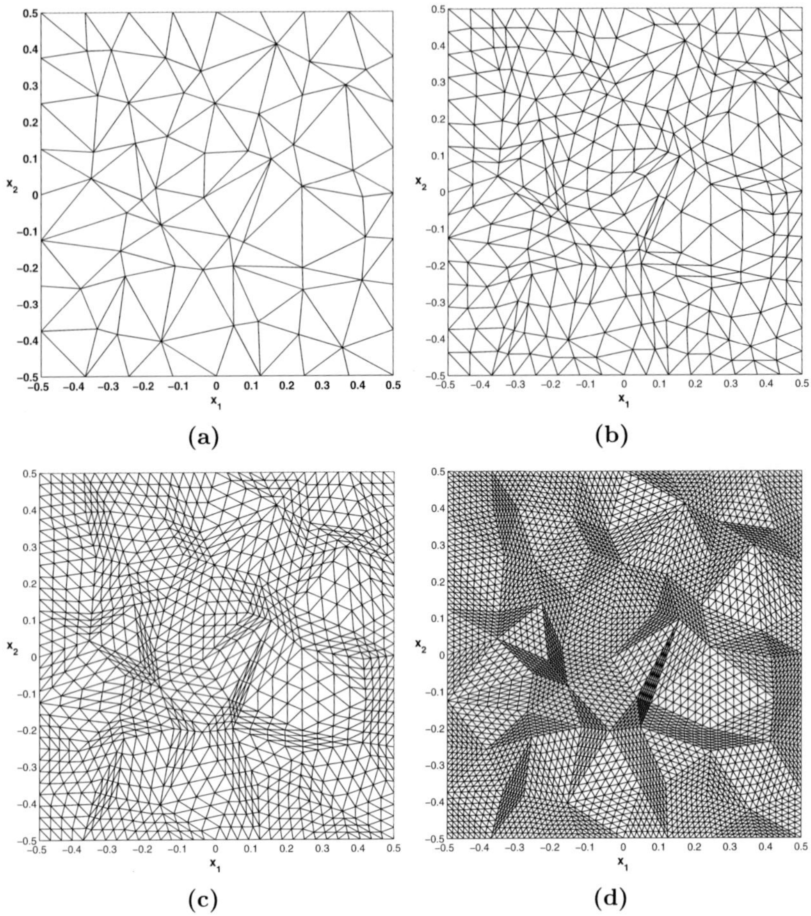

Fig. 10. The sequence of the four strongly irregular meshes with their mesh widths **(a)** $\mathbf{C}_0$ ($h = 0.125$), **(b)** $\mathbf{C}_1$ ($h = 0.0625$), **(c)** $\mathbf{C}_2$ ($h = 0.03125$), **(d)** $\mathbf{C}_3$ ($h = 0.015625$).

and to $\tau = 0.0125$ on mesh $\mathbf{C}_0$. With successive refinement levels the time step τ is halved accordingly. We remark, that the errors presented are those of the cell averages $\bar{u}$ of the solution u and the reference solution $\tilde{u}$, which are computed by a 7-point quadrature rule for triangles (see [9]). Table 1 shows the results for the approximation errors

$$E_p(h) = \|u - \tilde{u}\|_p \, ,$$

for the norms $\|\cdot\|_1$, $\|\cdot\|_2$, and $\|\cdot\|_\infty$, together with the corresponding convergence orders k_p,

Table 1. Results for the linear advection obtained by **(a)** ADER2, **(b)** ADER3, and **(c)** ADER4 schemes on the regular meshes $\mathbf{A}_0$ to $\mathbf{A}_4$.

h	$E_1(h)$	k_1	$E_2(h)$	k_2	$E_\infty(h)$	k_∞
1/8	$3.1024 \cdot 10^{-1}$	–	$3.5613 \cdot 10^{-1}$	–	$5.0293 \cdot 10^{-1}$	–
1/16	$8.9463 \cdot 10^{-2}$	1.79	$9.9535 \cdot 10^{-2}$	1.84	$1.4043 \cdot 10^{-1}$	1.84
1/32	$2.2632 \cdot 10^{-2}$	1.98	$2.5127 \cdot 10^{-2}$	1.99	$3.5492 \cdot 10^{-2}$	1.98
1/64	$5.6576 \cdot 10^{-3}$	2.00	$6.2828 \cdot 10^{-3}$	2.00	$8.8815 \cdot 10^{-3}$	2.00
1/128	$1.4139 \cdot 10^{-3}$	2.00	$1.5703 \cdot 10^{-3}$	2.00	$2.2205 \cdot 10^{-3}$	2.00

(a)

h	$E_1(h)$	k_1	$E_2(h)$	k_2	$E_\infty(h)$	k_∞
1/8	$3.4715 \cdot 10^{-1}$	–	$4.0503 \cdot 10^{-1}$	–	$5.8055 \cdot 10^{-1}$	–
1/16	$5.0290 \cdot 10^{-2}$	2.79	$5.7626 \cdot 10^{-2}$	2.81	$8.1474 \cdot 10^{-2}$	2.83
1/32	$6.1105 \cdot 10^{-3}$	3.04	$6.7352 \cdot 10^{-3}$	3.10	$9.1247 \cdot 10^{-3}$	3.16
1/64	$6.1757 \cdot 10^{-4}$	3.31	$6.8527 \cdot 10^{-4}$	3.30	$9.5297 \cdot 10^{-4}$	3.26
1/128	$7.1885 \cdot 10^{-5}$	3.10	$7.9826 \cdot 10^{-5}$	3.10	$1.1223 \cdot 10^{-4}$	3.09

(b)

h	$E_1(h)$	k_1	$E_2(h)$	k_2	$E_\infty(h)$	k_∞
1/8	$1.1148 \cdot 10^{-1}$	–	$1.3002 \cdot 10^{-1}$	–	$1.9584 \cdot 10^{-1}$	–
1/16	$3.9053 \cdot 10^{-3}$	4.84	$4.4563 \cdot 10^{-3}$	4.87	$8.8605 \cdot 10^{-3}$	4.47
1/32	$2.2444 \cdot 10^{-4}$	4.12	$2.4633 \cdot 10^{-4}$	4.18	$3.4535 \cdot 10^{-4}$	4.68
1/64	$1.4011 \cdot 10^{-5}$	4.00	$1.5064 \cdot 10^{-5}$	4.03	$2.0771 \cdot 10^{-5}$	4.06
1/128	$8.2222 \cdot 10^{-7}$	4.09	$9.1268 \cdot 10^{-7}$	4.04	$1.3338 \cdot 10^{-6}$	3.96

(c)

$$k_p = \frac{\log\Big(E_p(h) \,/\, E_p(h/2)\Big)}{\log(2)},$$

obtained by ADER2, ADER3, and ADER4 schemes on the sequence of regular meshes $\mathbf{A}_0$ to $\mathbf{A}_4$. Tables 2 and 3 show the corresponding results obtained on the sequences of the slighly and strongly irregular meshes **B** and **C**.

Note, that the ADER schemes reach the expected orders of convergence in all norms as well as on all meshes. However, a closer look at the individual approximation errors shows, that the slightly irregular meshes give the best results. This is because many of the triangles of the mesh sequence $\mathbf{B}_0$ to $\mathbf{B}_4$ are closer to being equilateral than in the other cases of meshes **A** and **C**. Indeed, as shown in [4, 20], the shapes of triangular cells play an important role for the accuracy in the sense that simulation results obtained on triangular meshes of equilateral triangles are more accurate than those obtained on other non-equilateral cells. These results are supported by our tests on ADER schemes. Moreover, it is obvious that even for the strongly distorted mesh **C**

Table 2. Results for the linear advection obtained by **(a)** ADER2, **(b)** ADER3, and **(c)** ADER4 schemes on the slightly irregular meshes $\mathbf{B}_0$ to $\mathbf{B}_4$.

h	$E_1(h)$	k_1	$E_2(h)$	k_2	$E_\infty(h)$	k_∞
1/8	$1.1265 \cdot 10^{-1}$	–	$1.2826 \cdot 10^{-1}$	–	$2.7656 \cdot 10^{-1}$	–
1/16	$4.2780 \cdot 10^{-2}$	1.40	$4.8948 \cdot 10^{-2}$	1.39	$1.0326 \cdot 10^{-1}$	1.42
1/32	$1.1288 \cdot 10^{-2}$	1.92	$1.2915 \cdot 10^{-2}$	1.92	$2.6589 \cdot 10^{-2}$	1.96
1/64	$2.6513 \cdot 10^{-3}$	2.42	$3.0153 \cdot 10^{-3}$	2.43	$1.1444 \cdot 10^{-2}$	1.41
1/128	$6.3234 \cdot 10^{-4}$	2.13	$7.1838 \cdot 10^{-4}$	2.14	$3.7882 \cdot 10^{-3}$	1.65

(a)

h	$E_1(h)$	k_1	$E_2(h)$	k_2	$E_\infty(h)$	k_∞
1/8	$1.4226 \cdot 10^{-1}$	–	$1.6078 \cdot 10^{-1}$	–	$2.7919 \cdot 10^{-1}$	–
1/16	$1.6160 \cdot 10^{-2}$	3.14	$1.8617 \cdot 10^{-2}$	3.11	$3.9276 \cdot 10^{-2}$	2.83
1/32	$1.5446 \cdot 10^{-3}$	3.39	$1.8346 \cdot 10^{-3}$	3.34	$4.2469 \cdot 10^{-3}$	3.21
1/64	$2.0259 \cdot 10^{-4}$	3.40	$2.2524 \cdot 10^{-4}$	3.51	$4.2128 \cdot 10^{-4}$	3.87
1/128	$2.4139 \cdot 10^{-5}$	3.17	$2.6835 \cdot 10^{-5}$	3.17	$5.1008 \cdot 10^{-5}$	3.14

(b)

h	$E_1(h)$	k_1	$E_2(h)$	k_2	$E_\infty(h)$	k_∞
1/8	$2.9912 \cdot 10^{-2}$	–	$3.4907 \cdot 10^{-2}$	–	$7.2935 \cdot 10^{-2}$	–
1/16	$1.1801 \cdot 10^{-3}$	4.66	$1.5787 \cdot 10^{-3}$	4.47	$5.2470 \cdot 10^{-3}$	3.80
1/32	$6.9519 \cdot 10^{-5}$	4.09	$8.9930 \cdot 10^{-5}$	4.13	$3.2150 \cdot 10^{-4}$	4.03
1/64	$6.4714 \cdot 10^{-6}$	3.97	$8.0984 \cdot 10^{-6}$	4.03	$3.1137 \cdot 10^{-5}$	3.91
1/128	$4.4070 \cdot 10^{-7}$	4.00	$5.5669 \cdot 10^{-7}$	3.99	$2.2974 \cdot 10^{-6}$	3.88

(c)

we still get very satisfying results, indicating that the proposed ADER scheme combined with the discussed WENO reconstruction technique seems to be a very useful and robust approach applicable to strongly distorted unisotropic meshes.

Nonlinear Advection. For the nonlinear problem we solve the two-dimensional inviscid Burgers equation

$$u_t + \left(\frac{1}{2}u^2\right)_{x_1} + \left(\frac{1}{2}u^2\right)_{x_2} = 0\,, \tag{22}$$

a nonlinear example of equation (1), with the initial condition

$$u_0(x) = u(0,x) = 0.3 + 0.7\sin\Big(2\pi(x_1 + x_2)\Big)$$

on the computational domain $\Omega = [-0.5, 0.5] \times [-0.5, 0.5]$. The computations are carried out for the time interval $I = [0, \frac{1}{4\pi}]$, such that no discontinuity has

Table 3. Results for the linear advection obtained by **(a)** ADER2, **(b)** ADER3, and **(c)** ADER4 schemes on the strongly irregular meshes $\mathbf{C}_0$ to $\mathbf{C}_4$.

h	$E_1(h)$	k_1	$E_2(h)$	k_2	$E_\infty(h)$	k_∞
1/8	$1.3924 \cdot 10^{-1}$	–	$1.6233 \cdot 10^{-1}$	–	$3.9986 \cdot 10^{-1}$	–
1/16	$3.2158 \cdot 10^{-2}$	2.11	$3.8800 \cdot 10^{-2}$	2.06	$1.4476 \cdot 10^{-1}$	1.47
1/32	$6.8809 \cdot 10^{-3}$	2.22	$8.3858 \cdot 10^{-3}$	2.21	$3.9424 \cdot 10^{-2}$	1.88
1/64	$1.6080 \cdot 10^{-3}$	2.10	$1.9787 \cdot 10^{-3}$	2.08	$1.0345 \cdot 10^{-2}$	1.93
1/128	$3.8924 \cdot 10^{-4}$	2.05	$4.8469 \cdot 10^{-4}$	2.03	$3.1769 \cdot 10^{-3}$	1.70

(a)

h	$E_1(h)$	k_1	$E_2(h)$	k_2	$E_\infty(h)$	k_∞
1/8	$2.7500 \cdot 10^{-1}$	–	$3.0955 \cdot 10^{-1}$	–	$4.9177 \cdot 10^{-1}$	–
1/16	$3.8493 \cdot 10^{-2}$	2.84	$4.4821 \cdot 10^{-2}$	2.79	$9.5172 \cdot 10^{-2}$	2.37
1/32	$4.5424 \cdot 10^{-3}$	3.08	$5.3011 \cdot 10^{-3}$	3.08	$1.1456 \cdot 10^{-2}$	3.05
1/64	$5.2333 \cdot 10^{-4}$	3.12	$6.0649 \cdot 10^{-4}$	3.13	$1.2106 \cdot 10^{-3}$	3.24
1/128	$6.1609 \cdot 10^{-5}$	3.09	$7.1088 \cdot 10^{-5}$	3.09	$1.4629 \cdot 10^{-4}$	3.05

(b)

h	$E_1(h)$	k_1	$E_2(h)$	k_2	$E_\infty(h)$	k_∞
1/8	$6.6326 \cdot 10^{-2}$	–	$7.9679 \cdot 10^{-2}$	–	$1.5932 \cdot 10^{-1}$	–
1/16	$3.9170 \cdot 10^{-3}$	4.08	$5.2793 \cdot 10^{-3}$	3.92	$1.3527 \cdot 10^{-2}$	3.56
1/32	$2.0676 \cdot 10^{-4}$	4.24	$2.7034 \cdot 10^{-4}$	4.29	$8.8686 \cdot 10^{-4}$	3.93
1/64	$1.3002 \cdot 10^{-5}$	3.99	$1.5726 \cdot 10^{-5}$	4.10	$5.3229 \cdot 10^{-5}$	4.06
1/128	$7.7907 \cdot 10^{-7}$	4.06	$9.5160 \cdot 10^{-7}$	4.05	$3.7559 \cdot 10^{-6}$	3.82

(c)

developed yet, i.e., the solution is still smooth at the end of the simulation time. Again, periodic boundary conditions are used. Note, that this initial condition leads to a transonic rarefaction.

Then, the same sequences of meshes **A**, **B**, and **C** as shown in Figures 8, 9, and 10 are used as in the linear case. The cell averages of the exact solution $\tilde{u}$ are calculated via the 7-point quadrature rule [9], where the values at each quadrature point is calculated via Newton's method.

Analogously to the linear case, Tables 4, 5, and 6 show the errors $E_p(h)$ of the cell averages at the end of the simulation together with the experimental orders of convergence k_p for the ADER2, ADER3, and ADER4 schemes on the meshes $\mathbf{A}_0$ to $\mathbf{A}_4$, $\mathbf{B}_0$ to $\mathbf{B}_4$, and $\mathbf{C}_0$ to $\mathbf{C}_4$, respectively. In analogy to the results for the linear problem the considered ADER schemes achieve the expected orders of convergence. Only the orders k_∞ of the ADER4 scheme on the strongly distorted mesh sequence $\mathbf{C}_0$ to $\mathbf{C}_4$ seem not to reach the expected order of 4 (see Table 6 **(c)**). We believe that this is due to the appearance of long and thin triangles, which in turn may lead to reconstruc-

Table 4. Results for Burgers equation obtained by **(a)** ADER2, **(b)** ADER3, and **(c)** ADER4 schemes on the regular meshes $\mathbf{A}_0$ to $\mathbf{A}_4$.

h	$E_1(h)$	k_1	$E_2(h)$	k_2	$E_\infty(h)$	k_∞
1/8	$1.3523 \cdot 10^{-2}$	–	$1.5406 \cdot 10^{-2}$	–	$4.5874 \cdot 10^{-2}$	–
1/16	$3.4015 \cdot 10^{-3}$	1.99	$4.2354 \cdot 10^{-3}$	1.86	$9.9464 \cdot 10^{-3}$	2.21
1/32	$8.1563 \cdot 10^{-4}$	2.06	$1.0681 \cdot 10^{-3}$	1.99	$2.6907 \cdot 10^{-3}$	1.89
1/64	$1.9851 \cdot 10^{-4}$	2.04	$2.6442 \cdot 10^{-4}$	2.01	$6.7492 \cdot 10^{-4}$	2.00
1/128	$4.8844 \cdot 10^{-5}$	2.02	$6.5561 \cdot 10^{-5}$	2.01	$1.6728 \cdot 10^{-4}$	2.01

(a)

h	$E_1(h)$	k_1	$E_2(h)$	k_2	$E_\infty(h)$	k_∞
1/8	$2.4460 \cdot 10^{-2}$	–	$2.6815 \cdot 10^{-2}$	–	$4.6138 \cdot 10^{-2}$	–
1/16	$3.2812 \cdot 10^{-3}$	2.90	$3.9929 \cdot 10^{-3}$	2.75	$8.3484 \cdot 10^{-3}$	2.47
1/32	$3.2445 \cdot 10^{-4}$	3.34	$4.4015 \cdot 10^{-4}$	3.18	$1.0869 \cdot 10^{-3}$	2.94
1/64	$3.3403 \cdot 10^{-5}$	3.28	$4.6285 \cdot 10^{-5}$	3.25	$1.2003 \cdot 10^{-4}$	3.18
1/128	$3.9009 \cdot 10^{-6}$	3.10	$5.4216 \cdot 10^{-6}$	3.09	$1.4117 \cdot 10^{-5}$	3.09

(b)

h	$E_1(h)$	k_1	$E_2(h)$	k_2	$E_\infty(h)$	k_∞
1/8	$6.4417 \cdot 10^{-3}$	–	$8.5263 \cdot 10^{-3}$	–	$2.1361 \cdot 10^{-2}$	–
1/16	$3.2915 \cdot 10^{-4}$	4.29	$4.9585 \cdot 10^{-4}$	4.10	$1.3426 \cdot 10^{-3}$	3.99
1/32	$2.0740 \cdot 10^{-5}$	3.99	$3.5896 \cdot 10^{-5}$	3.79	$1.2353 \cdot 10^{-4}$	3.44
1/64	$1.6025 \cdot 10^{-6}$	3.69	$2.6907 \cdot 10^{-6}$	3.74	$9.0486 \cdot 10^{-6}$	3.77
1/128	$1.1391 \cdot 10^{-7}$	3.81	$1.8737 \cdot 10^{-7}$	3.84	$5.7273 \cdot 10^{-7}$	3.98

(c)

tion polynomials of rather poor approximation quality, due to degenerate one-sided stencils. The stencil construction discussed in Subsection 3.2 uses a sectoral search of subsequent von Neumann neighbours of increasing levels. Therefore, a very thin triangular cell with a very small angle leads to a very narrow sectors $\mathcal{F}_j$ and $\mathcal{B}_j$, which in turn results in very elongated stencils with shapes preferring a particular direction. The resulting reconstruction of rather low approximation quality then influences the computation of the fluxes and finally causes errors that appear especially in the $\|\cdot\|_\infty$ norm.

This effect, however, was not seen in the linear case. We believe, that in the linear case, the applied mesh sequences were fine enough to discretise the smooth solution u to reach the expected convergence orders. In the nonlinear example, the simulation time $T = \frac{1}{4\pi}$ was chosen to keep the solution u smooth, but steep gradients are present. We think, that these gradients are not sufficiently well-resolved by the mesh sequence $\mathbf{C}_0$ to $\mathbf{C}_4$ and therefore, the expected order is not quite reached. Further mesh refinement should help to eliminate this effect.

Table 5. Results for Burgers equation obtained by **(a)** ADER2, **(b)** ADER3, and **(c)** ADER4 schemes on the regular meshes $\mathbf{B}_0$ to $\mathbf{B}_4$.

h	$E_1(h)$	k_1	$E_2(h)$	k_2	$E_\infty(h)$	k_∞
1/8	$1.4816 \cdot 10^{-2}$	–	$2.1592 \cdot 10^{-2}$	–	$8.9534 \cdot 10^{-2}$	–
1/16	$5.0152 \cdot 10^{-3}$	1.56	$6.8720 \cdot 10^{-3}$	1.65	$3.2865 \cdot 10^{-2}$	1.45
1/32	$1.3421 \cdot 10^{-3}$	1.90	$1.8877 \cdot 10^{-3}$	1.86	$1.0561 \cdot 10^{-2}$	1.64
1/64	$3.4067 \cdot 10^{-4}$	1.98	$4.8618 \cdot 10^{-4}$	1.96	$2.7014 \cdot 10^{-3}$	1.97
1/128	$8.3667 \cdot 10^{-5}$	2.03	$1.2018 \cdot 10^{-4}$	2.02	$7.0141 \cdot 10^{-4}$	1.95

(a)

h	$E_1(h)$	k_1	$E_2(h)$	k_2	$E_\infty(h)$	k_∞
1/8	$1.2429 \cdot 10^{-2}$	–	$1.5481 \cdot 10^{-2}$	–	$4.7784 \cdot 10^{-2}$	–
1/16	$1.6329 \cdot 10^{-3}$	2.93	$2.2922 \cdot 10^{-3}$	2.76	$1.0174 \cdot 10^{-2}$	2.23
1/32	$1.9838 \cdot 10^{-4}$	3.04	$3.0528 \cdot 10^{-4}$	2.91	$2.1328 \cdot 10^{-3}$	2.25
1/64	$2.7484 \cdot 10^{-5}$	3.31	$4.0679 \cdot 10^{-5}$	3.37	$2.8764 \cdot 10^{-4}$	3.35
1/128	$3.5762 \cdot 10^{-6}$	3.04	$5.1999 \cdot 10^{-6}$	3.06	$4.9262 \cdot 10^{-5}$	2.63

(b)

h	$E_1(h)$	k_1	$E_2(h)$	k_2	$E_\infty(h)$	k_∞
1/8	$2.9430 \cdot 10^{-3}$	–	$3.9772 \cdot 10^{-3}$	–	$1.6612 \cdot 10^{-2}$	–
1/16	$2.2322 \cdot 10^{-4}$	3.72	$3.5916 \cdot 10^{-4}$	3.47	$1.5177 \cdot 10^{-3}$	3.45
1/32	$1.9599 \cdot 10^{-5}$	3.51	$3.7513 \cdot 10^{-5}$	3.26	$2.7872 \cdot 10^{-4}$	2.44
1/64	$1.7003 \cdot 10^{-6}$	4.09	$2.9834 \cdot 10^{-6}$	4.24	$2.9170 \cdot 10^{-5}$	3.78
1/128	$1.3478 \cdot 10^{-7}$	3.78	$2.4466 \cdot 10^{-7}$	3.72	$2.6691 \cdot 10^{-6}$	3.56

(c)

6.2 Computational Efficiency

An important consideration, when applying numerical schemes to particular problems is their *computational efficiency*, which is a measure of reaching a desired accuracy in a particular computing time. In general, there are two possibilities to enhance the accuracy of a given scheme. First, the same scheme can be used on a finer discretisation, i.e., on smaller cells, or secondly a higher order version of the scheme can be used. Either possibility leads to an increase in computation time. Therefore, it is important to investigate which option provides the desired accuracy in less computational time. In other words the following question has to be answered: Is it more efficient to use a simple and fast low-order scheme on fine meshes or to use a more sophisticated and slower scheme of higher order on rather coarse meshes?

In this subsection, we numerically evaluate the proposed ADER schemes with respect to computing time and achieved accuracy which should help to answer the above question. Therefore, we record the CPU time used by

Table 6. Results for Burgers equation obtained by **(a)** ADER2, **(b)** ADER3, and **(c)** ADER4 schemes on the regular meshes $\mathbf{C}_0$ to $\mathbf{C}_4$.

h	$E_1(h)$	k_1	$E_2(h)$	k_2	$E_\infty(h)$	k_∞
1/8	$2.4789 \cdot 10^{-2}$	–	$3.5598 \cdot 10^{-2}$	–	$1.2987 \cdot 10^{-1}$	–
1/16	$8.1998 \cdot 10^{-3}$	1.60	$1.1486 \cdot 10^{-2}$	1.63	$6.5593 \cdot 10^{-2}$	0.99
1/32	$2.2506 \cdot 10^{-3}$	1.87	$3.2835 \cdot 10^{-3}$	1.81	$2.7181 \cdot 10^{-2}$	1.27
1/64	$5.5952 \cdot 10^{-4}$	2.01	$8.4517 \cdot 10^{-4}$	1.96	$9.1484 \cdot 10^{-3}$	1.57
1/128	$1.3480 \cdot 10^{-4}$	2.05	$2.0520 \cdot 10^{-4}$	2.04	$2.5284 \cdot 10^{-3}$	1.86

(a)

h	$E_1(h)$	k_1	$E_2(h)$	k_2	$E_\infty(h)$	k_∞
1/8	$2.1345 \cdot 10^{-2}$	–	$2.7487 \cdot 10^{-2}$	–	$8.6973 \cdot 10^{-2}$	–
1/16	$3.0335 \cdot 10^{-3}$	2.81	$4.4508 \cdot 10^{-3}$	2.63	$1.9878 \cdot 10^{-2}$	2.13
1/32	$3.8506 \cdot 10^{-4}$	2.98	$6.4792 \cdot 10^{-4}$	2.78	$5.4981 \cdot 10^{-3}$	1.85
1/64	$4.5916 \cdot 10^{-5}$	3.07	$7.6192 \cdot 10^{-5}$	3.09	$6.2541 \cdot 10^{-4}$	3.14
1/128	$5.5909 \cdot 10^{-6}$	3.04	$9.2328 \cdot 10^{-6}$	3.04	$8.5906 \cdot 10^{-5}$	2.86

(b)

h	$E_1(h)$	k_1	$E_2(h)$	k_2	$E_\infty(h)$	k_∞
1/8	$5.6973 \cdot 10^{-3}$	–	$8.1636 \cdot 10^{-3}$	–	$4.3144 \cdot 10^{-2}$	–
1/16	$5.1513 \cdot 10^{-4}$	3.47	$9.2607 \cdot 10^{-4}$	3.14	$5.0294 \cdot 10^{-3}$	3.10
1/32	$3.9238 \cdot 10^{-5}$	3.71	$7.8427 \cdot 10^{-5}$	3.56	$6.1687 \cdot 10^{-4}$	3.03
1/64	$2.7966 \cdot 10^{-6}$	3.81	$6.0176 \cdot 10^{-6}$	3.70	$5.2142 \cdot 10^{-5}$	3.56
1/128	$1.8851 \cdot 10^{-7}$	3.89	$4.4105 \cdot 10^{-7}$	3.77	$5.6319 \cdot 10^{-6}$	3.21

(c)

the different ADER schemes in order to compute the various steps necessary to complete one time step. As an example we choose the computation of the solution of the nonlinear advection equation (22) with the four ADER schemes on the slighly irregular mesh $\mathbf{B}_2$ shown in Figure 9 **(c)**, which consists of 2048 fixed cells. We remark, that all computations have been carried out with MATLAB 6 Release 13 on a PC (model: IBM 236623G) with processor type Intel Pentium(R) 4 1600MHz.

Table 7 shows the CPU times in seconds needed to complete the different steps of each ADERm scheme, $m = 1, ..., 4$. Here, t_s denotes the time in CPU seconds required to construct the stencils, t_r is the time to compute the reconstruction polynomials, and t_o is the time required for all other computations, such as the flux evaluation and the update of cell averages. Note, that t_o also includes the Lax-Wendroff procedure described in Subsection 4.2 to replace the time derivatives by space derivatives. The total time t_{tot} indicates the required CPU seconds in order to complete one time step. It is obvious, that with increasing order of accuracy m an ADERm scheme

Table 7. CPU seconds for the different computational step of ADER schemes.

	ADER1	ADER2	ADER3	ADER4
t_s	0	0.058	0.208	0.467
t_r	0	2.429	2.948	4.001
t_o	0.426	0.430	0.448	0.481
t_{tot}	0.426	2.917	3.604	4.949

Table 8. Factors indicating the slowdown of ADER schemes.

ADER1	ADER2	ADER3	ADER4
1	6.8	8.5	11.6
	1	1.2	1.7
		1	1.4

becomes more expensive. Note, that the main contribution to the increasing total CPU times t_{tot} is caused by t_s for the stencil construction and t_r for the reconstruction, whereas the increase in t_o is almost negligible. Table 8 shows the factors that indicate the slowdown of an ADER scheme, when increasing the order of accuracy. The factors represent the ratios of the times t_{tot} normalized to ADER1 (first row), ADER2 (second row), or ADER3 (third row). For example, we can read from Table 8 that the CPU time for an ADER4 scheme is 11.6 times larger than that of an ADER1 scheme, or 1.7 times larger than that of an ADER2 scheme, etc. However, due to the higher order of accuracy, the higher order ADER schemes can be applied to much coarser meshes, which reduces the computational time as discussed below.

Figure 11 shows, how the approximation error $E_1(h)$, obtained by the four different ADER schemes, decreases with decreasing mesh width h. The plot also shows, that for ADER schemes of higher order the errors decrease very rapidly, if the mesh is refined. Recall, that the number N of required mesh cells for a two-dimensional problem is $N \sim h^{-2}$. The time to compute new cell averages at the next time step depends linearly on N, i.e., $t_{\text{tot}} \sim N$. To give an example, let us look at an error of $E_1(h) = 10^{-3}$ in Figure 11, where the ADER2, ADER3, and ADER4 schemes require $h \approx 0.028$, $h \approx 0.055$ and $h \approx 0.1$ to reach the desired accuracy. Therefore, the ADER2 scheme needs about $(0.1/0.028)^2 \approx 12$ times more cells and the ADER3 requires about $(0.1/0.055)^2 \approx 3$ times more cells than ADER4. Now, as the CPU time for one time step for the ADER4 scheme is only 1.7 times larger than for ADER2 and only 1.4 times larger than for ADER3 (see Table 8), the ADER4 scheme turns out to be much more efficient. In other words, we have to combine the results of Figure 11 with those in Table 7 to decide, which scheme gives the desired accuracy at the lowest overall computational time. These results are shown in Figure 12. The CPU time is normalized by the CPU time of the ADER1 scheme on the coarsest mesh $\mathbf{B}_0$. Coming back to the above example of $E_1 = 10^{-3}$, Figure 12 indicates, that the ADER2, ADER3,

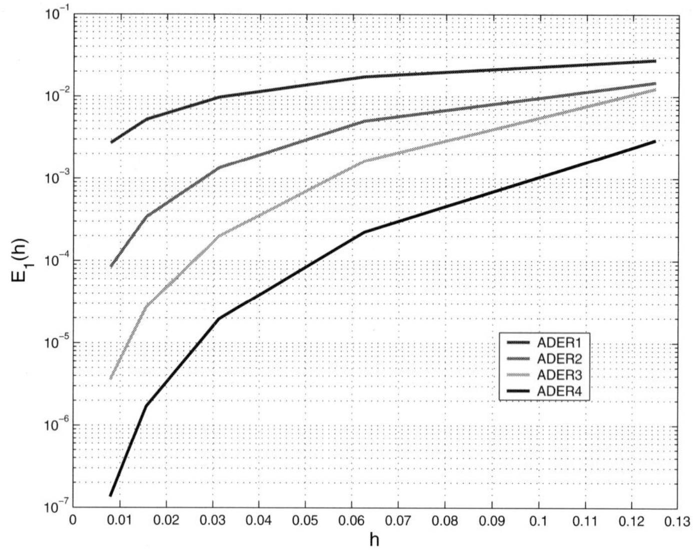

Fig. 11. Accuracy of ADER schemes for different mesh width h, for ADER1 (top), ADER2, ADER3, and ADER4 (bottom).

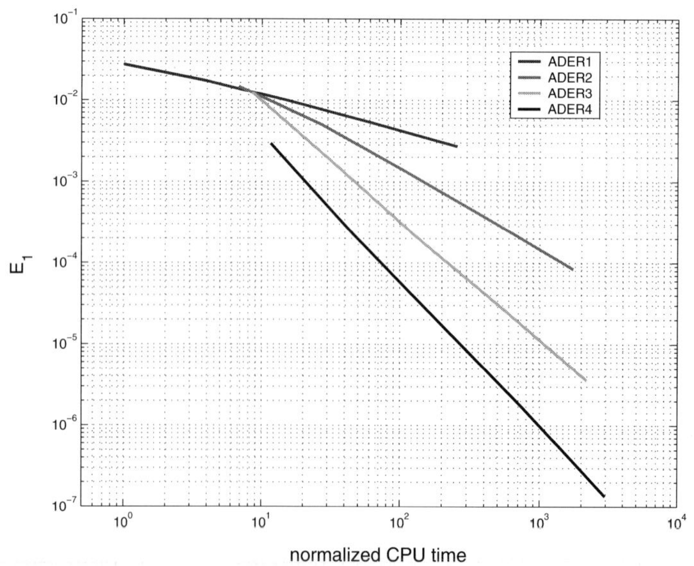

Fig. 12. Accuracy reached by different ADER schemes at their corresponding relative CPU time, for ADER1 (top), ADER2, ADER3, and ADER4 (bottom).

and ADER4 schemes require CPU times of about 150, 50, and 20 seconds, respectively. Note, that extrapolating the results the ADER1 scheme would require a CPU time larger than 10^3 to reach the desired error of $E_1 = 10^{-3}$. We also remark, that for results of low accuracy, i.e., of a rather large error $E_1 \approx 10^{-2}$, ADER1, ADER2, and ADER3 require roughly the same CPU time. There, the saving of mesh cells just balances the additional time used per time step.

In general, our results in Figure 12 show, that for problems with smooth solutions higher order schemes are more efficient than low order schemes with globally refined meshes, especially when highly accurate results are desired. In order to enhance the accuracy of the proposed ADER schemes even more, especially for solutions with discontinuities, we combine the ADER schemes with the ideas of adaptive mesh refinement in order to reduce numerical smearing. Details of our adaptive mesh strategy are discussed below.

7 Adaption Rules

One important feature of our ADER schemes on unstructured triangulations is the time dependent adaptive mesh. Adaptivity requires the modification of the triangulation $\mathcal{T}$ during the simulation in order to be able to balance the two conflicting requirements of good approximation quality and small computational costs. In fact, for the sake of reducing the computational complexity we wish to reduce the number of cells, whereas for the sake of good approximation quality we prefer to use a fine mesh and therefore increase the number of cells.

We have combined the proposed ADER schemes with the ideas of the adaption strategy, that has been discussed in previous work [5, 6, 18] and has proved to be efficient and robust. Here, we briefly review the basic ideas of the adaption strategy.

7.1 Error Indication

As described in [5, 6] we use a customized error indicator in order to adaptively modify the triangulation $\mathcal{T}$. A *significance* value η_ℓ for each cell $T_\ell \in \mathcal{T}$ is required to reflect the local approximation quality of the cell average $\bar{u}_\ell$. These significances η_ℓ, $\ell = 1, ..., \#\mathcal{T}$, are used in order to flag single triangles as *to be refined* or *to be coarsened.*

Definition 11. *Let* $\eta^* = \max_{1 \le \ell \le \#\mathcal{T}} \eta_\ell$, *and let* $\theta_{\text{crs}}, \theta_{\text{ref}}$ *be two tolerance values satisfying* $0 < \theta_{\text{crs}} < \theta_{\text{ref}} < 1$. *We say that a cell* $T \in \mathcal{T}$ *is* `to be refined`, *iff* $\eta_\ell > \theta_{\text{ref}} \cdot \eta^*$, *and* T *is* `to be coarsened`, *iff* $\eta_\ell < \theta_{\text{crs}} \cdot \eta^*$.

In our numerical experiments, we let $\theta_{\text{crs}} = 0.01$ and $\theta_{\text{ref}} = 0.05$ Note that a cell T cannot be refined and be coarsened at the same time; in fact, it may neither be refined nor be coarsened. In order to define the error indicator η_ℓ we first need to specify a set of neighbouring cells for each triangular cell $T \in \mathcal{T}$.

Definition 12. *Let $\mathcal{T}$ be a conforming triangulation. Then for any triangle $T_\ell \in \mathcal{T}$ the set*

$$K_M(T_\ell) = \{T \in \mathcal{T} \,:\, T \cap T_\ell \;\; \textit{is edge of } T_\ell \;\; \textit{or node of } T_\ell \textit{ and } T \neq T_\ell\}$$

is called `Moore neighbourhood`[6] *of T_ℓ and all triangles $T \in \mathcal{K}_M(T_\ell)$ are called* `Moore neighbours` *of T_ℓ.*

An example of a Moore neighbourhood is displayed in Figure 13. Following

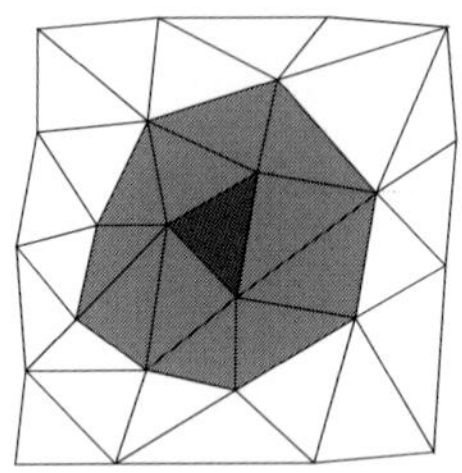

Fig. 13. A triangle T (dark shaded) with its Moore neighbours (light shaded).

along the lines of [13], and assuming that each cell average value $\bar{u}_\ell$ is assigned to the barycenter ξ_ℓ of cell T_ℓ, i.e., $\bar{u}_\ell \equiv \bar{u}(\xi_\ell)$, the error indicator is then given by

$$\eta_\ell = |\bar{u}(\xi_\ell) - s(\xi_\ell)|, \tag{23}$$

where for the Moore neighbourhood $\mathcal{K}_M(T_\ell)$ of T_ℓ the *thin plate spline* interpolant $s \equiv s_{\mathcal{K}_M}$ in (23), satisfying the interpolation conditions $s(\xi_\nu) = \bar{u}(\xi_\nu)$ for all $T_\nu \in \mathcal{K}_M(T_\ell)$, is of the form

$$s = \sum_{T_\nu \in \mathcal{K}_M} c_\nu \|\cdot - \xi_\nu\|^2 \log(\|\cdot - \xi_\nu\|) + p\,.$$

Here, p is a linear polynomial in two variables and $\|\cdot\|$ denotes the Euclidean norm. For more details concerning thin plate spline interpolation, due to Duchon [10], and related interpolation methods, we refer to the recent tutorial [17].

Hence, the thin plate spline interpolant s in (23) matches current cell average values of $\bar{u}_\nu$ in the Moore neighbourhood of the cell T_ℓ, but not at T_ℓ itself, i.e., we have $\bar{u}(\xi_\ell) \neq s(\xi_\ell)$ in general. Now the error indication η_ℓ for the cell T_ℓ is small whenever the reproduction quality of $\bar{u}_\ell$ by s around the

[6] This is a standard term used in the theory of cellular automata.

cell T_ℓ is good. In contrast, a high value of η_ℓ typically indicates that $\bar{u}_\ell$ is subject to strong variation locally around T_ℓ. Indeed, this observation relies on available local error estimates for thin plate spline interpolation (see the corresponding discussion on this in [5, 6]). We remark that the error indicator allows us to locate discontinuities of the solution u quite effectively. This is supported by the numerical results in the following Section 8.

7.2 Coarsening and Refinement

We balance the approximation quality and the required computational complexity by inserting new nodes at the barycenter of triangles $T \in \mathcal{T}$, whose error indicator η is large (refinement), whereas we remove nodes from the triangulation $\mathcal{T}$ in regions where η is small (coarsening).

Coarsening. A triangular cell $T_\ell \in \mathcal{T}$ is *coarsened* by the removal of its three vertices (nodes) from the current triangulation $\mathcal{T}$. This is done as follows. If *all* triangular cells adjacent to a node (vertex) ξ are flagged as *to be coarsened*, then the node ξ is removed from the triangulation $\mathcal{T}$. The triangulation is then updated by local Delaunay retriangulation.

Refinement. A cell $T_\ell \in \mathcal{T}$ is *refined* by the insertion of its barycenter ξ_ℓ into $\mathcal{T}$ and a subsequent local Delaunay retriangulation. This means that the number of nodes in the triangulation $\mathcal{T}$ is increased by one, ξ_ℓ, and the triangulation $\mathcal{T}$ is updated accordingly.

8 Reservoir Flow Simulation

The exploration and production of hydrocarbon reservoirs is still the most important technology to develop natural energy sources. Thereby, fluid flow simulators play a key role in order to help oil companies to make effective use of expensive data collected through field measurements, data processing and interpretation. In fact, simulation is one of the few tools available for modelling changes in a reservoir over time. Combined with other measurements it improves the degree of confidence in the understanding of reservoirs and heavily influences reservoir management decisions.

A central problem in petroleum reservoir simulation is to model the displacement of one fluid by another within a porous medium. A typical problem is characterized by the injection of a wetting fluid (e.g. water) into the reservoir at a particular location displacing the non-wetting fluid (e.g. oil), which is extracted or *produced* at another location. The nature of the front between the water and the oil is of primary importance and the goal is to withdraw as much oil as possible before water reaches the production location.

The physical phenomena that govern these enhanced oil recovery processes typically have important local properties. Thus, numerical schemes

Then, the equations describing reservoir fluid flow are given by the *Buckley-Leverett equation*

$$\frac{\partial}{\partial t}u + a \cdot \nabla f(u) = 0\,, \tag{27}$$

the *incompressibility relation*

$$\nabla \cdot a(t,x) = 0\,, \tag{28}$$

and *Darcy's law*

$$a(t,x) = -\mathbf{K}(x)M(u)\nabla p(t,x), \tag{29}$$

where $0 \leq u \leq 1$ is the water saturation.

In reservoir modelling, the function $f : u \rightarrow f(u)$ is monotonically increasing and satisfies $0 \leq f(u) \leq 1$ for all $u \in [0,1]$. In the following applications, all computations are based on the Corey model (cf. [2]) with quadratic relative permeabilities of the form

$$kr_w(u) = u^2\,, \qquad kr_o(u_o) = (1-u)^2.$$

This yields

$$M(u) = \frac{u^2}{\mu_w} + \frac{(1-u)^2}{\mu_o}$$

for the total mobility, and so in this case the fractional flow function f is of the form

$$f(u) = \frac{u^2}{u^2 + \frac{\mu_w}{\mu_o}(1-u)^2}\,, \tag{30}$$

also often referred to as the *Buckley-Leverett flux*, and used to model the displacement of oil by water.

8.2 The Five-Spot Problem

In the following, the computational domain $\Omega = [-0.5, 0.5] \times [-0.5, 0.5]$ represents a *homogeneous* medium, with $\phi(x) \equiv 1$ and $\mathbf{K}(x) \equiv 1$ for all $x \in \Omega$. Furthermore, we assume unitmobility, $M(u) \equiv 1$ in (29), and keep the reservoir pressure constant in time, i.e., $p(t,x) = p(x)$. Therefore, the total velocity field $a(t,x)$ in (29) has to be computed only once at the beginning of a simulation and is then independent of time, i.e., $a(t,x) = a(x)$. In general, the total velocity field will change during the simulation as the mobility $M(u)$ depends on the changes in the saturation u. Substituting equation (29) in (28) yields a set of elliptic equations, that would have to be solved for the pressure, which in turn provides an updated total velocity field $a(t,x)$ through (29). However, as we focus on solving the nonlinear conservation law (27), we separate the coupled differential equations for saturation and pressure and neglect the pressure equation. This can be justified by the fact, that for the

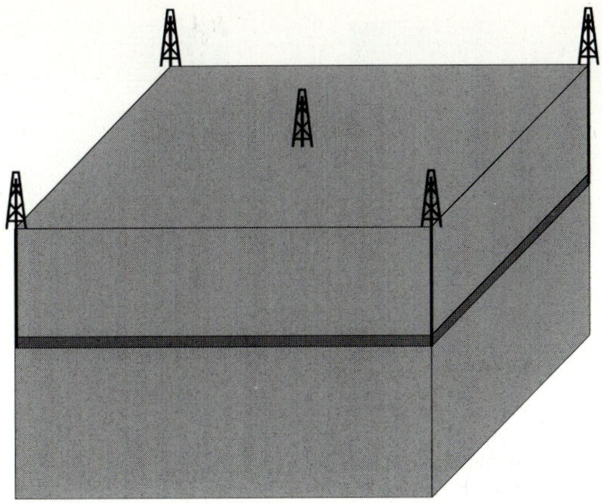

Fig. 14. A model of the five-spot problem with one production well in the center and four injection wells at the corners.

chosen five-spot problem the pressure changes have rather small effects on the solution [10]. The main idea of using this approximation is to evaluate the performance of the proposed ADER schemes of Chapter 3 as adaptive unstructured saturation solvers. We demonstrate how adaptive ADER schemes cope with velocity fields that typically arise in reservoir simulations.

Now, let us consider the standard test case of the five-spot problem. Here, one production well P in the center of the computational domain Ω is surrounded by four injection wells I_j, $j = 1, ..., 4$, located at the corners of the model as displayed in Figure 14. Assume a thin oil bearing layer trapped between two sealing layers as shown in Figure 14, which allows us to reduce the problem to two-dimensions by taking a $(x_1\text{-}x_2)$-slice of the model at the depth of the oil bearing layer. The distances between a particular location $x = (x_1, x_2)$ in the computational domain Ω and the wells are given by

$$
\begin{aligned}
r_{I_j}(x) &= \sqrt{(x_1 - x_{1,I_j})^2 + (x_2 - x_{2,I_j})^2}\,, \quad j = 1, \ldots, 4, \\
r_P(x) &= \sqrt{(x_1 - x_{1,P})^2 + (x_2 - x_{2,P})^2}\,,
\end{aligned}
$$

for the four injection wells I_j and the production well P. Then, we define a scalar pressure field through

$$
p(x) = \log\Big(r_P(x)\Big) - \sum_{j=1}^{4} \log\Big(r_{I_j}(x)\Big)\,.
$$

[10]Even in sophisticated, full reservoir simulators the pressure field and therefore the velocity field are recomputed infrequently compared to the saturation.

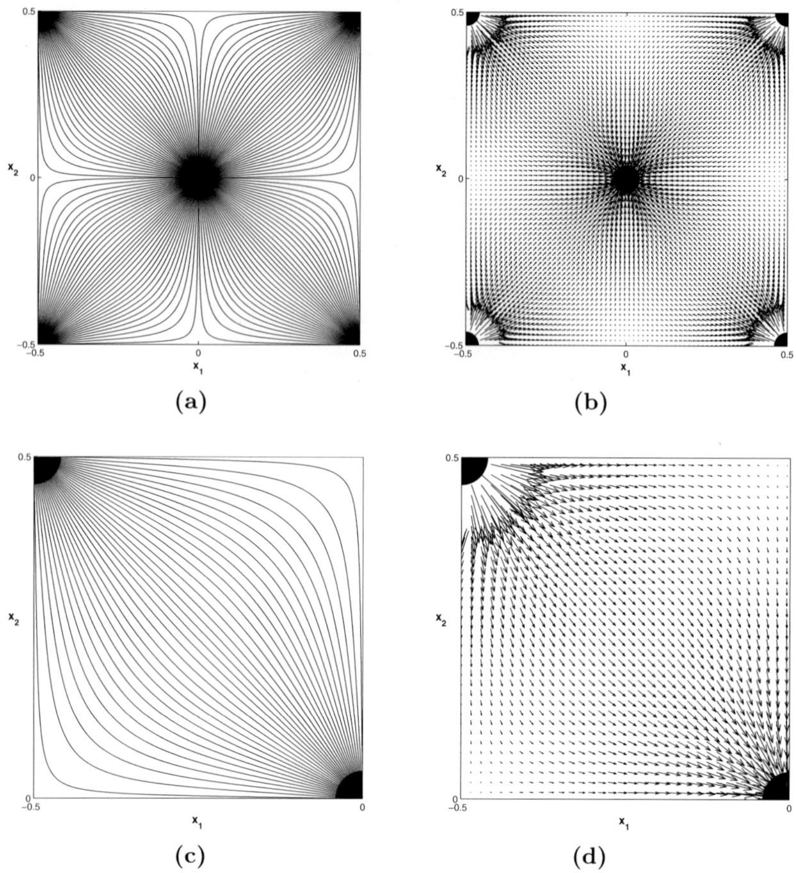

Fig. 15. The total velocity field is displayed as streamlines **(a)** and as velocity vectors **(b)** for the five-spot problem. The plots in **(c)** and **(d)** show the zoomed section of the top left quarter.

We then use Darcy's law (29) and the assumptions discussed above to compute the total velocity field. To be more precise the components of the velocity field $a(x) = a(x_1, x_2) = (a_1(x_1, x_2), a_2(x_1, x_2))$ are given by

$$\begin{aligned} a_1(x_1, x_2) &= -\frac{x_1 - x_{1,P}}{(x_1 - x_{1,P})^2 + (x_2 - x_{2,P})^2} + \sum_{j=1}^{4} \frac{x_1 - x_{1,I_j}}{(x_1 - x_{1,I_j})^2 + (x_2 - x_{2,I_j})^2} \\ a_2(x_1, x_2) &= -\frac{x_2 - x_{2,P}}{(x_1 - x_{1,P})^2 + (x_2 - x_{2,P})^2} + \sum_{j=1}^{4} \frac{x_2 - x_{2,I_j}}{(x_1 - x_{1,I_j})^2 + (x_2 - x_{2,I_j})^2} . \end{aligned} \tag{31}$$

Figure 15 displays the above velocity field $a(x)$ by showing its streamlines and the corresponding plots of the vector field. It can be shown by differentiation, that this velocity field $a(x)$ is divergence-free, as required by equation (28). Note, that it is not uncommon to have orders of magnitude differences in the absolute values of the total velocity a, with high velocities near the wells and lower velocities in places between the wells.

8.3 Results of ADER Schemes

Here, we solve the Buckley-Leverett equation (27) with the total velocity field a given in (31) and the fractional flow specified in (30). We set the radius of the injection wells to $R = 0.05$ and use the initial condition

$$u_0(x) = \begin{cases} 1 & \text{for } \|x - c_j\| \leq R, \quad j = 1, \ldots, 4, \\ 0 & \text{otherwise,} \end{cases}$$

where the points $c_j \in \mathbb{R}^2$ are the centers of the four injection wells. This way, we model pure water injection into an initially 100% oil saturated reservoir. Note, that periodic boundary conditions can be used for this model problem.

In order to see the differences between low and high order ADER schemes, we first show the color-coded water saturation obtained with an adaptive ADER1 scheme in Figure 16. The shocks representing the interface between pure oil and a mixture of oil and water are moving from the corners of the model reservoir towards its center. This way, oil in the porous medium is displaced by water, i.e., it is effectively pushed towards the production well. Before the shocks actually reach the production well at the center[11], the sucking effect of the production well becomes obvious, which is due to the increasing total velocity resulting from the pressure drop at the production well.

The underlying adaptive mesh is displayed in Figure 17, which clearly shows, how the locally refined mesh adaptively captures the shocks. After a shock has passed a particular location, the mesh is recoarsened, if the error indicator allows, in order to reduce the computational costs. However, the mesh behind the shocks remains finer than it was originally, which is due to the rarefaction following the moving shock, where the saturation of water slowly increases. In fact, in the zones of the rarefactions in Figure 16 one can recognize the shapes of some triangular cells emerged from the recoarsening.

Figure 18 shows the results (on the five-spot model problem) obtained by an adaptive ADER4 scheme. The shocks are resolved sharper in this case compared to the ADER1 results of Figure 16. Moreover, the rarefaction appears to be much smoother, there is no structure of an underlying triangular mesh visible. However, if we look at the corresponding adaptive mesh in Figure 19, we see that there is even a coarser mesh in the areas of the rarefactions. In fact, behind the shocks the error indicator allows the recoarsening of the mesh to its coarsest level. This is due to the increased approximation quality of the higher order ADER4 scheme, which uses piecewise cubic polynomials instead of piecewise constant functions to reconstruct the water saturation

[11] The time, when water arrives at the production well, is called the *breakthrough*.

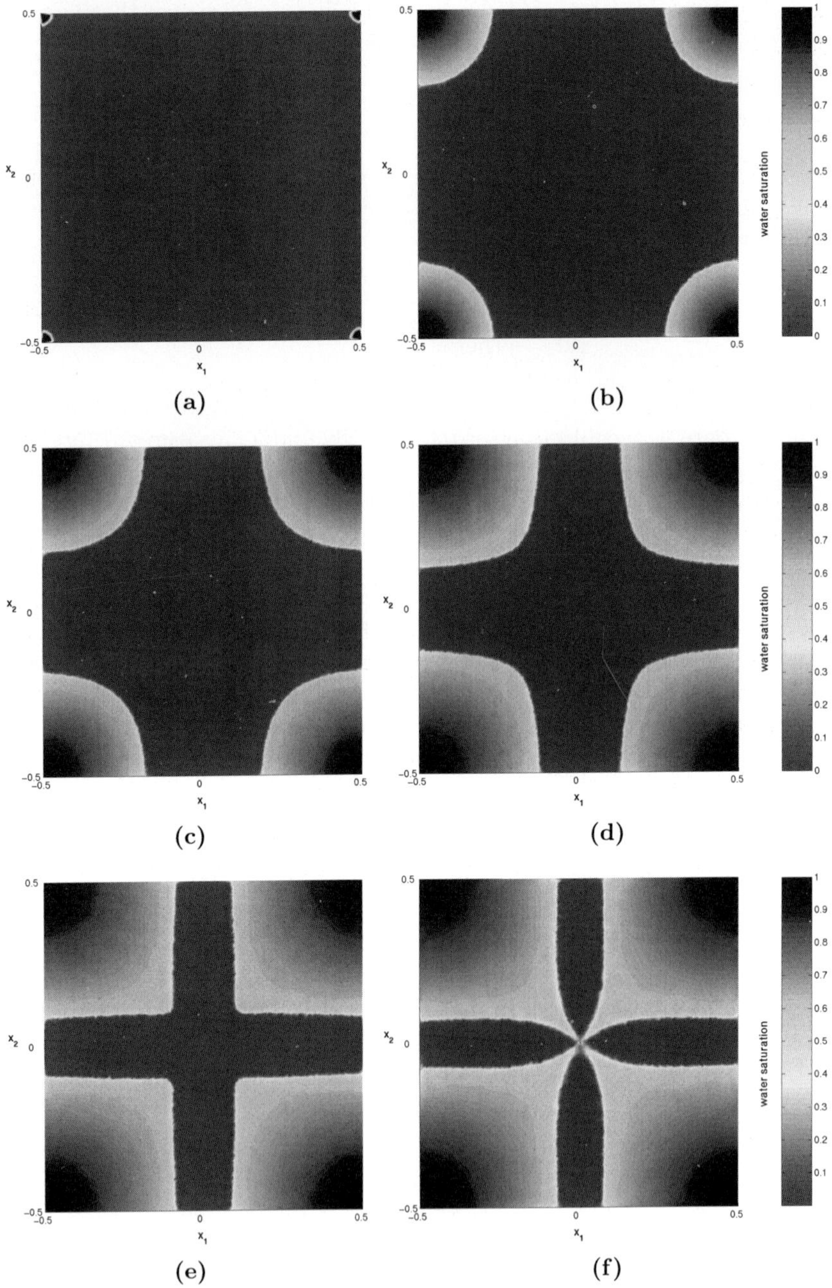

Fig. 16. [Reproduced in colour in Plate 31 on page 443.] Five-spot problem computed with ADER1. Color plots indicating the injection of water during the simulation at six different times, **(a)** $t = t_0$; **(b)** $t = t_{120}$; **(c)** $t = t_{240}$; **(d)** $t = t_{360}$; **(e)** $t = t_{480}$; and **(f)** $t = t_{600}$.

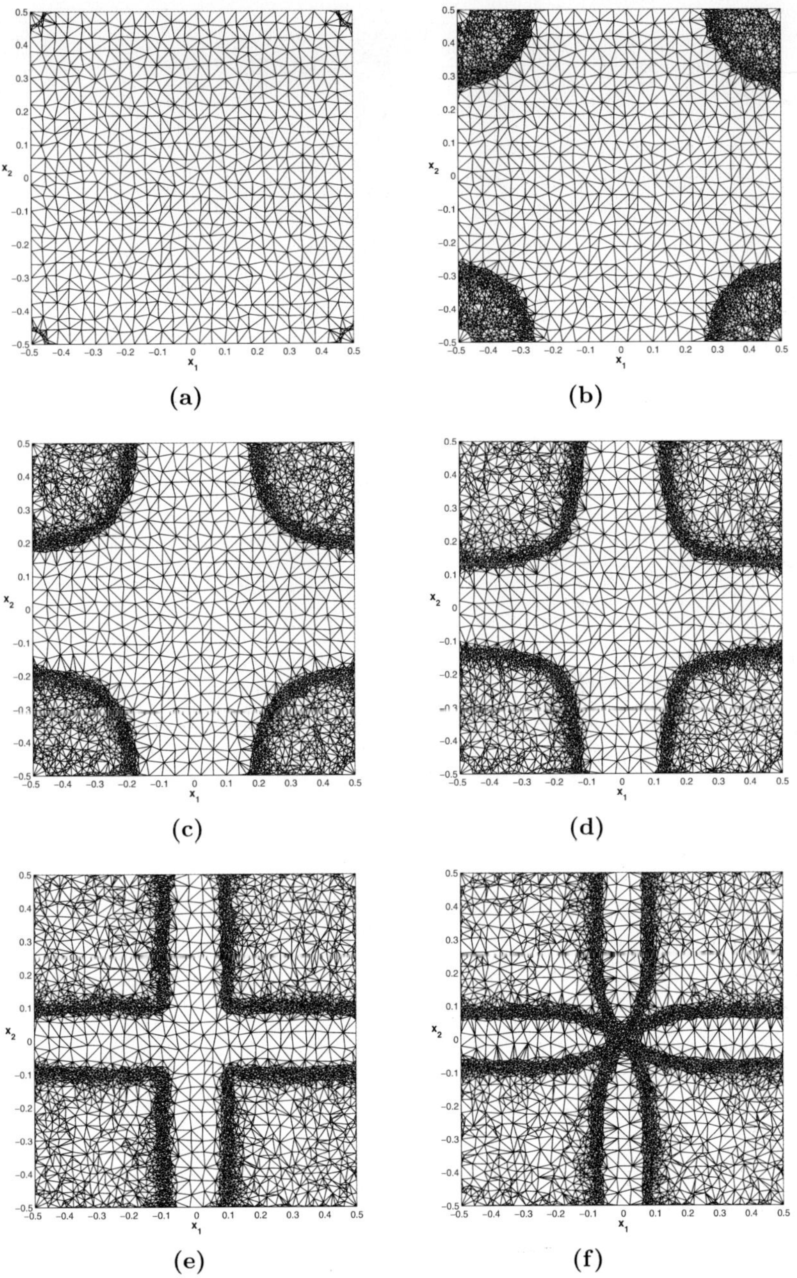

Fig. 17. Five-spot problem computed with ADER1. Adaptive triangulation during the simulation at six different times, **(a)** $t = t_0$; **(b)** $t = t_{120}$; **(c)** $t = t_{240}$; **(d)** $t = t_{360}$; **(e)** $t = t_{480}$; and **(f)** $t = t_{600}$.

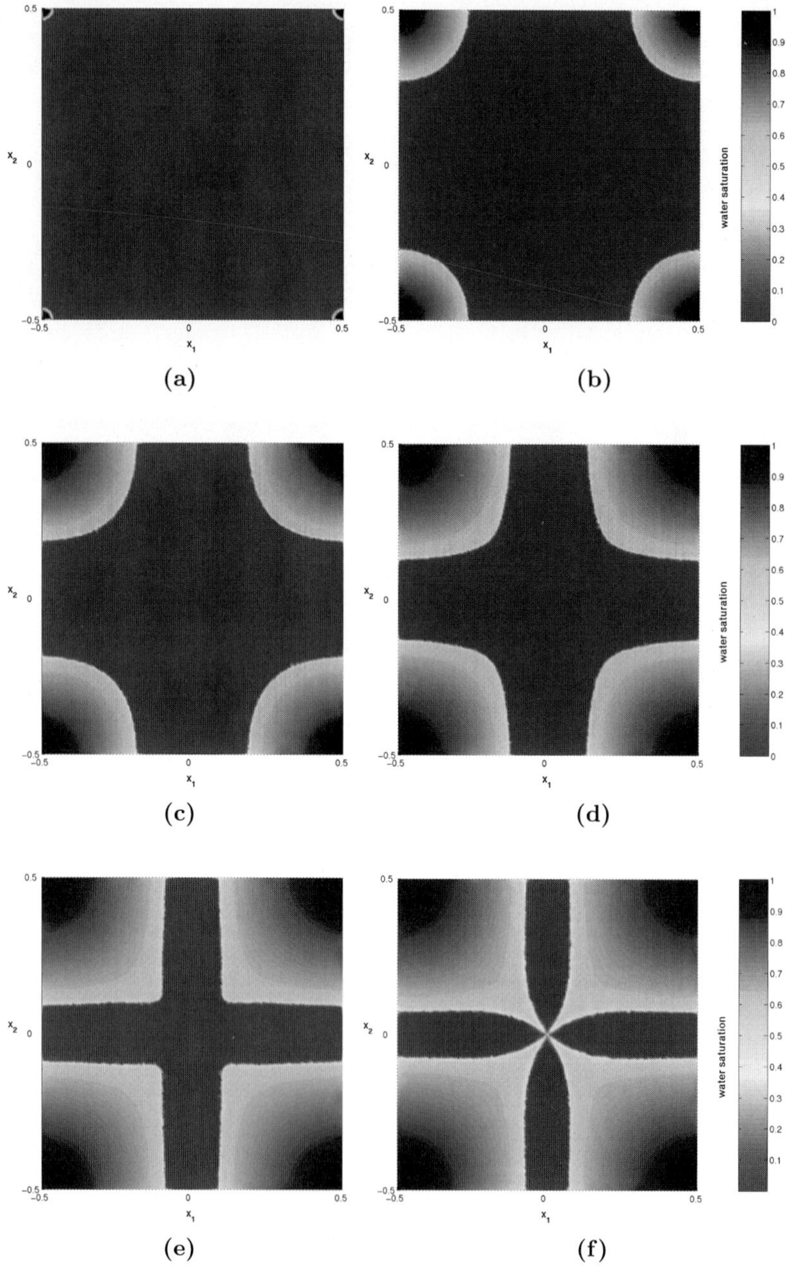

Fig. 18. [Reproduced in colour in Plate 32 on page 444.] Five-spot problem computed with ADER4. Color plots indicating the injection of water during the simulation at six different times, **(a)** $t = t_0$; **(b)** $t = t_{120}$; **(c)** $t = t_{240}$; **(d)** $t = t_{360}$; **(e)** $t = t_{480}$; and **(f)** $t = t_{600}$.

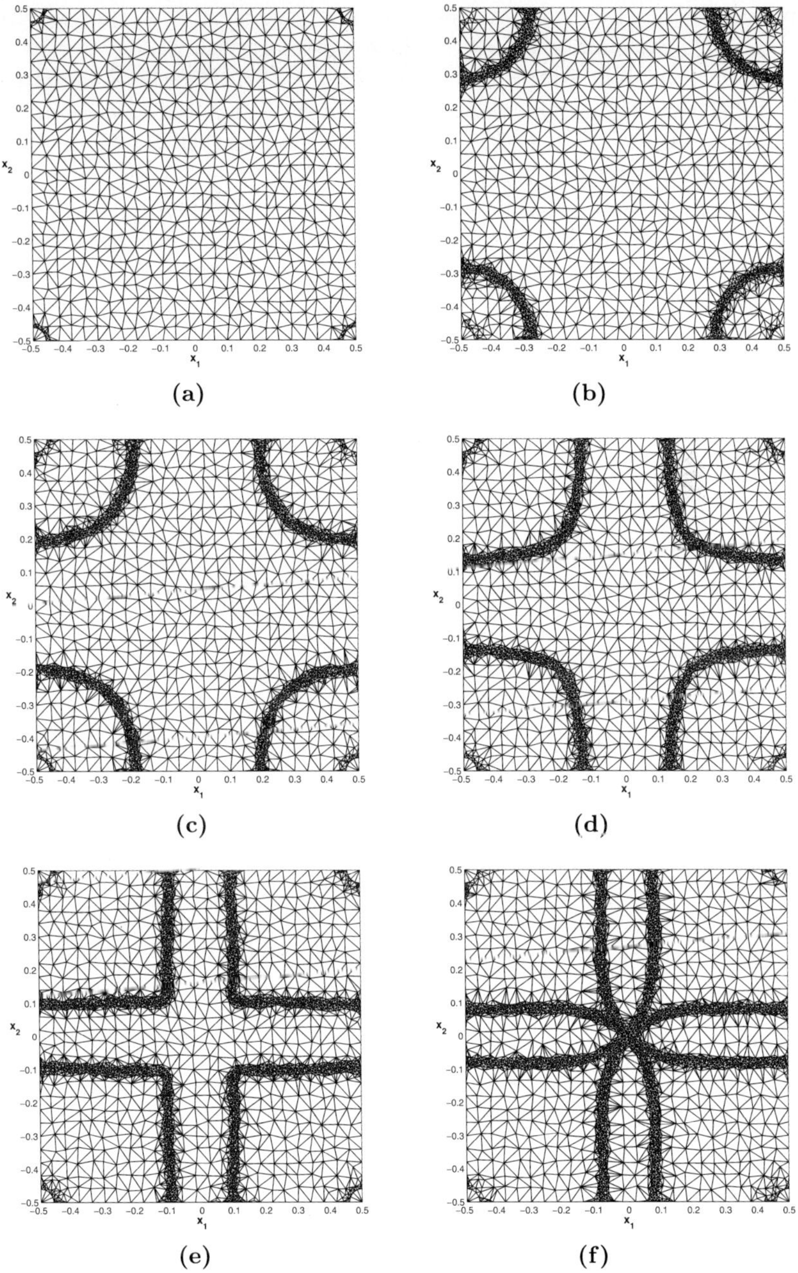

Fig. 19. Five-spot problem computed with ADER4. Adaptive triangulation during the simulation at six different times, **(a)** $t = t_0$; **(b)** $t = t_{120}$; **(c)** $t = t_{240}$; **(d)** $t = t_{360}$;**(e)** $t = t_{480}$; and **(f)** $t = t_{600}$.

function u. Therefore, coarser meshes can be used without loosing accuracy. In Figure 20 the number of cells of the adaptive mesh is plotted versus time for the four different ADERm schemes, $m = 1, ..., 4$. Clearly, if the order of the ADERm scheme is increased, the number of required cells is reduced. Note, that the reduction in mesh cells going from an ADER1 to an ADER2 scheme is significant, whereas by going to higher order ADER schemes this reduction is not very remarkable. This is due to the fact, that the shock is always resolved with a very fine mesh, whereas in the the regions of the rarefaction wave the error estimator already allows an ADER2 scheme to recoarsen the mesh almost to its initial state.

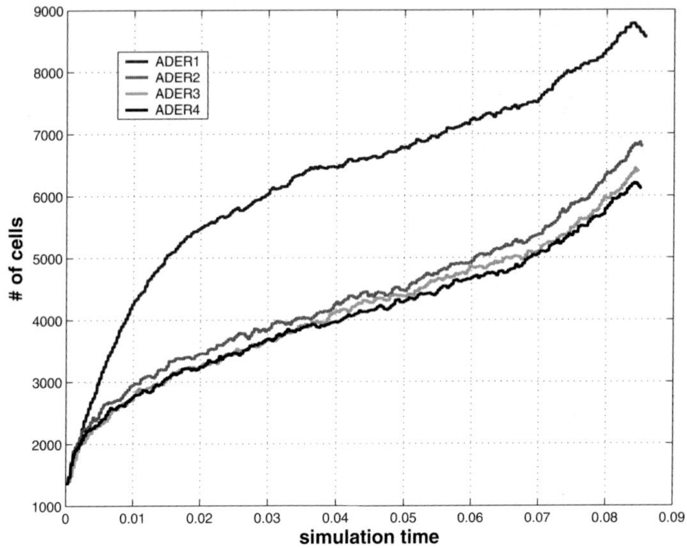

Fig. 20. Number of cells during the simulation for four different ADER schemes: ADER1 (top), ADER2, ADER3, and ADER4 (bottom).

8.4 Comparison with Reference Solutions

In order to confirm the good performance of the proposed ADER schemes, we compare their numerical results, on the five-spot problem, with two different reservoir simulators, whose results we use as reference solutions. The chosen simulators ECLIPSE and FrontSim are two commercial software packages used by the majority of reservoir simulation groups in oil industry. We remark, that these simulators solve the coupled system of the pressure and saturation equations (29)-(27) and therefore consider the effect of pressure changes due to changes in saturation.

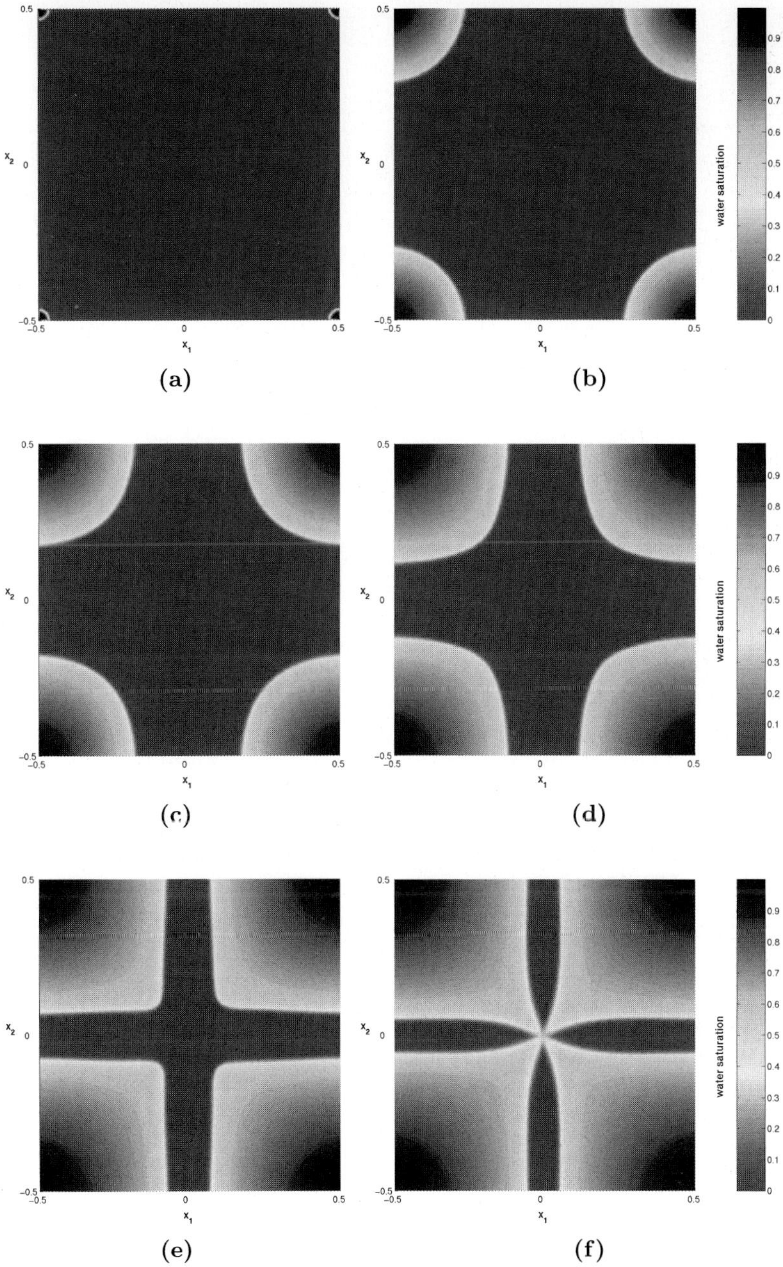

Fig. 21. [Reproduced in colour in Plate 33 on page 445.] Five-spot problem computed with ECLIPSE. Color plots indicating the injection of water during the simulation at six different times, **(a)** $t = t_0$; **(b)** $t = t_{120}$; **(c)** $t = t_{240}$; **(d)** $t = t_{360}$; **(e)** $t - t_{480}$; and **(f)** $t = t_{600}$.

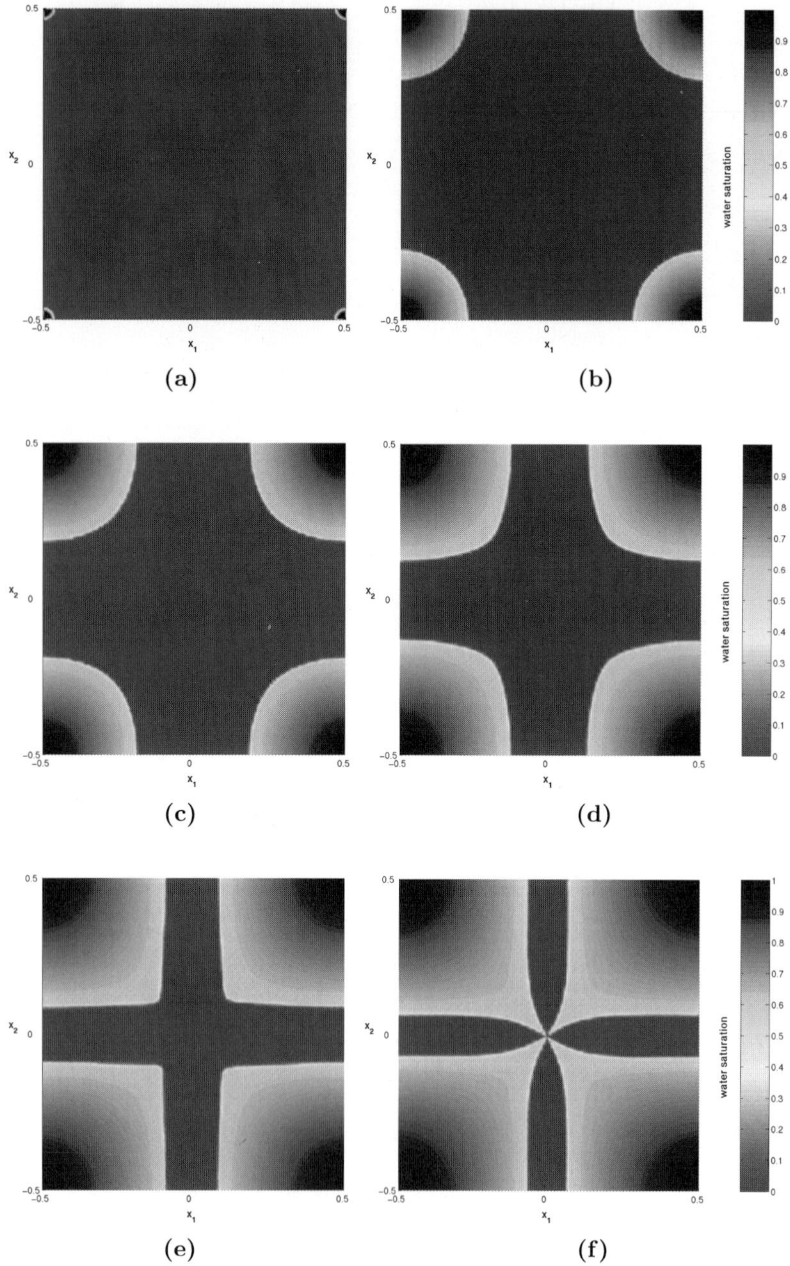

Fig. 22. [Reproduced in colour in Plate 34 on page 446.] Five-spot problem computed with FrontSim. Color plots indicating the injection of water during the simulation at six different times, **(a)** $t = t_0$; **(b)** $t = t_{120}$; **(c)** $t = t_{240}$; **(d)** $t = t_{360}$; **(e)** $t = t_{480}$; and **(f)** $t = t_{600}$.

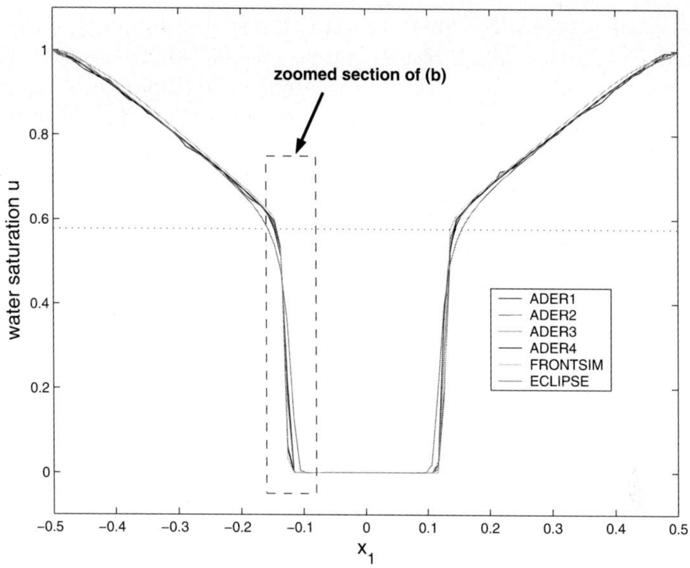

(a)

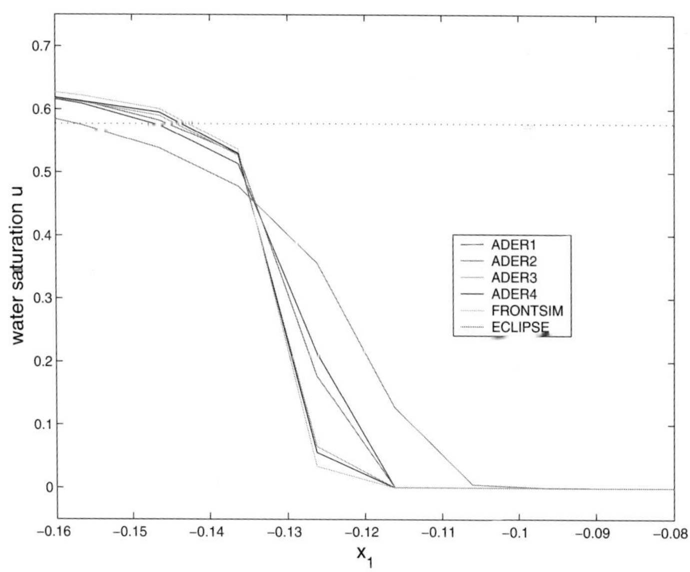

(b)

Fig. 23. A comparison of saturation profiles obtained by different ADER schemes together with the reference solutions computed with ECLIPSE and FrontSim.

This in turn typically causes changes in the total velocity field, but as confirmed by the following results, these effects can be neglected for the homogeneous five-spot problem. The reference solutions are both computed on a two-dimensional Cartesian mesh consisting of 100×100 rectangular cells, which are fixed throughout the simulation.

Figure 21 shows the water saturation as obtained with the simulator ECLIPSE. It is obvious, that the moving shocks are much more smeared than in the case of the ADER schemes. On the other hand, the general behaviour of the solutions is very similar as well as the jump height of the discontinuity. The results obtained by the simulator FrontSim are displayed in Figure 22. Here, the numerical diffusion is very small and the interface at the oil-water contact is resolved very sharply.

In order to get a better comparison of results of the different ADER schemes and the reference solutions of ECLIPSE and FrontSim, we look at a cross section of the water saturation u. Figure 23 shows the water saturation profile taken at the time step $t = t_{360}$ along the line $x_1 \in [-0.5, 0.5]$ cutting through the computational domain Ω at $x_2 = 0.45$. The plot in Figure 23 **(a)** displays an overlay of all saturation profiles obtained by the schemes for ADERm schemes, $m = 1, ..., 4$ together with the two reference solutions. In addition, the theoretical expected jump height of the shock is drawn as dotted line. We see, that all methods lead to very similar results and the expected jump at the oil-water contact is more or less reproduced by all schemes. To see the differences between the chosen methods in more detail, we zoom in at an area around the left shock as indicated in Figure 23 **(a)** and show the magnified plot in Figure 23 **(b)**.

Obviously, the reference solution computed by ECLIPSE is heavily smeared by numerical diffusion. This is because ECLIPSE is based on a finite volume scheme of (merely) first order. This is improved by the adaptive ADER1 scheme, which is also of first order, but resolves the shock much better due to the local mesh refinement in the vicinity of the shock. Increasing the order of the ADER schemes, the shock becomes sharper (see ADER2, ADER3 and ADER4 in Figure 23(**b**)). However, the highest resolution of the shock is obtained by FrontSim, which is based on a *front tracking* scheme which is well-known for its small numerical diffusion. Nevertheless, the saturation profile obtained by ADER4, which belongs to the class of *front capturing* schemes, is very close to the good results of FrontSim.

9 Conclusion

We presented an extension of the new ADER schemes on adaptive, unstructured triangulations in order to solve linear and nonlinear scalar conservation laws. Originally, the ADER approach based on Arbitrary high order DERivatives was introduced by Toro, Millington and Nejad in [38] for linear problems on Cartesian meshes and further expanded by Toro and Titarev to nonlinear problems in [36, 39, 41]. As ADER schemes belong to the class of finite volume methods, we discuss in detail the reconstruction of high order polynomials from cell average values on unstructured triangular meshes. We apply the WENO reconstruction technique in order to achieve a high order approximation quality while avoiding spurious oscillations of the solution. To this end, a customized stencil selection algorithm is introduced that relies on the idea of a sectoral search. The resulting piecewise polynomial approximation of the solution provides generalized Riemann problems at the cells interfaces, which can be solved by reducing them to a series of conventional derivative Riemann problems. Furthermore, the Lax-Wendroff procedure uses the solutions of these Riemann problems in order to determine high order fluxes across the cell interfaces and update the cell average values for each time step.

The performance of the proposed ADER schemes is evaluated with respect to experimental orders of convergence and their computational efficiency. The results show, that using higher order ADER schemes, although increasing the computational costs per time step, finally pay off due to the tremendous reduction in mesh cells required to reach a desired accuracy. We then optimize the computational efficiency by combining the high order ADER approach with the strategy of adaptive mesh refinement. Therefore, the corresponding error estimator and the customized adaption rules for refining and coarsening the triangular mesh are explained in detail. Finally, the adaptive ADER schemes are applied to the nonlinear Buckley-Leverett equation in the form of the five-spot problem, a well-known model problem in the oil industry describing two-phase fluid flow in petroleum reservoirs.

Further developments of the proposed ADER schemes are subject to current research and may include more elaborate mesh adaption methods, that work with isotropic refinement strategies or mesh alignment methods where edges of particular cells are aligned parallel to the orientation of the shock front. Encouraging results of such mesh alignment approaches are shown in [21]. Additionally, the coupling of mesh adaptivity with time adaptivity may be a further issue to use small time steps only in small cells and allow for larger time steps in larger cells in order to reduce numerical diffusion. The approach in [40] to increase the accuracy of ADER schemes by using a TVD flux instead of a first order monotonic flux as a building block for designing high order ADER schemes also appears very promising and deserves further investigation.

Acknowledgement

The work was partly supported by the European Union through the research and training network NetAGES, contract no. IST-1999-29034. Special thanks to Brice Vallès from Schlumberger Stavanger Research for computing the reference solutions with ECLIPSE and FrontSim. Thanks also to E. F. Toro from the University of Trento for the invitation to the *Workshop on Very High Order Numerical Methods* at Cambridge, UK, 2003. Discussions with him, R. J. LeVeque, C. W. Shu, and K.-A. Lie have been very fruitful and encouraging.

References

1. R. Abgrall (1994) On essentially non-oscillatory schemes on unstructured meshes: analysis and implementation. J. Comput. Phys. **144**, 45–58.
2. K. Aziz and A. Settari (1979) *Petroleum Reservoir Simulation*. Applied Science.
3. T.J. Barth and P.O. Frederickson (1990) Higher order solution of the Euler equations on unstructured grids using quadratic reconstruction. Technical Report no. 90-0013, AIAA Paper.
4. M. Batdorf, L.A. Freitag, and C.F. Ollivier-Gooch (1997) Computational study of the effect of unstructured mesh quality on solution efficiency. Proc. 13th AIAA Computational Fluid Dynamics Conference, Snowmass Village, Colorado, 1997.
5. J. Behrens, A. Iske, and M. Käser (2002) Adaptive meshfree method of backward characteristics for nonlinear transport equations. *Meshfree Methods for Partial Differential Equations*, M. Griebel and M.A. Schweitzer (eds.), Springer, Berlin, 21–36.
6. J. Behrens, A. Iske, and S. Pöhn (2001) Effective node adaption for grid-free semi-Lagrangian advection. *Discrete Modelling and Discrete Algorithms in Continuum Mechanics*, T. Sonar and I. Thomas (eds.), Logos, Berlin, 110–119.
7. J.M. Buckley and M.C. Leverett (1942) Mechanism of fluid displacement in sands. Trans. AIME **146**, 107–116.
8. B. Cockburn, G. Karniadakis, and C.-W. Shu (2000) The development of discontinuous galerkin methods. *Discontinuous Galerkin Methods: Theory, Computation and Applications*, B. Cockburn, G. Karniadakis, and C.-W. Shu (eds.), Springer, Berlin, 3–50.
9. G.R. Cowper (1973) Gaussian quadrature formulas for triangles. Int. J. Numer. Meth. Eng. **7**, 405–408.
10. J. Duchon (1977) Splines minimizing rotation-invariant semi-norms in Sobolev spaces. *Constructive Theory of Functions of Several Variables*, W. Schempp and K. Zeller (eds.), Springer, Berlin, 85–100.
11. O. Friedrich (1998) Weighted essentially non-oscillatory schemes for the interpolation of mean values on unstructured grids. J. Comput. Phys. **144**, 194–212.

12. S.K. Godunov (1959) A finite difference method for the computation of discontinuous solutions of the equations of fluid dynamics. Mat. Sb. **47**, 271–306.
13. T. Gutzmer and A. Iske (1997) Detection of discontinuities in scattered data approximation. Numerical Algorithms **16**(2), 155–170.
14. A. Harten and S. Chakravarthy (1991) Multi-dimensional eno schemes for general geometries. Technical Report 91-76, ICASE.
15. A. Harten, B. Engquist, S. Osher, and S. Chakravarthy (1987) Uniformly high order essentially non-oscillatory schemes, iii. J. Comput. Phys. **71**, 231–303.
16. C. Hu and C.W. Shu (1999) Weighted essentially non-oscillatory schemes on triangular meshes. J. Comput. Phys. **150**, 97–127.
17. A. Iske (2002) Scattered data modelling using radial basis functions. *Tutorials on Multiresolution in Geometric Modelling*, A. Iske, E. Quak, and M.S. Floater (eds.), Springer, Heidelberg, 205–242.
18. A. Iske and M. Käser (2004) Conservative semi-Lagrangian advection on adaptive unstructured meshes. To appear in Numer. Meth. Part. Diff. Eq.
19. G.S. Jiang and C.W. Shu (1996) Efficient implementation of weighted ENO schemes. J. Comput. Phys. **126**, 202–228.
20. M. Käser, H. Igel, M. Sambridge, and J. Brown (2001) A comparative study of explicit differential operators on arbitrary grids. J. Comput. Acoustics **9**, 1111–1125.
21. D. Kröner (1997) *Numerical Schemes for Conservation Laws.* Wiley & Teubner.
22. C.L. Lawson and R.J. Hanson (1995) *Solving Least Squares Problems.* SIAM, Philadelphia.
23. P. Lax and B. Wendroff (1960) Systems of conservation laws. Comm. Pure Appl. Math. **13**, 217–237.
24. R.L. LeVeque (2002) *Finite Volume Methods for Hyperbolic Problems.* Cambridge University Press, Cambridge.
25. X. Liu, S. Osher, and T. Chan (1994) Weighted essentially non-oscillatory schemes. J. Comput. Phys. **115**, 200–212.
26. A. Meister and J. Struckmeier, eds. (2002) *Hyperbolic Partial Differential Equations: Theory, Numerics and Applications.* Vieweg, Braunschweig.
27. C.D. Munz and R. Schneider. An arbitrary high order accurate finite volume scheme for the Maxwell equations in two dimensions on unstructured meshes. Forschungszentrum Karlsruhe, Germany, unpublished report.
28. C.F. Ollivier-Gooch (1997) Quasi-ENO schemes for unstructured meshes based on unlimited data-dependent least-squares reconstruction. J. Comput. Phys. **133**, 6–17.
29. T.J. Ruuth and R.J. Spiteri (2002) Two barriers on strong-stability-preserving time discretization methods. J. Sci. Comput. **17**, 211–220.
30. T. Schwartzkopff, C.D. Munz, and E.F. Toro (2002) ADER: A high-order approach for linear hyperbolic systems in 2D. J. Sci. Comput. **17**, 231–240.
31. T. Schwartzkopff, C.D. Munz, E.F. Toro, and R.C. Millington (2001) The ADER approach in 2D. *Discrete Modelling and Discrete Algorithms on Continuum Mechanics*, T. Sonar and I. Thomas (eds.), Logos, Berlin, 207–216.
32. C.W. Shu (1988) Total-variation-diminishing time discretizations. SIAM J. Sci. Stat. Comput. **9**, 1073–1084.
33. C.W. Shu and S. Osher (1988) Efficient implementation of essentially non-oscillatory shock-capturing schemes. J. Comput. Phys. **77**, 439–471.

34. T. Sonar (1997) On the construction of essentially non-oscillatory finite volume approximations to hyperbolic conservation laws on general triangulations: polynomial recovery, accuracy and stencil selection. Comput. Methods Appl. Mech. Engrg. **140**, 157–181.
35. Y. Takakura and E.F. Toro (2002) Arbitrarily accurate non-oscillatory schemes for a nonlinear conservation law. Computational Fluid Dynamics Journal **11**(1), 7–18.
36. V.A. Titarev and E.F. Toro (2002) ADER: Arbitrary high order Godunov approach. J. Sci. Comput. **17**, 609–618.
37. E.F. Toro (1999) *Riemann Solvers and Numerical Methods for Fluid Dynamics.* 2nd edition, Springer.
38. E.F. Toro, R.C. Millington, and L.A.M. Nejad (2001) Towards very high order Godunov schemes. *Godunov Methods; Theory and Applications*, E.F. Toro (ed.), Kluwer Academic Plenum Publishers, 907–940.
39. E.F. Toro and V.A. Titarev (2001) Very high order Godunov-type schemes for nonlinear scalar conservation laws. European Congress on Computational Methods in Applied Sciences and Engineering, ECCOMAS Computational Fluid Dynamics Conference, Swansea, Wales, September 2001.
40. E.F. Toro and V.A. Titarev (2003) TVD fluxes for the high-order ADER schemes (2003) Preprint, Laboratory of Applied Mathematics, University of Trento, Italy.
41. E.F. Toro and V.A. Titarev (2002) Solution of the generalized Riemann problem for advection-reaction equations. Proc. R. Soc. Lond. A, 271–281.

Optimal Multivariate Interpolation

Tobias Werther

University of Vienna, Department of Mathematics, Austria

Summary. In this chapter, we are concerned with the problem of multivariate data interpolation. The main focus lies on the concept of minimizing a quadratic form which, in practice, emerges from a physical model, subject to the interpolation constraints. The approach is a natural extension of the one-dimensional polynomial spline interpolation. Besides giving a basic outline of the mathematical framework, we design a fast numerical scheme and analyze the performance quality. We finally show that optimal interpolation is closely related to standard linear stochastic estimation methods.

1 Introduction

In many geophysical applications we have a finite data set consisting of a scalar vector $\gamma = (\gamma_1, \dots, \gamma_n)^T$ of observations measured at corresponding known locations $\mathbf{z}_1, \dots, \mathbf{z}_N$, and we want to find an analytical or discrete representation of a surface f which interpolates or approximates these sampling values at the given nodes. The finite node set $Z = \{\mathbf{z}_1, \dots, \mathbf{z}_n\}$ contains points of some prescribed domain $\Omega \subset \mathbb{R}^d$ where $d = 1, 2, 3, \dots$ denotes the dimension.

In a mathematical setting, the multivariate interpolation problem consists of seeking a function

$$f : \Omega \to \mathbb{R}$$

that satisfies the interpolation condition

$$f(\mathbf{z}_j) = \gamma_j\,, \qquad j = 1, \dots, n,$$

or, in short,

$$f|_Z = \gamma\,. \tag{1}$$

Obviously, one can easily draw infinitely many solutions for this problem. In order to pick out a single one, we argue by following an isotropic physical model originated from the theory of elasticity.

Consider an ideal thin plate whose thickness can be neglected. If no forces are acting, the plate is assumed to be flat. Relative to a given coordinate system, this implies that the flat surface can be represented by a linear polynomial

$$f(\mathbf{x}) = d_0 + d_1 x_1 + d_2 x_2\,, \qquad \mathbf{x} = (x_1, x_2) \in \Omega \subset \mathbb{R}^2\,.$$

If we put point forces to the plate by, for example, attaching lead weights at the prescribed nodes, the plate will be deformed into a bending surface. The corresponding physical model is described by the potential energy $\tilde{U}$ of the deformed plate which is a quadratic homogeneous function of the reciprocal curvature [16]. If we approximate the reciprocal curvature by derivatives of second order, we arrive at the mathematical model

$$-U = \int \left|\frac{\partial^2 f}{\partial x_1^2}(\mathbf{x})\right|^2 + 2\left|\frac{\partial^2 f}{\partial x_1 x_2}(\mathbf{x})\right|^2 + \left|\frac{\partial^2 f}{\partial x_2^2}(\mathbf{x})\right|^2 d\mathbf{x}\,. \tag{2}$$

By applying the Hamilton principle of least action [16], it follows that the surface f satisfies the inhomogeneous biharmonic plate equation

$$\Delta^2 f = w \tag{3}$$

where w stands for the acting forces and Δ is the *Laplace operator* defined by

$$\Delta = \frac{\partial^2}{\partial x_1^2} + \frac{\partial^2}{\partial x_2^2}\,.$$

The *biharmonic equation* (3) constitutes the mathematical model of the deformation of thin plates. The right-hand side of (3) is a weighted sum $\sum_{j=1}^{n} c_j\, \delta_{\mathbf{z}_j}$ of point-measures at the given nodes, where the scalar c_j denotes the amplitude of the point-force acting on f at $\mathbf{z}_j$. The derivatives are to be understood in the distributional sense.

From this model, we see that if a fundamental solution of Δ^2 is given by a function ϕ, i.e.,

$$\Delta^2 \phi = \delta_0$$

then the function

$$f(\mathbf{x}) = \sum_{j=1}^{n} c_j \phi(\mathbf{x} - \mathbf{z}_j) + p(\mathbf{x}) \tag{4}$$

satisfies equation (3) for any polynomial $p(\mathbf{x}) = d_0 + d_1 x_1 + d_2 x_2$ that is in the nullspace of Δ^2, i.e., $\Delta^2 p = 0$.

It is well known that the continuous function

$$\phi(\mathbf{x}) = (x_1^2 + x_2^2) \log(x_1^2 + x_2^2)^{1/2}\,, \qquad \mathbf{x} \in \mathbb{R}^2,$$

is (up to some constant) a fundamental solution of Δ^2 on $\mathbb{R}^2$. The function f is called *thin plate spline interpolant.*

The condition $\Delta^2 p = 0$ implies that whenever p is the possible deformed surface, all amplitudes c_j must vanish which, by hypothesis on the thin plate, only leaves the flat solution

$$p(\mathbf{x}) = d_0 + d_1 x_1 + d_2 x_2\,.$$

We have now come very close to the solution of a surface exposed to point forces. In order to find the unknowns $c_1, \ldots, c_n$ and d_0, d_1, d_2, recall that we originally want the surface to attain the sampling values $\gamma_1, \ldots, \gamma_n$ at the corresponding nodes $\mathbf{z}_1, \ldots, \mathbf{z}_n$. This requirement combines the forces and the sampling values through the linear equations

$$\gamma_i = \sum_{j=1}^{n} c_j \phi(\mathbf{z}_i - \mathbf{z}_j) + d_0 + d_1 z_{i1} + d_2 z_{i2}\,, \qquad i = 1, \ldots, n. \tag{5}$$

Note that the linear system (5) is under-determined. For covering the remaining degrees of freedom, we add constraints in terms of moment conditions. The zero moment and the first moments of the amplitudes of the forces are assumed to vanish, i.e.,

$$\sum_{j=1}^{N} c_j = \sum_{j=1}^{N} c_j\, z_{j1} = \sum_{j=1}^{N} c_j\, z_{j2} = 0\,. \tag{6}$$

The linear system (5) together with (6) induces a linear system that is non-singular, see, for example, [9]. Inverting the corresponding matrix yields explicit coefficients for (4) which, in turn, provide an analytical representation of the interpolating surface, cf. Figure 1.

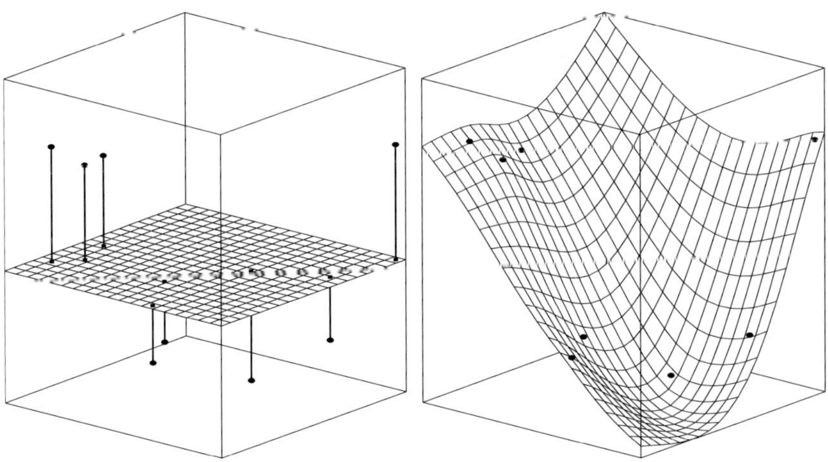

Fig. 1. Interpolating thin plate

In one dimension there is the analogue of the bending beam which leads to cubic splines when applying the same arguments, cf. [1].

The aim of this chapter is to build a rigorous mathematical framework on the top of this model and to extend the approach to a generalized minimum curvature principle in the context of multivariate data interpolation.

2 The Mathematical Model

The mathematical model is valid in any dimension, in particular, for two and three dimensional data. For a dimension independent notation, it is convenient to use the multi-index notation. A d-tuple of non-negative integers

$$\alpha = (\alpha_1, \ldots, \alpha_d)$$

is called a multi-index. We define

- the order of α by $|\alpha| = \alpha_1 + \cdots + \alpha_d$,
- the monomial of order $|\alpha| + 1$ by $\mathbf{x}^\alpha = x_1^{\alpha_1} \cdots x_d^{\alpha_d}$, $\mathbf{x} \in \mathbb{R}^d$,
- $\alpha! = \alpha_1! \cdots \alpha_d!$,
- the Euclidean norm for $\mathbf{x} \in \mathbb{R}^d$ by $|\mathbf{x}| = \sqrt{\mathbf{x}^T\mathbf{x}} = (x_1^2 + \cdots + x_d^2)^{1/2}$,
- the differential operator $D^\alpha = \frac{\partial^{\alpha_1}}{\partial x_1^{\alpha_1}} \cdots \frac{\partial^{\alpha_d}}{\partial x_d^{\alpha_d}}$.

We will only consider function spaces that are embedded in the space of distributions $\mathcal{D}'$ which is defined as the topological dual of $\mathcal{D}$, the space of all infinitely differentiable functions with compact support, see [11].

Let m be a positive integer. For the mathematical description of the multivariate interpolation model, we restrict the class of possible signals to the linear space

$$\mathcal{B}_m = \left\{ f \in \mathcal{D}' \;\middle|\; D^\alpha f \in L^2(\mathbb{R}^d),\ |\alpha| = m \right\} \tag{7}$$

where L^2 denotes the space of square integrable functions.

In what follows, we assume

$$m > d/2\,. \tag{8}$$

Then, the Sobolev Lemma implies that all elements of $\mathcal{B}_m$ are continuous functions on $\mathbb{R}^d$, see [12]. As a consequence, the interpolation problem (1) is well-defined.

We endow $\mathcal{B}_m$ with the semi-inner product

$$(f\,|\,g)_m = \sum_{|\alpha|=m} \frac{m!}{\alpha!} \int_{\mathbb{R}}^{d} D^\alpha f(\mathbf{x})\, D^\alpha g(\mathbf{x}) d\mathbf{x} \tag{9}$$

and the corresponding semi-norm

$$|f|_m = \sqrt{(f\,|\,f)_m} \tag{10}$$

with nullspace $\mathcal{P}_m^d$, the linear space of d-variate polynomials of order $\leq m$ whose dimension is

$$Q = \binom{m-1+d}{d}.$$

The optimal solution of the interpolation problem (1) is the function s in $\mathcal{B}_m$ that solves the optimization problem

$$\min_{f|_Z=\gamma} |f|_m \,. \tag{11}$$

If the node set Z is *unisolvent*, i.e.,

$$p \in \mathcal{P}: \quad p(\mathbf{z}_i) = 0\,, \quad i = 1, \ldots, n, \qquad \Longrightarrow \qquad p = 0\,,$$

it is known that (11) admits a unique solution, cf. [6].

For an analytical representation of s, we need the notion of a *kernel function.* Following [14], there exists a function

$$\phi(\mathbf{x}, \mathbf{y}) = \phi(\mathbf{x} - \mathbf{y})\,, \qquad (\mathbf{x}, \mathbf{y} \in \mathbb{R}^d)\,,$$

with the property that

$$\left(f\,, \sum_{i=1}^{n} c_i \phi(\mathbf{x} - \mathbf{x}_i) \right)_m = \sum_{i=1}^{n} c_i f(\mathbf{x}_i) \tag{12}$$

for all coefficients $c_1, \ldots, c_n$ and nodes $\mathbf{x}_1, \ldots, \mathbf{x}_n$ that satisfy

$$\sum_{i=1}^{n} c_i \mathbf{x}_i^\alpha = 0\,, \qquad |\alpha| < m\,.$$

Such a function ϕ is often referred to as kernel or semi-kernel of $\mathcal{B}_m$. Note the similarity of the last condition with the vanishing moments (6).

The optimal interpolation s is then of the form

$$s(\mathbf{x}) = \sum_{i=1}^{n} c_i \phi(\mathbf{x} - \mathbf{z}_i) + p(\mathbf{x})$$

where p is a polynomial in $\mathcal{P}_m^d$ and the coefficients satisfy

$$\sum_{i=1}^{n} c_i \mathbf{x}_i^\alpha = 0\,, \qquad |\alpha| < m\,. \tag{13}$$

Both the interpolation constraints and requirement (13) yield a quadratic system of linear equations which provide the coefficients $c_1, \ldots, c_n$ and the polynomial p.

Duchon has shown that possible kernel functions are the radial thin plate splines

$$\begin{aligned} \phi(\mathbf{x}, \mathbf{y}) &= |\mathbf{x} - \mathbf{y}|^{2m-d}, && \text{if } d \text{ is odd,} \\ \phi(\mathbf{x}, \mathbf{y}) &= |\mathbf{x} - \mathbf{y}|^{2m-d} \log |\mathbf{x} - \mathbf{y}|, && \text{if } d \text{ is even.} \end{aligned}$$

The name thin plate splines refers to the problem of elasticity as discussed in the introduction. Duchon's idea to identify these functions is based on the concept of fundamental solutions of the iterated Laplace operator

$$(-\Delta)^m\,.$$

Indeed, up to a constant, radial thin plate splines are fundamental solutions of this operator.

3 A Generalized Model

We now generalize the setting of the previous section in order to cover a wide range of optimal interpolation which is more flexible for application purposes.

Let Ω be an open and bounded domain in $\mathbb{R}^d$ with smooth boundaries. Let us consider the space

$$\mathcal{B}_m = \left\{ f \in \mathcal{D}'(\Omega) \,\middle|\, D^\alpha f \in L^2(\Omega),\ |\alpha| = m \right\}.$$

Again, for $m > d/2$, $\mathcal{B}_m$ is a linear space of continuous functions, and the interpolation problem (1) is well defined.

For optimal interpolation, we need to specify a semi-norm. Let us consider piecewise continuous weight functions w_α, $|\alpha| = m$, that are positive and bounded. The bilinear form

$$\left(f \,\middle|\, g \right)_{m,w} = \sum_{|\alpha|=m} \int_\Omega D^\alpha f(\mathbf{x}) D^\alpha g(\mathbf{x}) w_\alpha(\mathbf{x}) d\mathbf{x} \tag{14}$$

defines a semi-inner product on $\mathcal{B}_m$ with nullspace $\mathcal{P}_m^d$.

Following [14], there exists a function ϕ on $\Omega \times \Omega$ satisfying the kernel property (12) with respect to (14). As a consequence, the unique minimal semi-norm interpolation for a data vector γ and a corresponding unisolvent node set Z enjoys the representation

$$s(\mathbf{x}) = \sum_{i=1}^{n} c_i \phi(\mathbf{x}, \mathbf{z}_i) + p(\mathbf{x}). \tag{15}$$

The coefficients and the polynomial can be computed as above.

For numerical simulations, there is, however, a significant drawback of the analytical representation (15). In general, we can not expect an explicit closed form of the kernel function ϕ.

In the sequel, we propose a numerical scheme that approximates the solution s without any a-priori knowledge of the kernel function.

4 A Finite Dimensional Model

A widely used interpolation method for two- and three-dimensional data in geoscience is the minimum curvature interpolation proposed by Briggs [2]. Briggs' gridding technique is derived from the discretization of a continuous model. Replacing integrals by finite sums and derivatives by finite differences leads to a finite dimensional model whose solution approximates the minimal semi-norm solution constrained to the interpolation condition. This method needs to transfer sampling nodes to a regular grid causing some inaccuracy for the interpolation. Altogether, it is related to the concept of finite difference methods.

The approach that we describe in this section, is more in the spirit of finite elements which are successfully applied in the numerical treatment of elliptic differential equations, see [3]. Instead of discretizing an infinite dimensional model, we start with a finite dimensional function space $\mathcal{V}$ in which we can solve the constrained optimization problem exactly.

Let us consider a semi-Hilbert space $\mathcal{S}$ of continuous real-valued functions defined on a bounded domain $\Omega \subset \mathbb{R}^d$ with semi-norm $|\cdot|_{\mathcal{S}}$ and associated nullspace $\mathcal{P}$. The most well-known example of this setting is the homogeneous Sobolev space $\mathcal{B}_m$ with semi-norm $|\cdot|_m$ and nullspace $\mathcal{P}_m^d$ as introduced above. Let $\mathcal{V}$ be a finite-dimensional subspace of $\mathcal{S}$ with $\mathcal{P} \subset \mathcal{V}$.

Given data consisting of nodes $\mathbf{z}_1, \ldots, \mathbf{z}_n$ in Ω and corresponding scalar measurements $\gamma_1, \ldots, \gamma_n$, we look for an interpolant of the data which has minimal semi-norm. We transfer the problem from $\mathcal{S}$ to the finite dimensional semi-Hilbert subspace $\mathcal{V}$.

Let $b_1, \ldots, b_r$ be a basis of $\mathcal{V}$, i.e., every function $v \in \mathcal{V}$ has a unique expansion

$$v = \sum_{i=1}^{r} c_i \, b_i$$

with $\mathbf{c} = (c_1, \ldots, c_r)^T \in \mathbb{R}^r$. The squared semi-norm of v is simply

$$|v|_{\mathcal{S}}^2 = \left(v \,\middle|\, v \right)_{\mathcal{S}} = \mathbf{c}^T \, \mathbf{B} \, \mathbf{c} \, ,$$

where $\left(\cdot \,\middle|\, \cdot \right)_{\mathcal{S}}$ denotes the semi-inner product of $\mathcal{S}$, and $\mathbf{B}$ is the Gramian matrix with entries

$$B_{i,j} = \left(b_i \,\middle|\, b_j \right)_{\mathcal{S}} , \qquad i, j = 1, \ldots, r \, .$$

Then, the minimal semi-norm interpolation problem in $\mathcal{V}$ turns into an quadratic optimization problem with linear constraints of the form

$$\min_{\substack{\mathbf{c} \in \mathbb{R}^m \\ \mathbf{P}\,\mathbf{c} = \boldsymbol{\gamma}}} \mathbf{c}^T \mathbf{B} \, \mathbf{c} \, , \tag{16}$$

where $\mathbf{P}$ is the $n \times r$ matrix with entries $P_{ij} = b_i(\mathbf{z}_j)$ and $\boldsymbol{\gamma} = (\gamma_1, \ldots, \gamma_n)^T$.

Assuming $\mathbf{P}$ to be surjective, which guarantees that there exists, at least, one function in $\mathcal{V}$ interpolating the given data, we know that (16) has a unique solution in $\mathcal{V}$ as long as Z is unisolvent which is not a strong restriction and will always be assumed.

The standard approach in optimization theory for solving (16) makes use of a vector of so-called Lagrange multipliers $\boldsymbol{\mu} = (\mu_1, \ldots, \mu_r)^T$, see [5]. In this way, (16) can be transformed into a system of linear equations of the form

$$\begin{pmatrix} \mathbf{B} & \mathbf{P} \\ \mathbf{P}^T & \mathbf{0} \end{pmatrix} \begin{pmatrix} \mathbf{c} \\ \boldsymbol{\mu} \end{pmatrix} = \begin{pmatrix} \mathbf{0} \\ \boldsymbol{\gamma} \end{pmatrix} . \tag{17}$$

Solving (17) yields the coefficient vector $\mathbf{c}$ for the minimal semi-norm interpolant in $\mathcal{V}$ of the given data.

The system (17) provides the computational scheme of the general concept of solving the minimal semi-norm interpolation problem in finite dimensional spaces. In practice, numerical stability and low complexity impose requirements on the choice of the basic functions $b_1, \ldots, b_r$ which, in turn, depends on the semi-norm. Both, numerical stability and low complexity are often achieved when the system matrix is sparse. In the following, we will use compactly supported basis functions in order to satisfy the stated requirements.

In practice, a semi-norm has to be specified. Instead of the minimum curvature principle used by Briggs, we consider the second order homogeneous Sobolev semi-norm

$$|g|_2 = \left(\int_\Omega \sum_{|\alpha|=2} \frac{2}{\alpha!} \left| D^\alpha g(\mathbf{x}) \right|^2 d\mathbf{x} \right)^{1/2} . \tag{18}$$

We restrict our attention to $d = 1, 2, 3$. In this way, we make sure that the considered functions are continuous. The multi-index notation is convenient for having only one description covering all three cases.

Let Ω denote the open unit cube in $\mathbb{R}^d$, i.e.,

$$\Omega = (0, 1)^d .$$

Then, the linear space of all function on Ω for which (18) is finite, is a semi-Hilbert space $\mathcal{S}$ with nullspace $\mathcal{P}_2^d$, the space of d-variate polynomials of order ≤ 2.

We will use uniform translates of the tensor product of cubic B-splines that are dilated according to a resolution parameter N. These translates constitute a basis for a finite dimensional subspace $\mathcal{V}$ of $\mathcal{S}$. Since these basis functions have finite support the Gramian matrix $\mathbf{B}$ is sparse. Moreover, due to fundamental properties of cubic B-splines, $\mathcal{P}_2^d$ is included in $\mathcal{V}$.

In the next section we discuss the bivariate case. The one- and three-dimensional cases are analogue, though complexity increases with the dimension.

5 The Bivariate Cubic B-Spline Model

Consider the bivariate homogeneous Sobolev semi-norm

$$|g|_2 = \left(\int_0^1 \int_0^1 \left(\frac{\partial^2 g(x,y)}{\partial x^2} \right)^2 + 2 \left(\frac{\partial^2 g(x,y)}{\partial x \partial y} \right)^2 + \left(\frac{\partial^2 g(x,y)}{\partial y^2} \right)^2 dx dy \right)^{1/2}$$

with a three-dimensional nullspace $\mathcal{P}_2^2$, spanned by the polynomials $1, x, y$.

For a resolution parameter N with corresponding lattice constant

$$h = 1/(N-1) ,$$

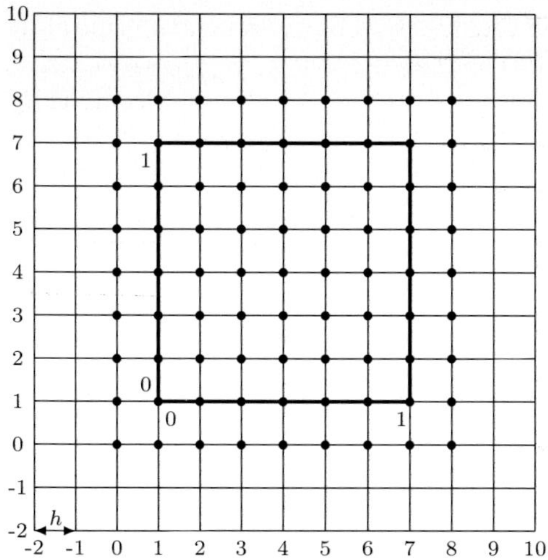

Fig. 2. Mesh covering $[0,1]^2$ with $\bullet$ indicating centers of B-splines ($N = 7$).

we cover the unit square by a uniform mesh with nodes

$$x_i = (i-1)h, \quad (i = -2, -1, \ldots, N+2, N+3),$$

$$y_j = (j-1)h, \quad (j = -2, -1, \ldots, N+2, N+3).$$

We further consider the tensor product of the centered cubic B-spline

$$b(x, y) = b(x)\, b(y), \tag{19}$$

where

$$b(x) = \frac{1}{h^3} \begin{cases} (x+2h)^3, & x \in [-2h, h], \\ h^3 + 3h^2(x+h) + 3h(x+h)^2 - 3(x+h)^3, & x \in [-h, 0], \\ h^3 + 3h^2(h-x) + 3h(h-x)^2 - 3(h-x)^3, & x \in [0, h], \\ (2h-x)^3, & x \in [h, 2h], \\ 0, & \text{otherwise}. \end{cases}$$

We define the finite-dimensional space $\mathcal{V}$ as the linear span of the basis functions

$$b_{i,j}(x, y) = b(x - x_i, y - y_j), \qquad i, j = 0, \ldots, N+1,$$

with $\text{supp}(b_{i,j}) = [x_{i-2}, x_{i+2}] \times [y_{j-2}, y_{j+2}]$. The non-zero part of the basis functions that lies outside the unit square is not considered. From the properties of the one-dimensional B-spline basis, it follows that $\mathcal{P}_3^2 \subset \mathcal{V}$.

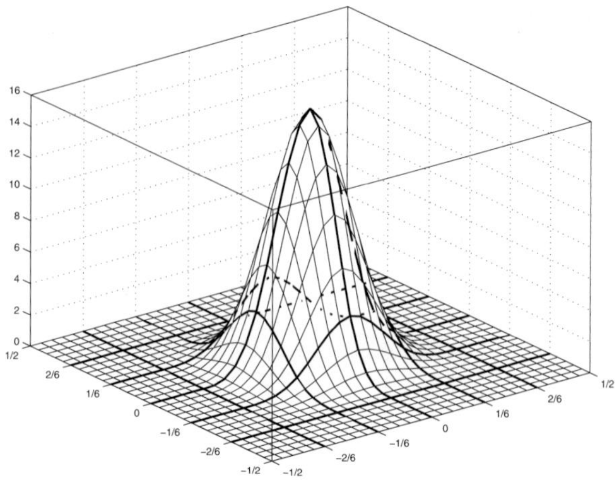

Fig. 3. Centered cubic B-spline tensor-product in two dimensions, $N = 7$.

In the constrained optimization problem (16) we look for a function of the form

$$v(x,y) = \sum_{i=0}^{N+1} \sum_{j=0}^{N+1} c_{i,j}\, b_{i,j}(x,y)$$

that minimizes $|\cdot|_s$ subject to interpolating data $\gamma_1, \ldots, \gamma_n$ at the corresponding nodes $\mathbf{z}_1 = (x_1, y_1), \ldots, \mathbf{z}_n = (x_n, y_n)$.

For practical purposes, it is convenient to reshuffle the two-dimensional index into a one-dimensional index in the following way:

$$(i,j) \to k = i + (N+2)j\,.$$

Then, we have

$$v(x,y) = \sum_{k=0}^{N^2+2N} c_k\, b_k(x,y)\,.$$

We also need the backward transformation

$$k \to (i_k, j_k)$$

with $j_k = \lfloor k/(N+2) \rfloor$ and $i_k = k - (N+2)j_k$.

The entries of the symmetric Gram matrix $\mathbf{B}$ of size $(N+2)^2$ are given by

$$B_{k,l} = \underbrace{\int_0^1\int_0^1 \frac{\partial^2 b_k(x,y)}{\partial x^2}\frac{\partial^2 b_l(x,y)}{\partial x^2}dxdy}_{=B^I_{k,l}}$$

$$+\,2\underbrace{\int_0^1\int_0^1 \frac{\partial^2 b_k(x,y)}{\partial x\partial y}\frac{\partial^2 b_l(x,y)}{\partial x\partial y}dxdy}_{=B^{II}_{k,l}}$$

$$+\underbrace{\int_0^1\int_0^1 \frac{\partial^2 b_k(x,y)}{\partial y^2}\frac{\partial^2 b_l(x,y)}{\partial y^2}dxdy}_{=B^{III}_{k,l}}\,.$$

Applying the above index transformation and using symmetry properties of the tensor product, we can simplify $B^I_{k,l}$ according

$$B^I_{k,l} = \int\limits_{x_{i_k-2}}^{x_{i_k+2}}\int\limits_{y_{j_k+2}}^{y_{j_k-2}} \frac{\partial^2 b_{i_k}(x)b_{j_k}(y)}{\partial x^2}\frac{\partial^2 b_{i_l}(x)b_{j_l}(y)}{\partial x^2}dxdy$$

$$= \int\limits_{(i_k-3)h}^{(i_k+1)h}\int\limits_{(j_k-3)h}^{(j_k+1)h} b_{j_k}(y)\frac{d^2 b_{i_k}(x)}{dx^2}b_{j_l}(y)\frac{d^2 b_{i_l}(x)}{dx^2}dxdy$$

$$= \int\limits_{(i_k-3)h}^{(i_k+1)h}\frac{d^2 b_{i_k}(x)}{dx^2}\frac{d^2 b_{i_l}(x)}{dx^2}dx\int\limits_{(j_k-3)h}^{(j_k+1)h} b_{j_k}(y)b_{j_l}(y)dy \tag{20}$$

and likewise $B^{II}_{k,l}$ and $B^{III}_{k,l}$. It follows that the value vanishes for

$$|i_k - i_m| > 3 \qquad \text{or} \qquad |j_k - j_m| > 3\,.$$

Hence, for every basis functions $b_k(x,y)$ there are only 49 possible contributions. Since the setting is shift invariant with respect to the lattice constant h, we can compute one example and then shift the entries accordingly. Moreover, the symmetry of the tensor product reduces the number of different cases. Special attention has to be paid for basis functions near the boundaries.

The integrants in (20) are piecewise polynomials. Therefore, $B_{k,l}$ can be computed exactly. Since

$$|v|_S^2 = \mathbf{c}^T\mathbf{B}\,\mathbf{c}$$

where $\mathbf{c}$ is the coefficient vector of the function v with the indexing described above, the only case where the quadratic form $\mathbf{c}^T\mathbf{B}\,\mathbf{c}$ vanishes is for those coefficient vectors that come from bivariate linear polynomials. In fact, we can easily verify numerically, that the kernel (nullspace) of $\mathbf{B}$ has dimension 3 and that the kernel of $\mathbf{B}$ consists of bivariate polynomials of order ≤ 2.

The second matrix $\mathbf{P} = (b_k(\mathbf{z}_l))$ is also sparse. When discarding sampling nodes such that in each cell of the lattice of size h there is at most one node, it is most likely that the matrix $\mathbf{P}$ will be surjective.

Under the assumption that $\mathbf{P}$ is surjective, it easily follows that the system matrix of

$$\begin{aligned} \mathbf{Bc} + \mathbf{P}\boldsymbol{\mu} &= \mathbf{0} \\ \mathbf{P}^T\mathbf{c} \qquad &= \mathbf{0} \end{aligned} \tag{21}$$

is non-singular and the constrained optimization problem (16) has a unique solution.

In order to increase the stability of the matrix of (21) for practical purposes, it is recommended to equalize the magnitude of the entries of both matrices $\mathbf{B}$ and $\mathbf{P}$. If we divide, for example, the matrix $\mathbf{B}$ by the factor 10^5, the condition number of (17) usually decreases significantly. Note that the optimization problem remains unchanged.

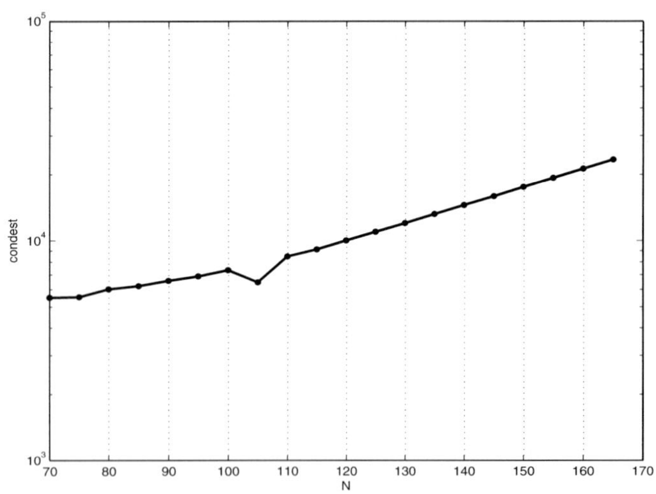

Fig. 4. Typical behavior of the condition number of (17) in 2D.

Whenever we want to resolve small features, we have to increase the resolution parameter N. This increases the size of the system matrix (17). Limits to the resolution are set by the complexity of the system (17) which, in the two-dimensional case, is of order $\mathcal{O}(N^2(\log N)^2)$ where N is the resolution parameter. This has been verified numerically, cf. Figure 5.

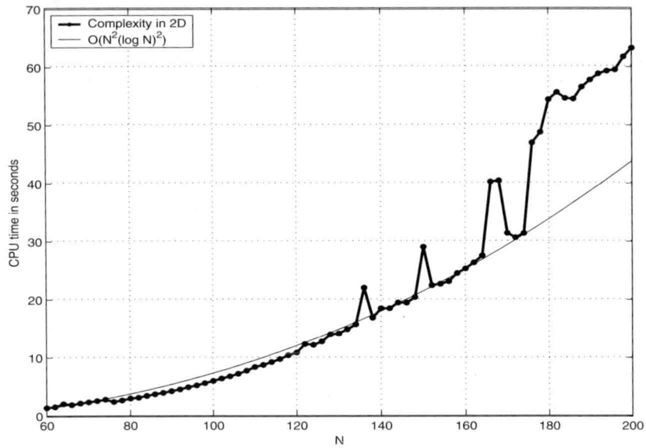

Fig. 5. Complexity of (17) in 2D.

In what follows, we compare the bivariate cubic B-spline interpolation with Duchon's surface spline interpolation.

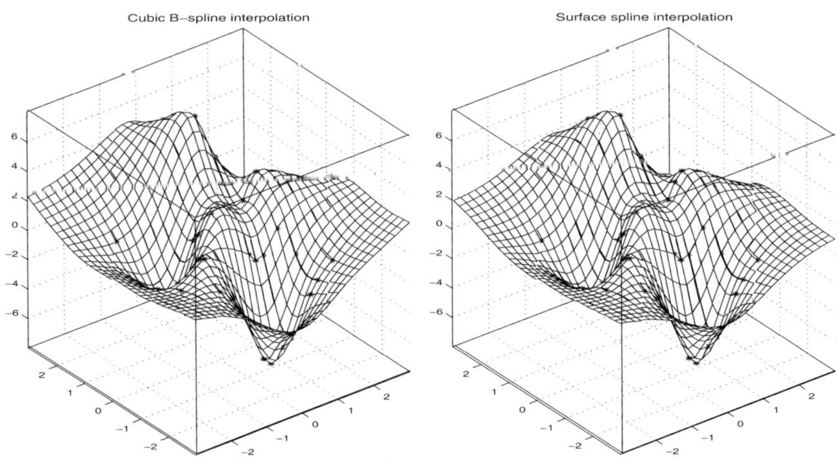

Fig. 6. Cubic B-spline interpolation versus surface spline interpolation.

The following Figure 7 shows the convergence of the relative error between successive B-spline solutions when increasing the resolution parameter.

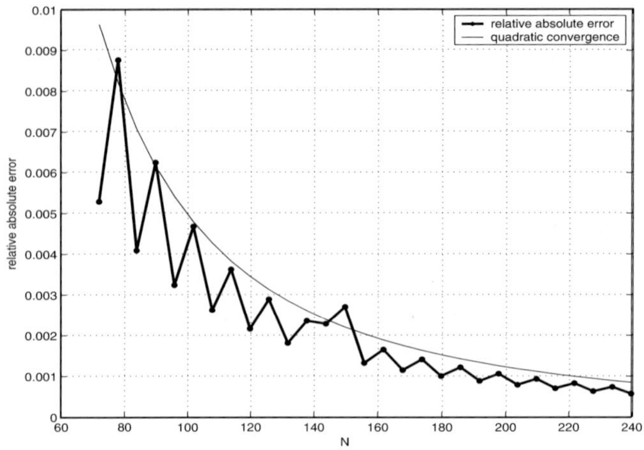

Fig. 7. Convergence of the relative absolute error in 2D.

An explicit convergence proof for Sobolev norms can be found in [15].

Let us fix the sampling nodes and increase the domain Ω. What we observe is that, similar to the surface spline, the bicubic B-spline interpolant tends to a linear polynomial.

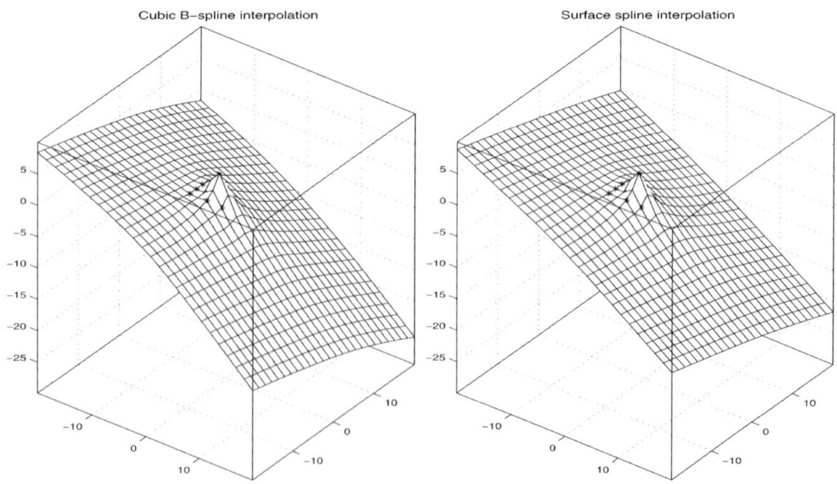

Fig. 8. Cubic B-spline versus surface spline interpolation in the far-field.

The B-spline model provides a method to compute an approximate of the minimal semi-norm interpolation on Ω. We emphasize, that in contrast to the global surface spline solution, no kernel function is explicitly used and that unlike Briggs' method, the B-spline solution is an exact interpolation and does not involve boundary conditions.

6 Best Linear Unbiased Prediction

Linear models of random fields are represented through the first two moments, namely the mean and the covariance. The mean function is usually not known but assumed to be a low dimensional linear parametric model. Suppose that we have a random field $Z(\mathbf{x})$ of the form

$$Z(\mathbf{x}) = \sum_{i=1}^{r} \beta_i \, m_i(\mathbf{x}) + \epsilon(\mathbf{x})$$

where $\epsilon(\mathbf{x})$ is a zero-mean random field with known covariance function C_ϵ, $m_1(\mathbf{x}), \ldots, m_r(\mathbf{x})$ are given real-valued functions, and $\boldsymbol{\beta} = (\beta_1, \ldots, \beta_r)^T$ is a vector of p unknown parameters. Often, the mean is a linear combination of first order polynomials indicating the trend or drift of the random field.

We observe the data vector

$$\gamma = (z(\mathbf{z}_1), \ldots, z(\mathbf{z}_n))^T$$

for some scalar function z, and wish to predict $Z(\mathbf{z}_0)$. We may look for the minimal squared error prediction among all predictors of the form

$$\tilde{Z}(\mathbf{z}_0) = \mathbf{c}^T \mathbf{Z} + d_0,$$

where $\mathbf{Z} = (Z(\mathbf{z}_1), \ldots, Z(\mathbf{z}_n))^T$, subject to the constraint that the prediction be unbiased, i.e.,

$$E\big[\tilde{Z}(\mathbf{z}_0)\big] = E\big[Z(\mathbf{z}_0)\big]$$

for all $\boldsymbol{\beta}$. The unbiasedness constraint is equivalent to

$$\mathbf{c}^T \mathbf{M} \boldsymbol{\beta} + d_0 = \mathbf{m}^T \boldsymbol{\beta}$$

for all $\boldsymbol{\beta}$, or

$$d_0 = 0 \qquad \text{and} \qquad \mathbf{M}^T \mathbf{c} = \mathbf{m},$$

where $\mathbf{M} = (m_i(\mathbf{z}_j))^T$ is an $n \times r$ matrix and $\mathbf{m} = (m_1(\mathbf{z}_0), \ldots, m_r(\mathbf{z}_0))^T$. The goal is then to minimize the mean-squared error

$$E\big[(Z(\mathbf{z}_0) - \mathbf{c}^T \mathbf{Z})^2\big]$$

subject to the condition $\mathbf{M}^T \mathbf{c} = \mathbf{m}(\mathbf{z}_0)$. If $\mathbf{c}$ solves this constrained minimization problem, then $\mathbf{c}^T \gamma$ is called a *best linear unbiased predictor for* $Z(\mathbf{z}_0)$.

Recall that the mean-squared error is just the mean of the prediction error plus its variance. Because of the unbiasedness condition the mean of the prediction vanishes and we obtain

$$E\big[(Z(\mathbf{z}_0) - \mathbf{c}^T\mathbf{Z})^2\big] = C(\mathbf{z}_0, \mathbf{z}_0) + \sum_{i=1}^{n}\sum_{j=1}^{n} c_i c_j C(\mathbf{z}_i, \mathbf{z}_j) - 2\sum_{i=1}^{n} c_i C(\mathbf{z}_0, \mathbf{z}_i)$$
$$= k_0 + \mathbf{c}^T\mathbf{K}\mathbf{c} - 2\mathbf{c}^T\mathbf{k}\,,$$

where

$$k_0 = C(\mathbf{z}_0, \mathbf{z}_0),$$
$$\mathbf{k} = (C(\mathbf{z}_0, \mathbf{z}_1), \dots, C(\mathbf{z}_0, \mathbf{z}_n))^T,$$
$$\mathbf{K} = \big(C(\mathbf{z}_i, \mathbf{z}_j)\big)_{1 \le i,j \le n}\,.$$

According to classic optimization techniques, the optimal solution of the minimization problem subject to the unbiasedness constraint must realize the minimum of the *Lagrange function*

$$L(c_1, \dots, c_n, \mu_1, \dots, \mu_r) = E\big[(Z(\mathbf{z}_0) - \mathbf{c}^T\mathbf{Z})^2\big] - 2\boldsymbol{\mu}^T(\mathbf{M}^T\mathbf{c} - \mathbf{m})$$

where $\boldsymbol{\mu} = (\mu_1, \dots, \mu_r)^T$ are the so-called Lagrange multipliers [5]. After differentiating the Lagrange function L with respect to $c_1, \dots, c_n, \mu_1, \dots, \mu_r$ and equating the result at zero we arrive at the linear system

$$\begin{pmatrix} \mathbf{K} & \mathbf{M} \\ \mathbf{M}^T & \mathbf{0} \end{pmatrix} \begin{pmatrix} \mathbf{c} \\ \boldsymbol{\mu} \end{pmatrix} = \begin{pmatrix} \mathbf{k} \\ \mathbf{m} \end{pmatrix}. \tag{22}$$

If $\mathbf{K}$ and $\mathbf{M}$ are of full rank, then the system matrix of (22) is invertible, and we obtain

$$\begin{pmatrix} \mathbf{c} \\ \boldsymbol{\mu} \end{pmatrix} = \begin{pmatrix} \mathbf{K} & \mathbf{M} \\ \mathbf{M}^T & \mathbf{0} \end{pmatrix}^{-1} \begin{pmatrix} \mathbf{k} \\ \mathbf{m} \end{pmatrix}.$$

Since

$$\begin{pmatrix} \mathbf{K} & \mathbf{M} \\ \mathbf{M}^T & \mathbf{0} \end{pmatrix}^{-1} = \begin{pmatrix} \mathbf{U} & \mathbf{V} \\ \mathbf{V}^T & \mathbf{W} \end{pmatrix}$$

with

$$\mathbf{U} = \mathbf{K}^{-1} - \mathbf{K}^{-1}\mathbf{M}(\mathbf{M}^T\mathbf{K}^{-1}\mathbf{M})^{-1}\mathbf{M}^T\mathbf{K}^{-1}\,,$$
$$\mathbf{V} = \mathbf{K}^{-1}\mathbf{M}(\mathbf{M}^T\mathbf{K}^{-1}\mathbf{M})^{-1}\,,$$
$$\mathbf{W} = -(\mathbf{M}^T\mathbf{K}^{-1}\mathbf{M})^{-1}\,,$$

we have $\mathbf{c} = \mathbf{U}\mathbf{k} + \mathbf{V}\mathbf{m}$, so that the resulting predictor is

$$\mathbf{c}^T\mathbf{Z} = \mathbf{k}^T\mathbf{K}^{-1}(\mathbf{Z} - \mathbf{M}\hat{\boldsymbol{\beta}}) + \mathbf{m}^T\hat{\boldsymbol{\beta}}\,,$$

where $\hat{\boldsymbol{\beta}} = \mathbf{V}^T\mathbf{Z}$ is the generalized least square estimator of $\boldsymbol{\beta}$, see [13].

If $m(\mathbf{x}) \equiv 1$, so that the mean of the process is assumed to be an unknown constant, then best linear unbiased prediction is called *ordinary kriging*. Best linear unbiased prediction for a more general mean function m is known as *universal kriging* and best linear prediction with the mean assumed to be zero is a special case of *simple kriging.*

When $Z(\mathbf{x})$ is Gaussian, then the best linear unbiased prediction has a Bayesian interpretation, see [7] and [10].

We now turn our attention to the dual formulation of kriging which leads us to linear equations of the same type as those that arise from interpolation with *radial basic functions* such as thin plate splines. Some of the basic ideas can be found in [4]. The first studies go back to [8].

Recall that the coefficient vector $\mathbf{c}$ of the best linear unbiased prediction at $\mathbf{z}_0$ depends obviously on the location $\mathbf{z}_0$. Often, one is not only interested in a single location but in many locations usually forming a regular grid. It seems that in this case we have to compute a coefficient vector for every grid node. However, the matrix to invert does not depend on the prediction nodes but only on the given locations. The dual formulation of kriging is an elementary transcription of the linear model such that the alternative coefficients are independent of the nodes of interest.

The estimate $\tilde{z}(\mathbf{z}_0)$ is given by the inner product of the vector $\mathbf{c} = \mathbf{c}(\mathbf{z}_0)$ and the observed data vector $\boldsymbol{\gamma}$, i.e.,

$$\tilde{z}(\mathbf{z}_0) \;=\; \mathbf{c}^T\boldsymbol{\gamma}\,.$$

With the above notation, we obtain

$$\begin{aligned}
\tilde{z}(\mathbf{z}_0) &= \boldsymbol{\gamma}^T\,\mathbf{c} \\
&= (\boldsymbol{\gamma}^T, \mathbf{0}^T)\begin{pmatrix} \mathbf{K} & \mathbf{M} \\ \mathbf{M}^T & \mathbf{0} \end{pmatrix}^{-1}\begin{pmatrix} \mathbf{k} \\ \mathbf{m} \end{pmatrix} \\
&= (\mathbf{v}^T, \mathbf{w}^T)\begin{pmatrix} \mathbf{k} \\ \mathbf{m} \end{pmatrix},
\end{aligned}$$

where the coefficient vectors $\mathbf{v}$ and $\mathbf{w}$ are given by

$$\begin{pmatrix} \mathbf{v} \\ \mathbf{w} \end{pmatrix} = \begin{pmatrix} \mathbf{K} & \mathbf{M} \\ \mathbf{M}^T & \mathbf{0} \end{pmatrix}^{-1}\begin{pmatrix} \boldsymbol{\gamma} \\ \mathbf{0} \end{pmatrix}$$

or implicitly,

$$\begin{pmatrix} \mathbf{K} & \mathbf{M} \\ \mathbf{M}^T & \mathbf{0} \end{pmatrix}\begin{pmatrix} \mathbf{v} \\ \mathbf{w} \end{pmatrix} = \begin{pmatrix} \boldsymbol{\gamma} \\ \mathbf{0} \end{pmatrix}. \tag{23}$$

Note that this linear system is independent of the node $\mathbf{z}_0$. The estimator $\tilde{z}(\mathbf{z}_0)$ has now the form

$$\tilde{z}(\mathbf{z}_0) = \sum_{i=1}^{n} v_i\, C(\mathbf{z}_0, \mathbf{z}_i) + \sum_{j=1}^{r} w_j\, m_j(\mathbf{z}_0)\,.$$

This is what is referred to as the *dual formulation of kriging*, see [4]. Instead of a fixed location $\mathbf{z}_0$, any $\mathbf{x} \in \Omega$ can be inserted in the formula and we obtain the global estimator

$$\tilde{z}(\mathbf{x}) = \sum_{i=1}^{n} v_i\, C(\mathbf{x}, \mathbf{z}_i) + \sum_{j=1}^{p} w_j\, m_j(\mathbf{x})\,. \tag{24}$$

Similar equations can be obtained when replacing the covariance function by the variogram, see [4]. Even though this estimator seems purely deterministic the basic model functions $C(\mathbf{x}, \mathbf{z}_i)$ and $m_j(\mathbf{x})$ are derived from a probabilistic model.

Observing that the only computational challenge is the invertibility of the system matrix in (23), gives a different viewpoint of the interpolation method. Indeed, one could ignore the stochastic interpretation and simply look for deterministic functions $C(\mathbf{x}, \mathbf{y})$ and $m_1(\mathbf{x}), \ldots, m_r(\mathbf{x})$ with the proviso that the system matrix be non-singular. Then, the function $\tilde{z}(\mathbf{x})$ interpolates the data $z(\mathbf{z}_1), \ldots, z(\mathbf{z}_n)$ at the given nodes $\mathbf{z}_1, \ldots, \mathbf{z}_n$. This is the common viewpoint in radial basic function interpolation such as surface spline interpolation [6].

Acknowledgement

The work was partly supported by the European Union through the research and training network NetAGES, contract no. IST-1999-29034. Special thanks to Hans Georg Feichtinger for the excellent support during the author's PhD studies.

References

1. J.H. Ahlberg, E.N. Nilson, and J.L. Walsh (1967) *The Theory of Splines and Their Applications.* Academic Press, New York.
2. I.C. Briggs (1974) Machine contouring using minimum curvature. Geophysics **39**(1), 39–48.
3. P.G. Ciarlet (1978) *The Finite Element Method for Elliptic Problems.* SIAM Classics in Applied Mathematics.
4. N.A.C. Cressie (1991) *Statistics for Spatial Data.* John Wiley.
5. R. Fletcher (1987) *Practical Methods in Optimization.* 2nd edition, John Wiley and Sons.

6. A. Iske (2002) Scattered data modelling using radial basis functions. *Tutorials on Multiresolution in Geometric Modelling*, A. Iske, E. Quak, and M. S. Floater (eds.), Springer, Heidelberg, 205–242.
7. P.K. Kitanidis (1986) Parameter uncertainty in estimation of spatial functions: Bayesian analysis. Water Resource Research **19**, 499–507.
8. G. Matheron (1981) Splines and kriging: their formal equivalence. *Down-to-Earth Statistics: Solutions Looking for Geological Problems*, D.F. Merriam (ed.), Syracuse University Geology Contributions, 77–95.
9. C.A. Micchelli (1986) Interpolation of scattered data: distance matrices and conditionally positive definite functions. Constr. Approx. **2**, 11–22.
10. H. Omre (1987) Bayesian kriging-merging observations and qualified guesses in kriging. Math. Geol. **19**, 25–39.
11. W. Rudin (1991) *Functional Analysis*. 2nd edition, Mc Graw-Hill.
12. L. Schwartz (1966) *Théorie des Distributions*. Hermann, Paris.
13. M.L. Stein (1999) *Interpolation of Spatial Data*. Springer.
14. T. Werther (2003) Characterization of semi-Hilbert spaces with application in scattered data interpolation. *Curve and Surface Fitting: Saint-Malo 2002*, A. Cohen, J.-L. Merrien, and L.L. Schumaker (eds.), Nashboro Press, Brentwood, 374–383.
15. T. Werther (2003) *Optimal Interpolation in Semi-Hilbert Spaces*. PhD thesis, University of Vienna.
16. F. Ziegler (1998) *Technische Mechanik der festen und flüssigen Körper*. Springer.

A Method for Ranking CO_2 Flow Models Using Seismic Modeling and Time-Lapse Data

Magne Lygren[1], Erik Lindeberg[2], Per Bergmo[2], Geir Vaaland Dahl[1], Kristine Årland Halvorsen[1], Trygve Randen[1], and Lars Sønneland[1]

[1] Schlumberger Stavanger Research, P.O. Box 8013, N-4068 Stavanger, Norway
[2] SINTEF Petroleum Research, Trondheim, Norway

Summary. A method for discriminating between different reservoir flow models using forward modeling and time-lapse seismic is presented. A rock-physical model is used in order to generate synthetic time-lapse acoustic responses based on flow model predictions. From the acoustic properties a pull-down caused by modifications in acoustic velocity is calculated and compared to real measurements. Full synthetic seismograms are also generated. The method has been applied to the Sleipner CO_2 sequestration project where time-lapse seismic is used to monitor the injected gas. Different vertical migration processes of the CO_2 may explain the observed time-lapse response. In this chapter, this new methodology is used in order to discriminate between these processes.

1 Introduction

One of the few options for large-scale reduction of anthropogenic CO_2 emissions is to sequester CO_2 into underground formations. Since September 1996, Statoil has been applying this technology by injecting about 1 million tons of CO_2 per year into a highly-permeable, highly-porous saline aquifer located in the Sleipner area in the North Sea at about 1,000 m depth. At the prevailing reservoir condition the CO_2 will be a low-viscosity fluid with a density of about 650 kg/m^3. The formation, the Utsira Sand, is homogeneous and isotropic except for a set of almost horizontal, discontinuous shale layers [5]. The gas is injected near the base of the Utsira Sand which is about 200 m thick near the injection site.

In order to monitor the CO_2, repeated seismic surveys have been acquired. The first seismic survey was acquired 1994, i.e., before the injection had started. Repeated 3D surveys took place in 1999 and 2001 after three and five years of injection, respectively. The seismic showed a clear response due to an impedance-contrast caused by the injected CO_2. As shown in Figure 1, nearly horizontal reflectors were observed in the repeated seismic. These reflectors were interpreted as accumulations of CO_2 below the different shale layers, which act as barriers to the vertical migration of the CO_2.

A reservoir flow model was developed by Lindeberg et al. [4] in order to study the dominant flow processes. The predicted CO_2 distribution was

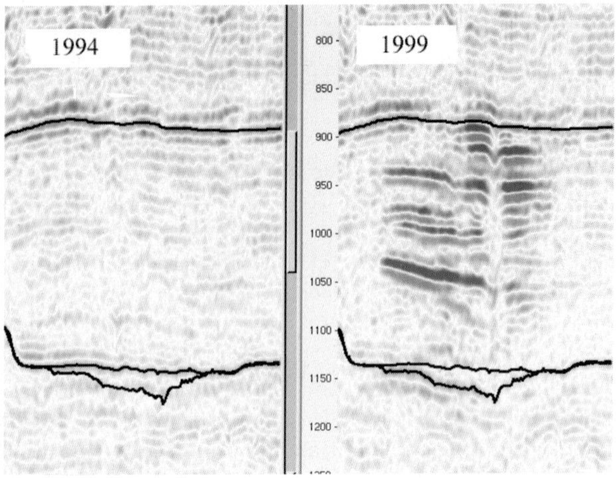

Fig. 1. Inline sections from the 1994 and 1999 seismic, showing reflections caused by the injected CO_2 in the 1999 data. The pull-down is seen on the 1999 seismic below the CO_2.

matched with the seismic in an iterative process. The matching consisted in predicting the same number of CO_2 accumulations and approximately the same extension of the individual accumulations as observed from the time-lapse seismic.

In [4], mechanisms for the transport of CO_2 through the shale layers are investigated. These shale layers can either be semi-permeable, leading to a dispersed flow between the layers, or the layers can be sealing with localized holes conducting the CO_2 in columns. In the first case, more water will be in contact with the CO_2 giving rise to a larger portion of the CO_2 being dissolved in the water than in the second case. The question of how much CO_2 is dissolved in water needs essentially to be addressed in order to predict the long-term fate of the CO_2. However, due to tuning effects and uncertainties in the acoustic properties of the injected CO_2, it is a great challenge to properly invert the time-lapse seismic, and consequently there are inherent uncertainties in the matching between the seismic and the flow model.

In this chapter, additional observations are therefore used in order to distinguish between different possible flow models. The presence of CO_2 introduces a reduction in compressional wave velocity, which in turn results in an increase of travel-time for reflections below the CO_2. This *pull-down* can be observed and quantified from the time-lapse seismic. A forward modeling of acoustic properties is performed from the flow model, and the modeled pull-down is compared to the observed pull-down. This evaluation will reduce both the uncertainties associated with the flow model and the uncertainties in the rock-physical transforms. From the matched flow model and rock-physical model, a full synthetic seismic response is calculated and compared to the real seismic by a *consistency-check*.

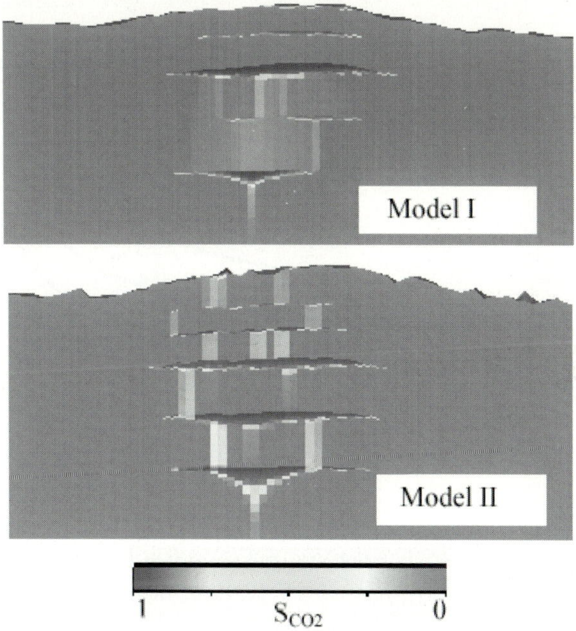

Fig. 2. Cross-sections of the predicted CO_2 saturation in 1999, from Model I and Model II.

2 Methodology

In this section, the flow model and rock-physical model are described. In addition, the method for generating the synthetic response is outlined and the way to quantify the pull-down is explained.

Flow models usually exhibit non-unique solutions, i.e., different models may give rise to the same simulation results. In this case, we have considered two different versions of the 3D flow model Lindeberg [4], each consisting of 428,400 grid cells. Each model simulates equal numbers and approximately equal sizes of the high-concentration CO_2 accumulations. Transmissibility multipliers have been used to model the effect of the shale layers. In one model, hereafter denoted as *Model I*, the shale layers are semi-permeable, and in addition the layers contain one or two high-permeable holes. In the second model, denoted as *Model II*, the layers are impermeable except for a higher number of holes in each layer (in total about 300 holes). The position of the holes in Model II is stochastically distributed. In each model, the CO_2 gets trapped below the shale layers. However, the distribution of CO_2 in the sand *between* the layers is very different for the two models, as shown in Figure 2.

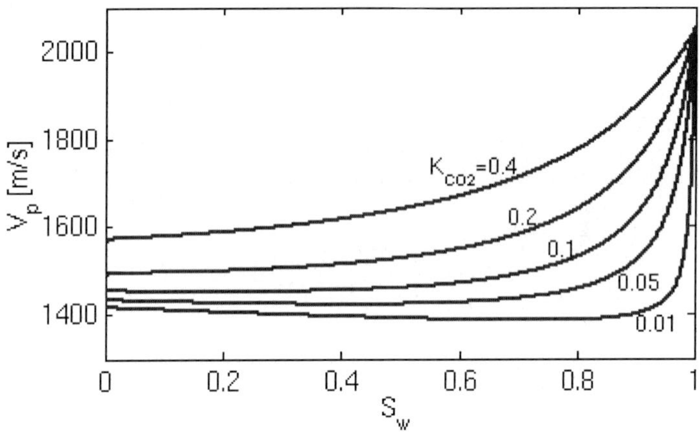

Fig. 3. Gassmann-modeled seismic velocity V_p in a CO_2-water system as function of the water saturation S_w. The density of CO_2 is 650 kg/m^3. The bulk modulus of the CO_2 varies from 0.01 to 0.4 GPa.

Since it is not possible to distinguish between these models solely by considering the high-concentration accumulations, we exploit a different observation from the 4D seismic, namely the pull-down. In order to derive pull-down values from the simulation model, rock-physics modeling is needed. The rock-physical modeling is based on the *Gassmann equation.* The input parameters used have been determined from well log data by [1]. The density of the CO_2 under reservoir conditions is about 650 kg/m^3. The acoustic velocity is highly sensitive to variations in the CO_2 bulk modulus K_{CO_2}, see Figure 3. Laboratory experiments indicate that the value of K_{CO_2} is most likely $\ll 0.675$ GPa. In our study, the value of the CO_2 bulk modulus is varied according to the uncertainties in this parameter. There are uncertainties associated with other parameters in the rock-physical model as well, but the variation of the bulk modulus has the largest impact on the present results. From Figure 3 it is seen that when K_{CO_2} is low, even small values of the CO_2 saturation lead to a large drop in the velocity. Consequently, a flow situation with low-saturated CO_2 between the accumulations can lead to a very strong pull-down if K_{CO_2} is low.

From the rock properties and the fluid properties in the flow model, acoustic velocities and impedances are calculated for each grid cell and each time step in the model. The irregular simulation grid is then resampled to a regular, equidistant Cartesian grid, see [3]. While simulation grids may be tilted, the new grid lines in the depth direction are vertical. This grid can then be converted from depth to travel-time, using the modeled velocities. The effect of this depth to time conversion is illustrated in Figure 4. Using the acoustic properties in the regular sampled grid, calculations of the synthetic pull-down are straightforward. From the impedance, reflectivities can be calculated and a full synthetic seismogram can be generated from a convolutional model.

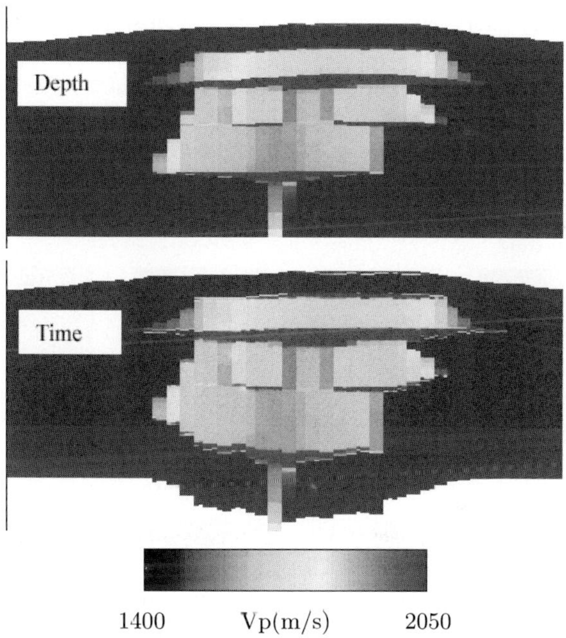

Fig. 4. [Reproduced in colour in Plate 35 on page 447.] Model I. P-velocity in depth and time after a depth-to-time conversion.

There are different ways to measure the pull-down from seismic signals. One method is to cross-correlate the seismic signals between the seismic surveys, using only the signals from below the injected CO_2 [2]. Also, direct interpretations of seismic horizons below the CO_2 can be compared, as indicated in Figure 1. These two methods give the integrated time-delay, caused by total volume of injected CO_2.

3 Results

Maps showing the interpreted pull-down (two-way-time) measured by the seismic in 1999 and 2001 are shown in Figures 5 and 6. It is shown that the basic shape of the pull-down is similar for the two measurements, but the pull-down has increased and the 3D volume effected by the pull-down has been extended, due to the additional CO_2 injections between 1999 and 2001. Moreover, Figures 5 and 6 show synthetically generated maps of the pull-down. In order to incorporate uncertainties associated with the rock-physical transformations, three different values of K_{CO_2} are used. In order to evaluate the migration mechanisms of the CO_2, data from the two models are used. Observe from Figures 5 and 6 that neither of the synthetically generated pull-down-maps captures the exact shape of the map based on real data. The pull-down of Model II varies quite rapidly laterally in contrast to the real pull-down. The high-valued peaks of the pull-down of Model II are caused by the vertical columns of CO_2 above the high-permeable holes in the shale layer. In Model I, the variation of the pull-down smoother, due to the presence of low CO_2-saturated rocks in the region between the shale layers.

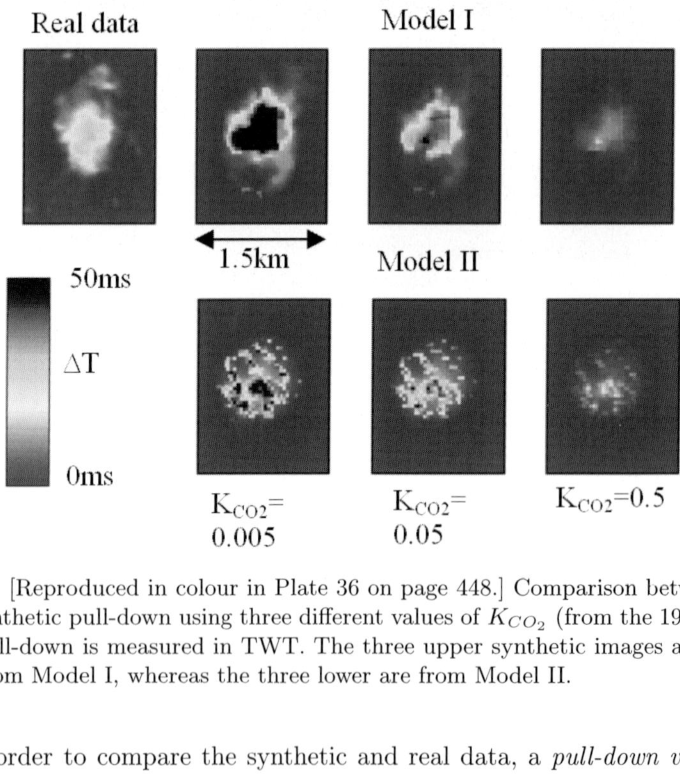

Fig. 5. [Reproduced in colour in Plate 36 on page 448.] Comparison between real and synthetic pull-down using three different values of K_{CO_2} (from the 1999 data). The pull-down is measured in TWT. The three upper synthetic images are generated from Model I, whereas the three lower are from Model II.

In order to compare the synthetic and real data, a *pull-down volume* is calculated. This volume is the area integral of the pull-down, and hence it is a quasi-volume with unit m^2s. The results are shown in Figure 7 where a continuous range of K_{CO_2} is used. Based on both the 1999 and 2001 data, it is seen that for most values of K_{CO_2} the pull down from Model II is underestimated. Only for very low values (< 0.005) of K_{CO_2} the observed pull-down volume is reproduced.

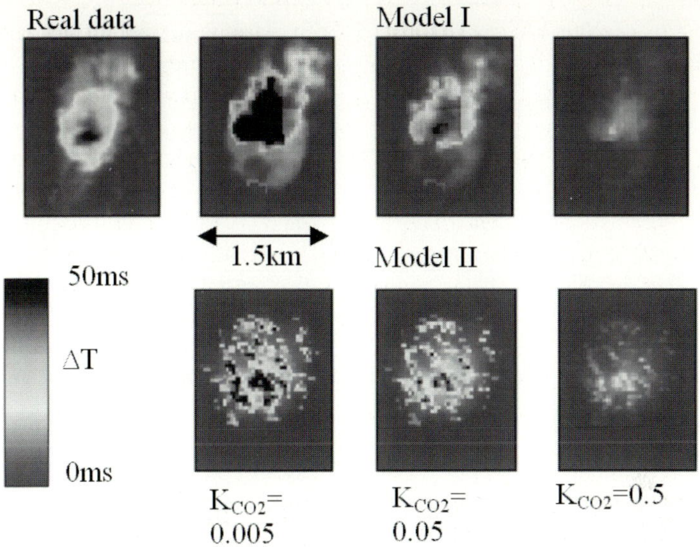

Fig. 6. [Reproduced in colour in Plate 37 on page 448.] Comparison between real and synthetic pull-down from the 2001 data.

Inline sections of synthetic seismograms are given in Figure 8. Even though there is not an exact match, the seismic from Model I reproduce the dominant features of the real seismic. In contrast, Model II gives a more noisy pattern due to the strong lateral variations caused by the vertical columns of CO_2. The strong variations in the seismic from Model II indicate that more advanced modeling, incorporating diffraction effects, should be used in generating synthetic seismic from Model II.

4 Discussion and Conclusion

Based on our results, Model II, which transports the CO_2 in distinct vertical columns between the shale layers, give too small pull-down. The CO_2 columns are also strongly reflected in both the pull-down and the synthetic seismic. This is not observed in the real seismic. The low CO_2-saturated rock between the shale in Model I, on the other hand, give a stronger pull-down. The lateral variation of the pull-down and the synthetic seismic complies with the real observations.

We therefore conclude that it is unlikely that the CO_2 is solely transported through a set of distinct holes in the shale layers having the size of the simulation grid cell (about $30 \times 30m$). Therefore, the shale layers seem to be either semi-permeable, or they may have a more dense distribution of holes of a smaller scale than the simulation cells.

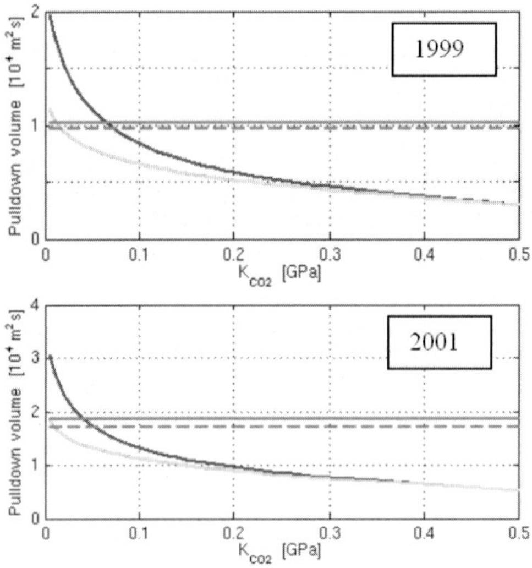

Fig. 7. 1999 and 2001 data. Pull-down volume from Model I (solid dark), Model II (solid bright) compared to observed values from direct interpretation (solid horizontal) and from cross-correlation (dotted horizontal [1]).

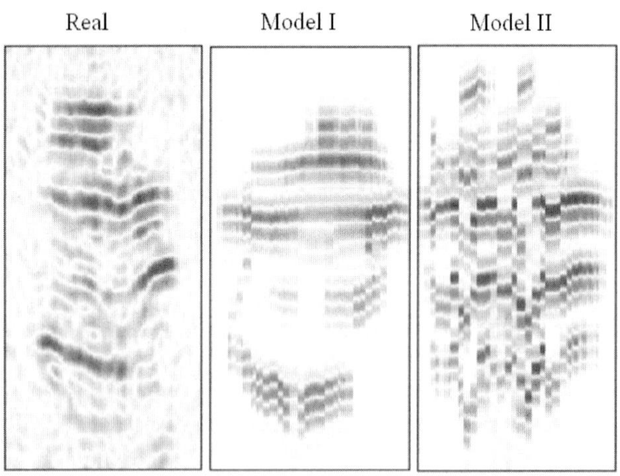

Fig. 8. Seismic difference images from real and synthetic data.

Acknowledgement

Den norske stats oljeselskap (Statoil), Sleipner partners Esso Norge, Norsk Hydro, TotalFinaElf and SACS partners BP, Norsk Hydro, Mobil, Saga Petroleum and Vattenfall, BGS (British Geological Survey), BRGM (Bureau de Recherches Gologiques et Minires), GEUS (Geological Survey of Denmark and Greenland), IFP (Institut Franais du Ptrole), NITG-TNO (Netherlands Institute of Applied Geoscience TNO-National Geological Survey) and Sintef Petroleum Research are acknowledged for permission to publish this paper. The European Union R&D programme Thermie is acknowledged for financial support.

References

1. R. Arts, A. Chadwick, O. Eiken, G.A. Kirby, E. Lindeberg, and P. Zweigel (2003) 4d seismic imaging of a CO2 bubble at the Sleipner field, central North Sea. *3D Seismic Data: Advances in the Understanding of Stratigraphic and Structural Architecture*, R. Davies, J. Cartwright, S. Stewart, R. Underhill, and M. Lappin (eds.), Special Memoir of the Geological Society, London.
2. R. Arts, R. Esayed, L. van der Meer, O. Eiken, S. Ostmo, A. Chadwick, G. Kirby, and B. Zinszner (2002) Estimation of the mass of injected CO2 at Sleipner using time-lapse seismic data. 64th EAGE Annual Conference & Exhibition, Florence, Italy, 2002.
3. K. Fagervik, M. Lygren, T.S. Valen, A. Hetlelid, G. Berge, G.V. Dahl, L. Sønneland, H.E. Lie, and I. Magnus (2001) A method for performing history matching of reservoir flow models using 4d seismic. Society of Exploration Geophysicists 71th Annual Meeting, San Antonio, Texas, USA, 2001.
4. E. Lindeberg, P. Zweigel, P. Bergmo, A. Ghaderi, and A. Lothe (2000) Prediction of CO2 dispersal pattern improved by geology and reservoir simulation and verified by time-lapse seismic. 5th International Conference on Greenhouse Gas Control Technologies, Cairns, Australia, 2000.
5. P. Zweigel, M. Hamborg, R. Arts, A. Lothe, and A. Tømmeras (2000) Prediction of migration of CO2 injected into an underground depository: reservoir geology and migration modelling in the Sleipner case (North Sea). 5th International Conference on Greenhouse Gas Control Technologies, Cairns, Australia, 2000.

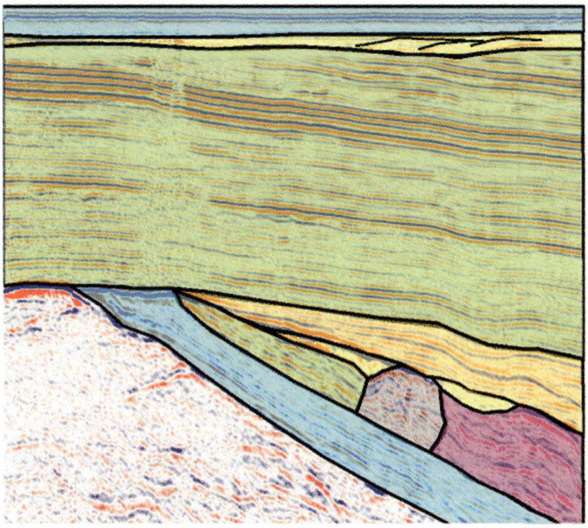

Plate 3. [Figure 14 on page 12] Overlay of different colours onto the same inline as for Figure 13. This represents a vision how a seismic data set should be segmented into shapes and textures. Different colours correspond to different shapes and textures. Compare Figure 13 for legend.

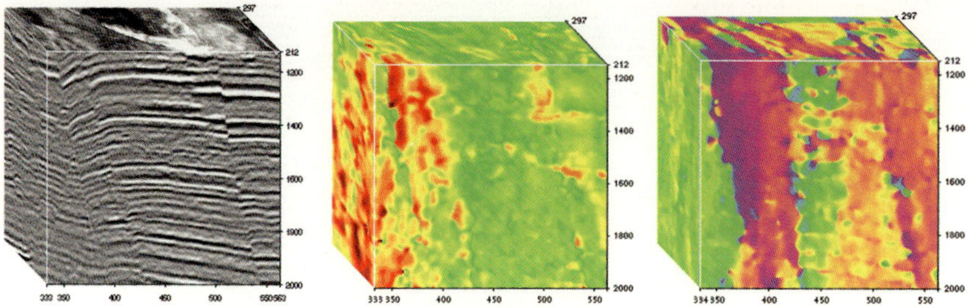

(a) (b) (c)

Plate 4. [Figure 3 on page 27] Examples of the dip (b) and azimuth (c) attributes computed from the seismic cube (a). For the dip cube, increasing dip is here encoded from green through red to darker red. For the azimuth cube, yellow/green indicates that the layers are dipping to the right, whereas pink indicate layers dipping to the left.

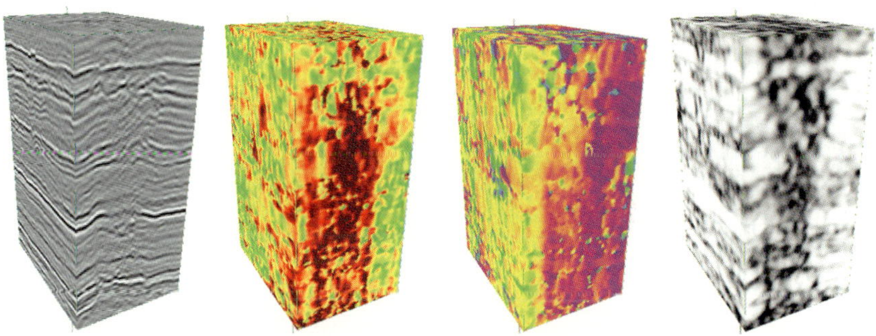

(a) (b) (c) (d)

Plate 5. [Figure 4 on page 29] An example of a seismic cube with a chaotic pattern representing a gas chimney is shown in (a), its dip and azimuth attributes in (b) and (c) and a chaos texture attribute highlighting the gas migration path in (d).

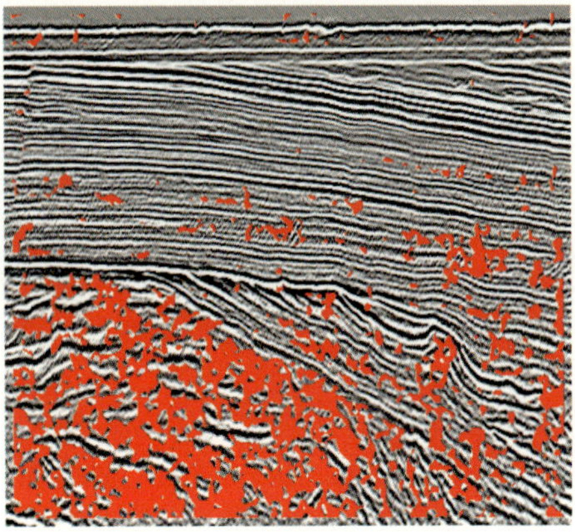

Plate 6. [Figure 5 on page 68] Corner pixels detected by the Harris corner detector for $k = 0.2$, $\sigma = 2$.

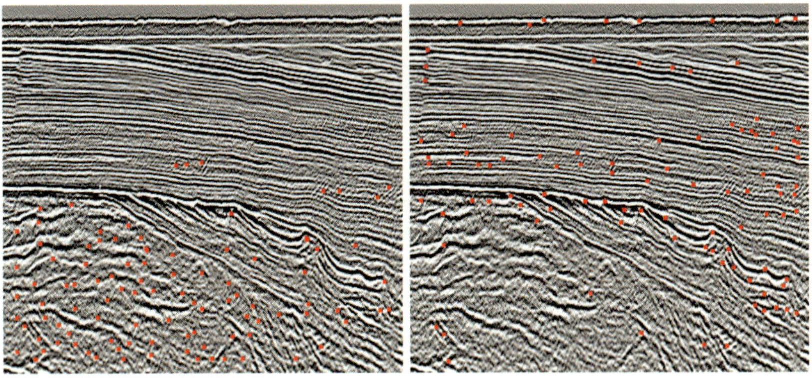

(a) (b)

Plate 7. [Figure 6 on page 69] (a) The most significant 100 corner points for $k = 0.2$, $\sigma = 5$ and (b) for $k = 0.04$, $\sigma = 4$.

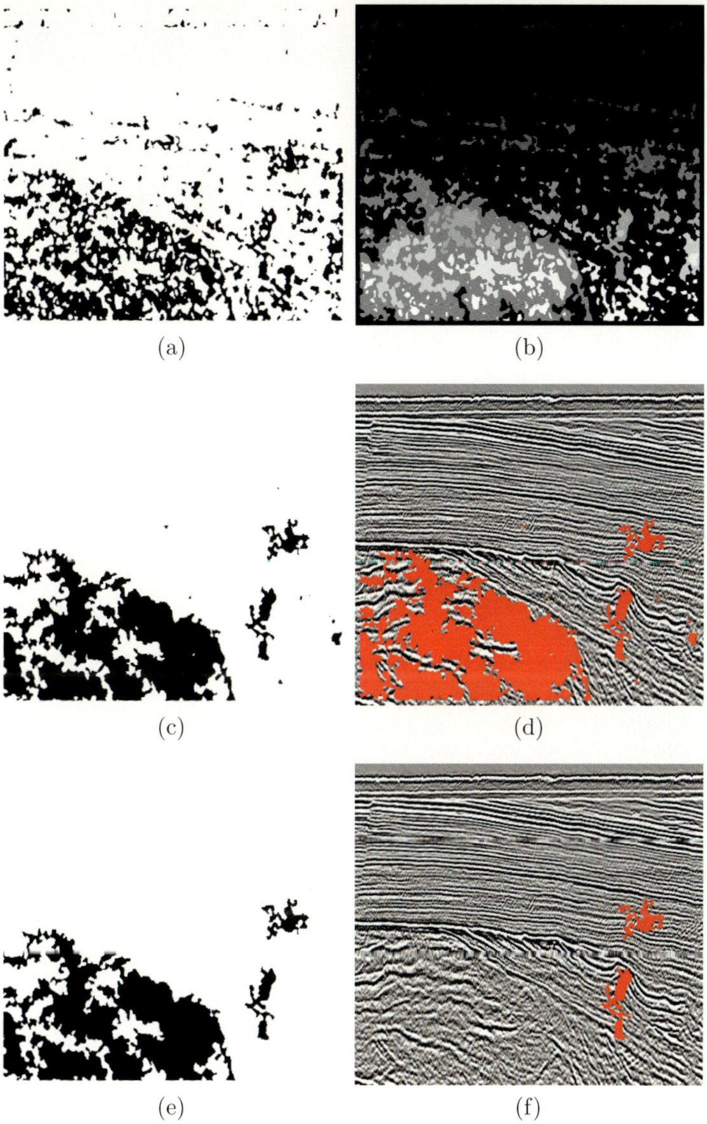

Plate 8. [Figure 7 on page 71] (a) Binary image with corner points, (b) different labels of each connected component, (c) binary image after small regions have been thresholded away, (d) final result superimposed on the original image, (e) binary image after very small regions have been thresholded away, (f) areas of interest.

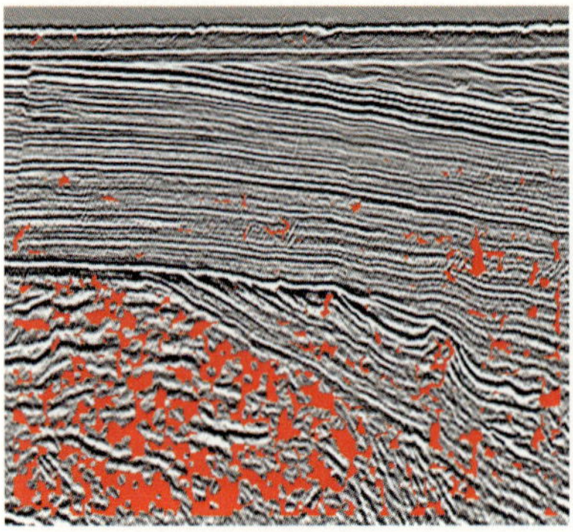

(a) Erosion

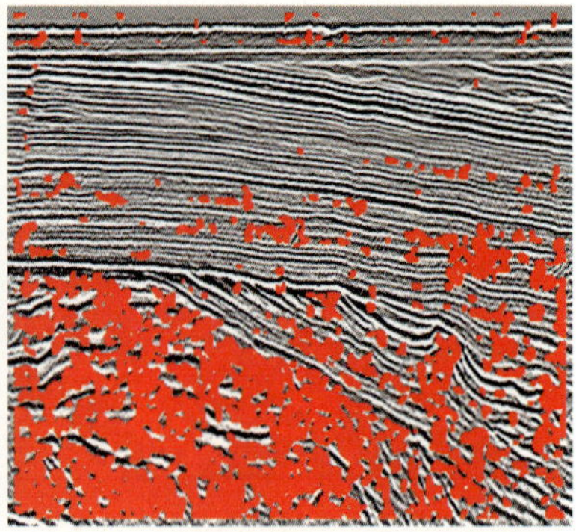

(b) Dilation

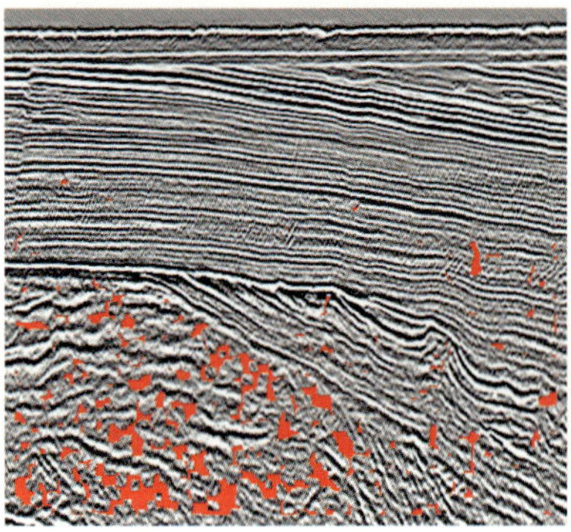

(c) Erosion-Erosion

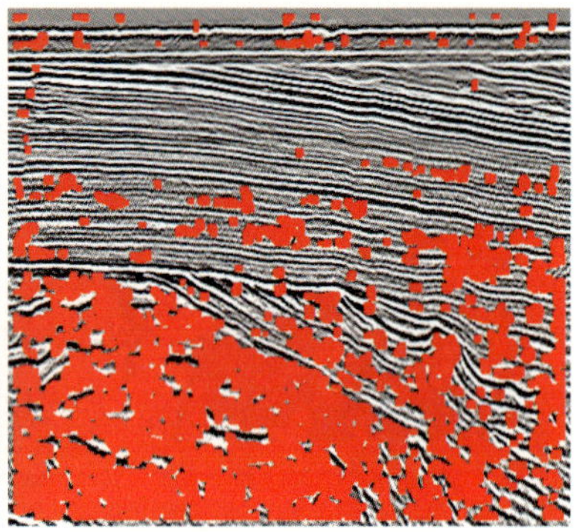

(d) Dilation-Dilation

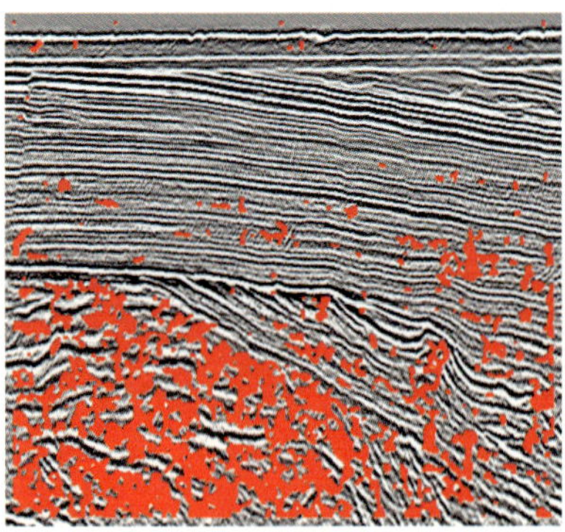

(e) Opening

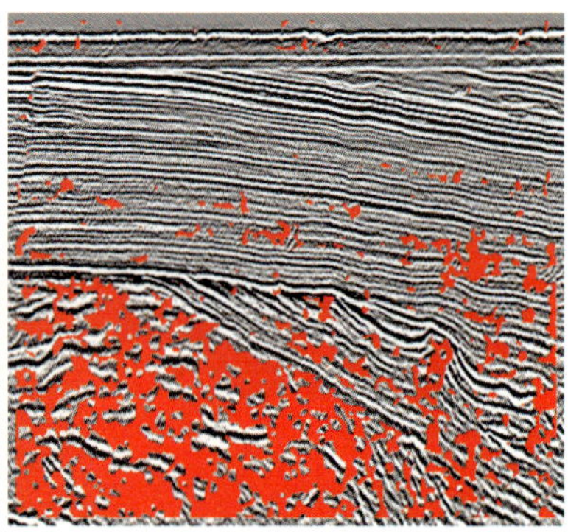

(f) Closing

Plate 9. [Figure 8 on page 72] Several morphological transformations.

(a)

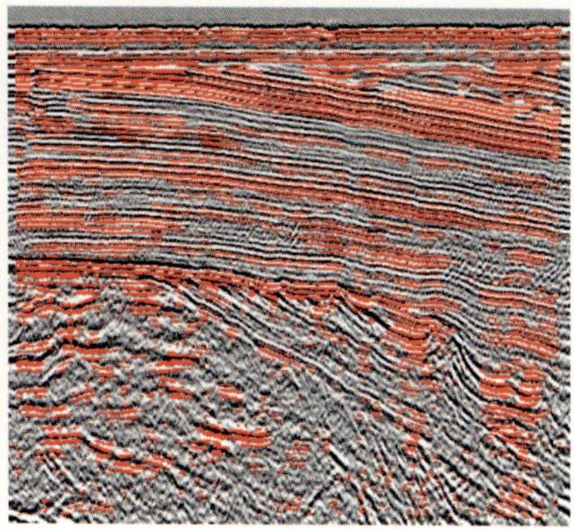

(b)

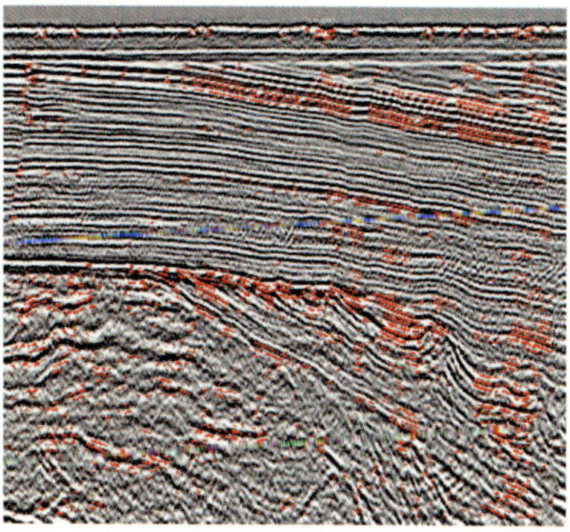

(c)

Plate 10. [Figure 12 on page 77] (a) Image after thresholding the determinant of tensor A with $T_A = 4000$, (b) image after thresholding the determinant of tensor B with $T_B = 3000$, and (c) combination of results (a) and (b) with an AND operator.

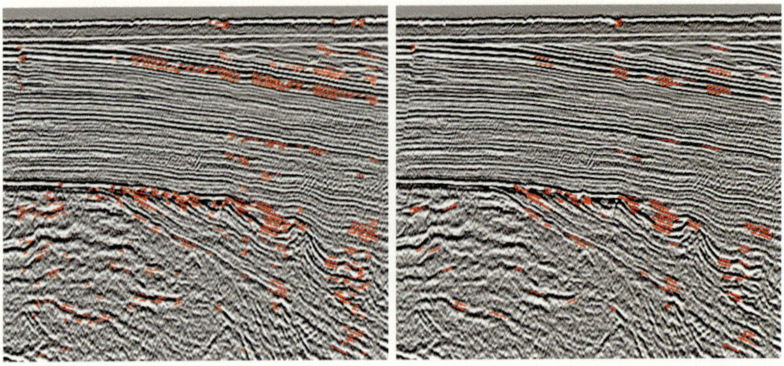

(a) (b)

Plate 11. [Figure 14 on page 79] (a) Final image of chaotic and mixed areas after connected component analysis with $T_A = 4000$, $T_B = 3000$ and $T_C = 10$, (b) final image of chaotic and mixed areas after connected component analysis with $T_A = 6000$, $T_B = 2000$ and $T_C = 20$.

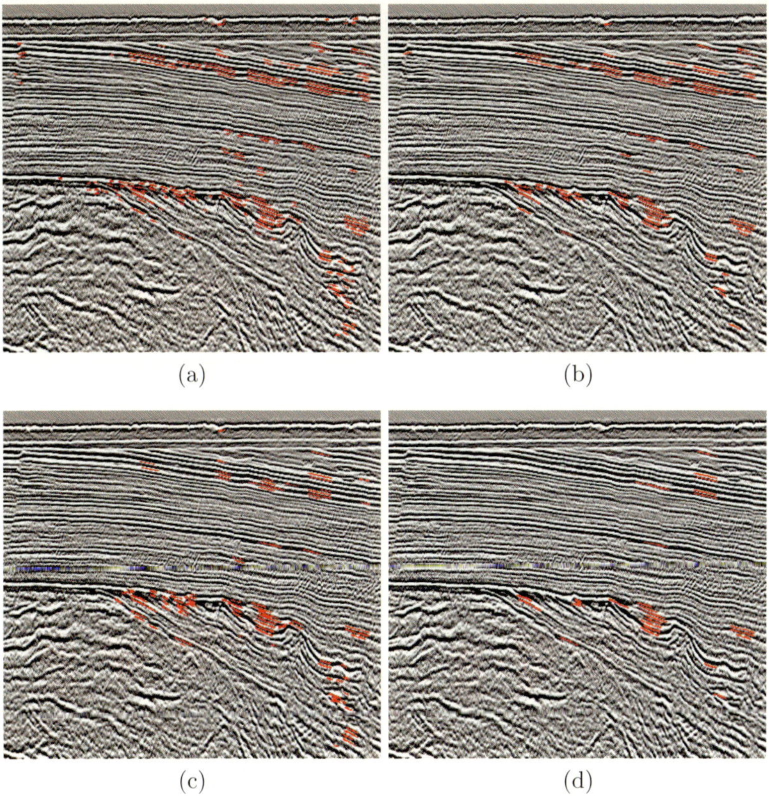

Plate 12. [Figure 16 on page 81] (a),(c) Combination of both determinant and magnitude methods, (b),(d) combination of both determinant and magnitude methods after connected component analysis and region thresholding.

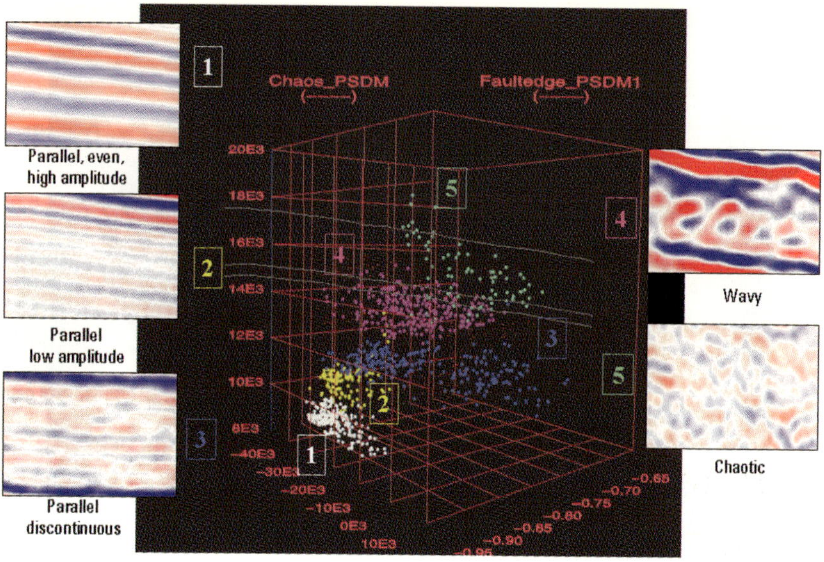

Plate 18. [Figure 5 on page 311] Training data distribution in attribute space. Each class defines a cluster in the attribute space, where training data can be checked and edited.

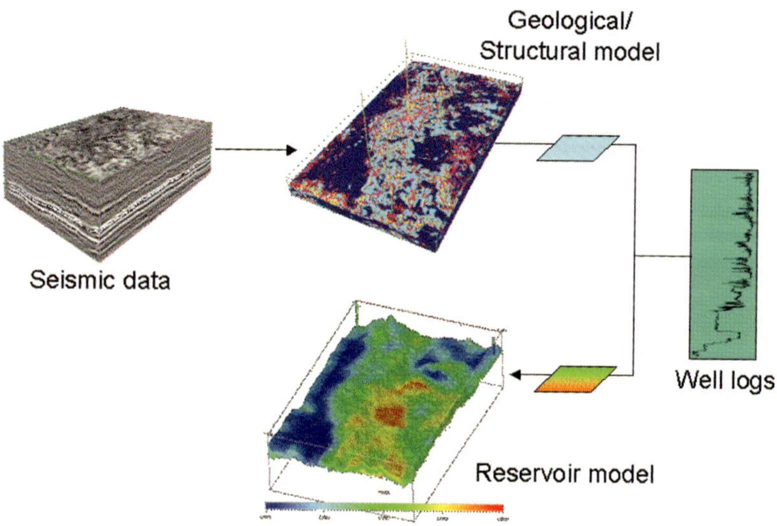

Plate 19. [Figure 6 on page 311] Classification of seismic facies based on texture attributes defines a geological/structural model. Using iterative and hierarchical classification capability, seismic facies can be calibrated and assigned to lithology and fluids using well data or another set of attributes such as amplitude-based, or AVO data.

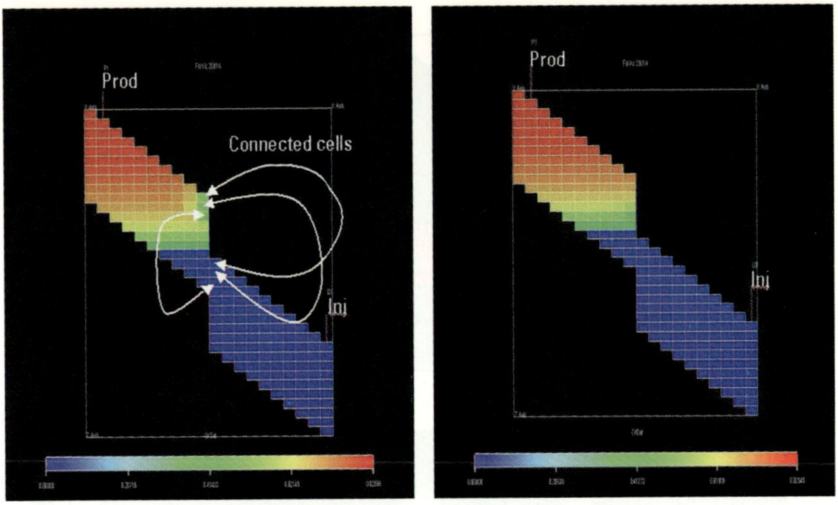

Plate 20. [Figure 10 on page 316] Cell connection across fault plane. Left: Block Centred (BC) grid. Right: Corner Point (CP) grid. Colour scale given by oil saturation values. Fault plane is not taken into account with the BC grid. Cells are connected according to their grid indices and not their physical location in space.

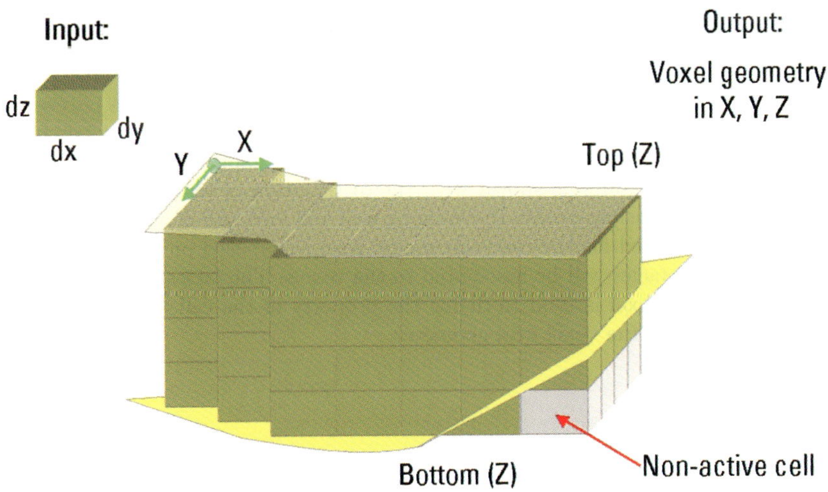

Plate 21. [Figure 11 on page 318] Automated voxel grid extraction. The inputs are: voxel size (dX, dY, dZ), the Top and Bottom horizons (given in depth) and a reference frame. The output is a voxel grid geometry in physical coordinates (X, Y, Z).

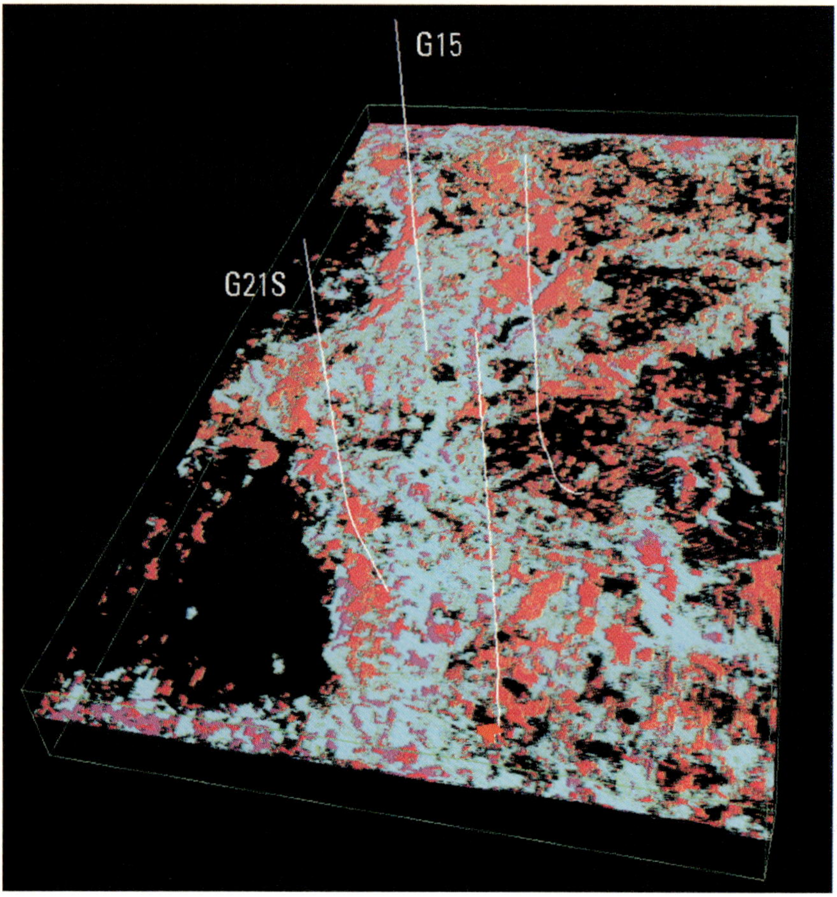

Plate 24. [Figure 14 on page 324] Distribution of potential sand injections mapped by the red seismic facies from PS data above the top of the reservoir (some facies are set to transparent on the display). The injected sands are mostly visible along the margins of the main sand lobe and locally above it. The four wells displayed show evidence of injected sands above the main reservoir body.

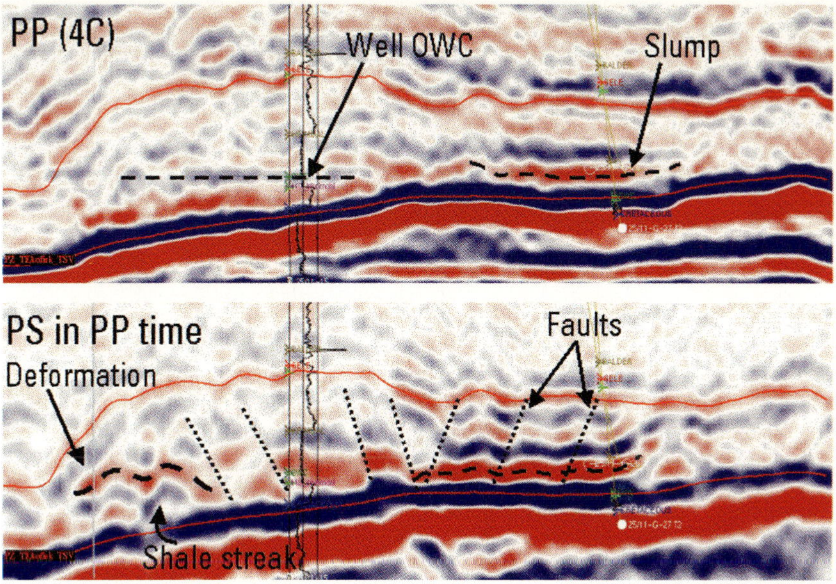

Plate 25. [Figure 15 on page 325] Seismic cross-sections of multicomponent data: PP data (upper section) and PS data in PP time (lower section).

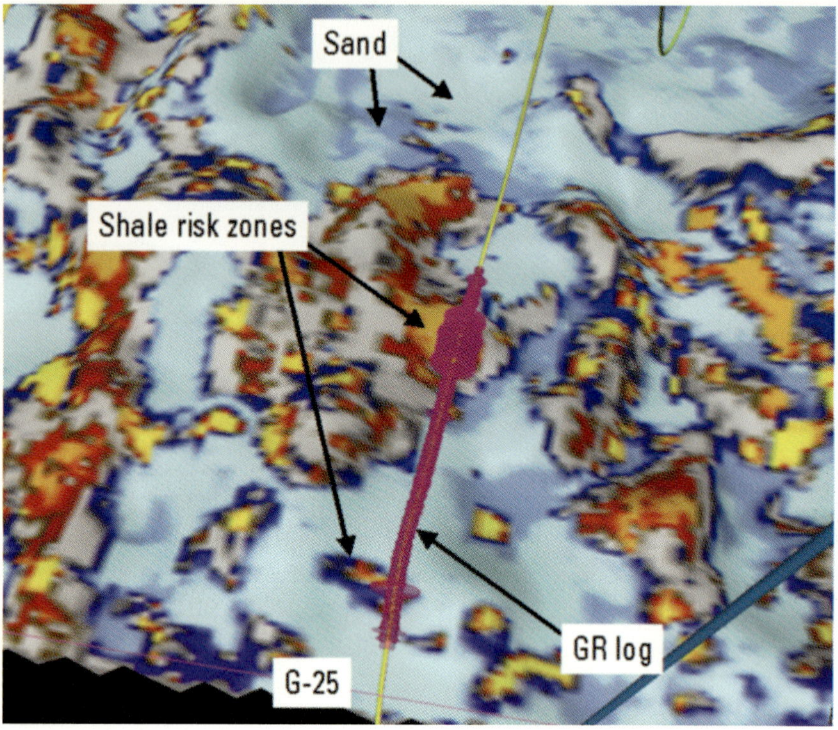

Plate 26. [Figure 16 on page 326] Stratal slice of the seismic facies model 8 ms above the base of the Grane reservoir. The dark blue and light blue facies represent flat continuous seismic facies and are calibrated as sands. The yellow and red facies represent discontinuous and dipping seismic facies and are calibrated as shale prone deformation features as indicated by the Gamma Ray log along a horizontal well.

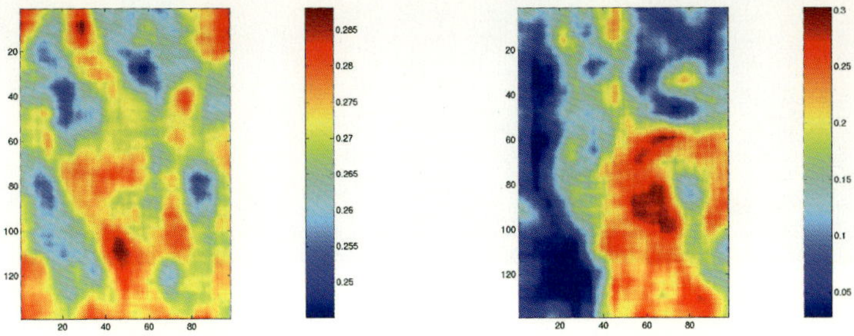

(a) (b)

Plate 27. [Figure 18 on page 331] Comparison of porosity at the top of the reservoir for the two voxel models. (a) Voxel model based on acoustic impedance only. (b) Voxel model based on acoustic impedance and three-dimensional seismic facies results.

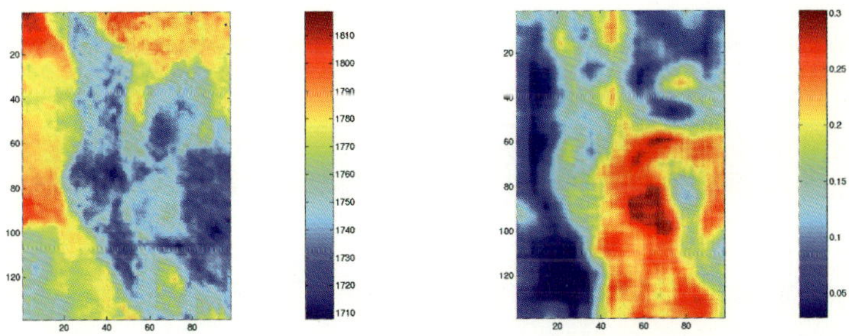

(a) (b)

Plate 28. [Figure 19 on page 331] (a) Illustration of the top reservoir horizon (in ms). (b) Porosity distribution along the top of the reservoir for the voxel model based on acoustic impedance and three-dimensional seismic facies results.

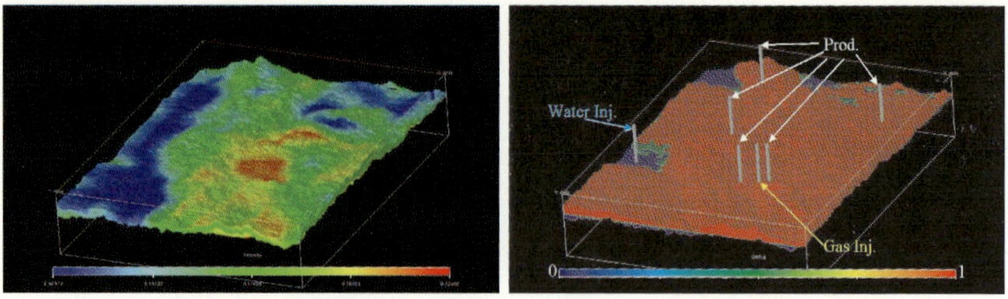

(a) (b)

Plate 29. [Figure 20 on page 332] Voxel model based on acoustic impedance and three-dimensional seismic facies results. (a) Porosity distribution; (b) initial oil saturation distribution together with the locations of the wells.

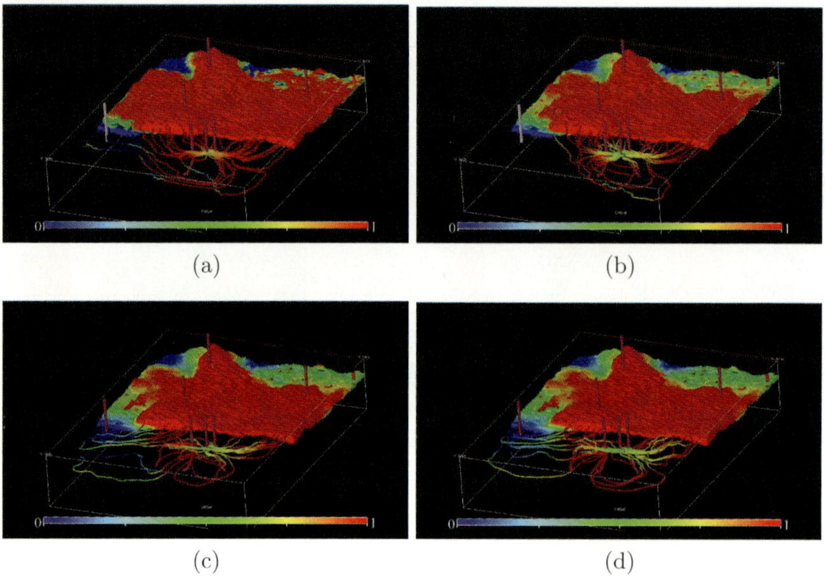

Plate 30. [Figure 21 on page 332] Voxel model based on acoustic impedance and three-dimensional seismic facies results. Simulation results for the oil saturation at day (a) 300, (b) 1800, (c) 3300, and (d) 4800 of the production.

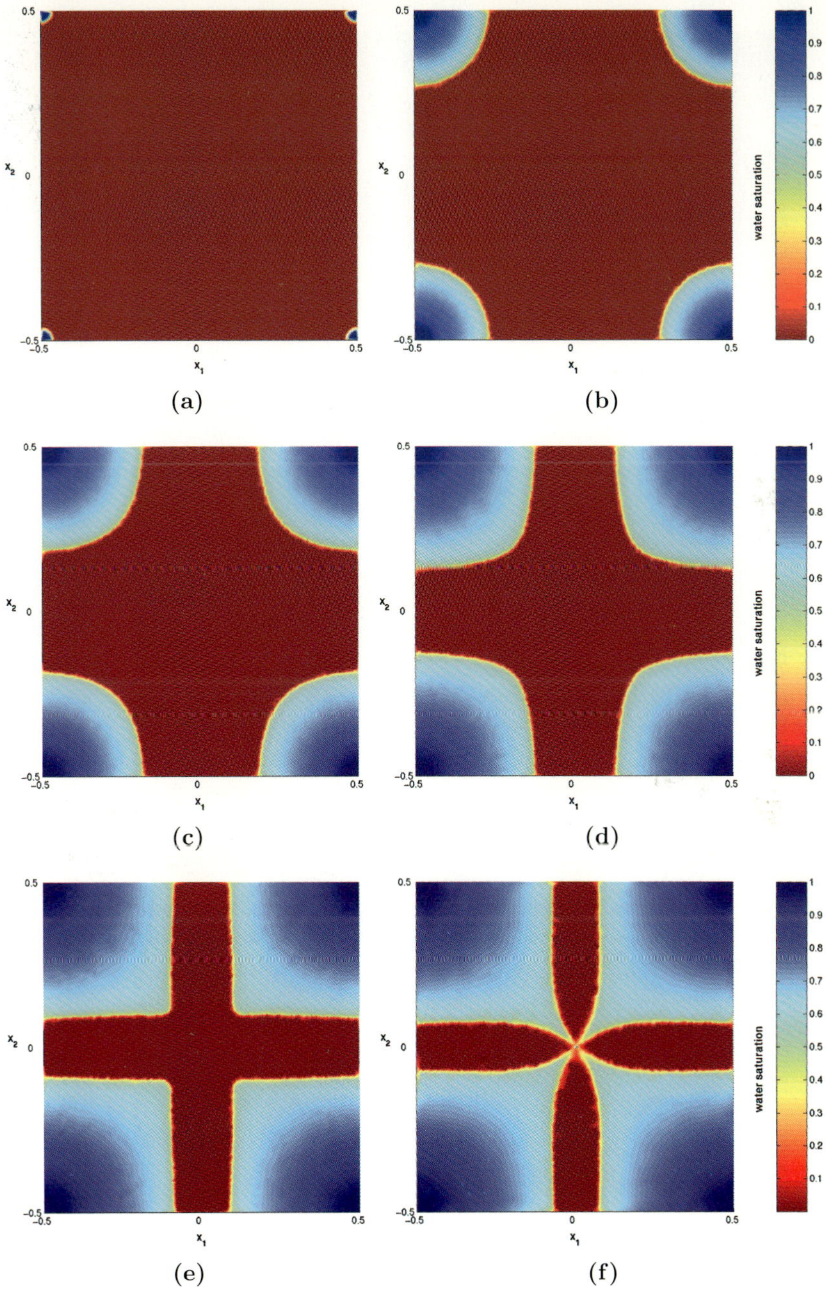

Plate 31. [Figure 16 on page 376] Five-spot problem computed with ADER1. Color plots indicating the injection of water during the simulation at six different times, **(a)** $t = t_0$; **(b)** $t = t_{120}$; **(c)** $t = t_{240}$; **(d)** $t = t_{360}$; **(e)** $t = t_{480}$; and **(f)** $t = t_{600}$.

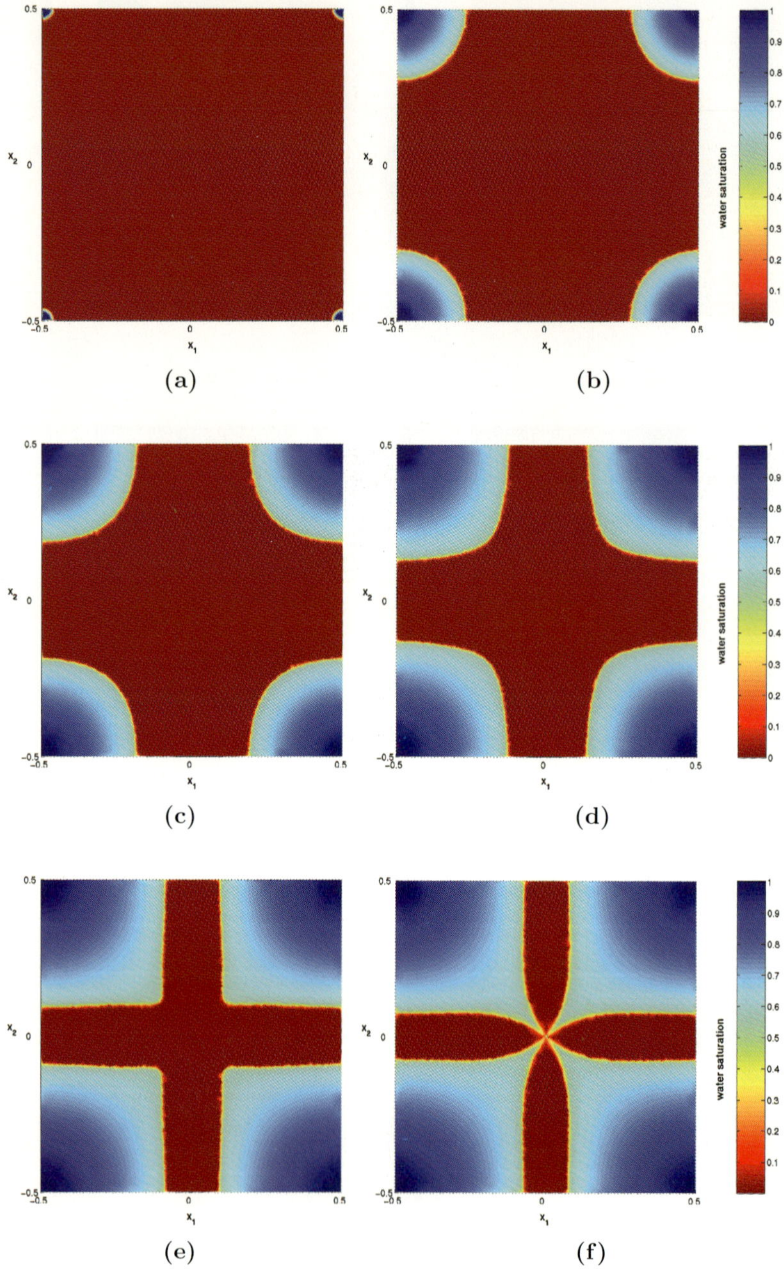

Plate 32. [Figure 18 on page 378] Five-spot problem computed with ADER4. Color plots indicating the injection of water during the simulation at six different times, **(a)** $t = t_0$; **(b)** $t = t_{120}$; **(c)** $t = t_{240}$; **(d)** $t = t_{360}$; **(e)** $t = t_{480}$; and **(f)** $t = t_{600}$.

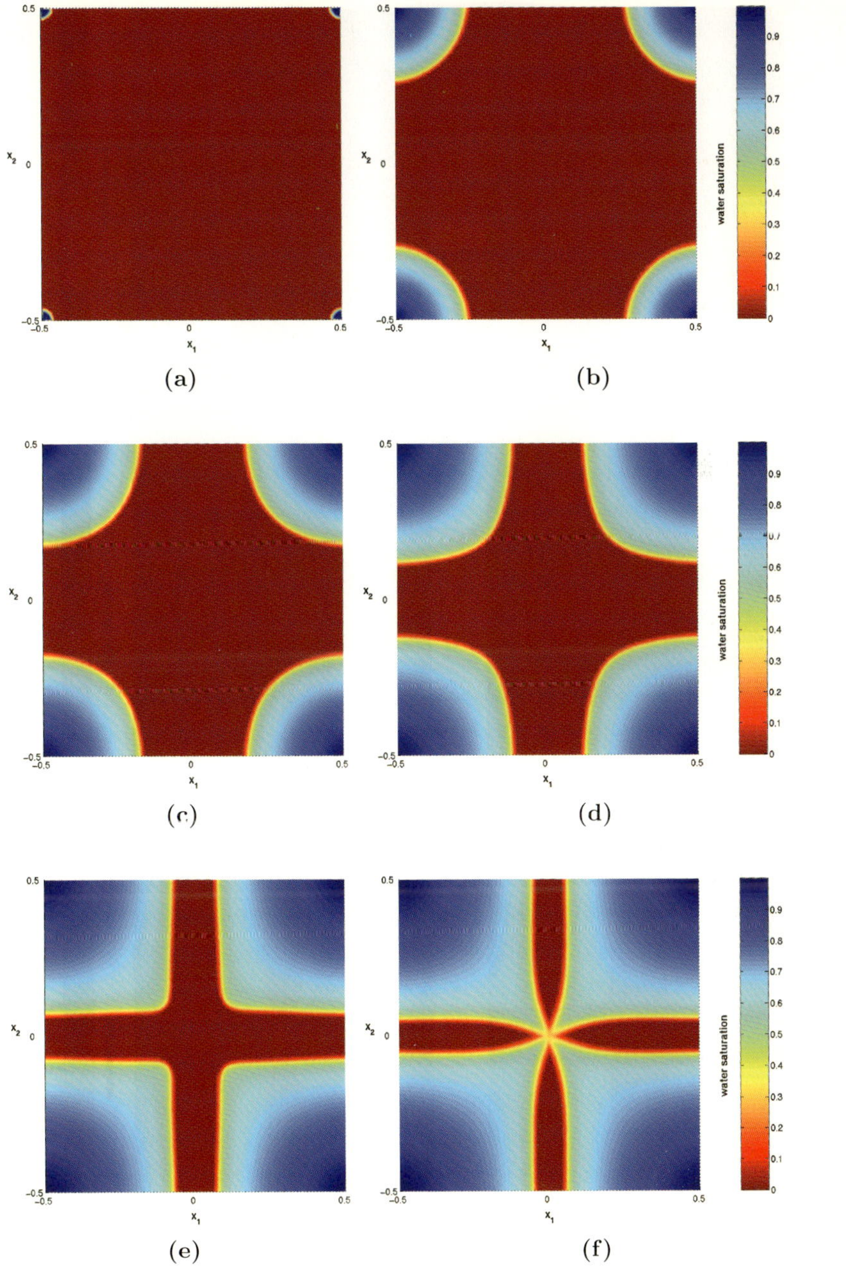

Plate 33. [Figure 21 on page 381] Five-spot problem computed with ECLIPSE. Color plots indicating the injection of water during the simulation at six different times, **(a)** $t = t_0$; **(b)** $t = t_{120}$; **(c)** $t = t_{240}$; **(d)** $t = t_{360}$; **(e)** $t = t_{480}$; and **(f)** $t = t_{600}$.

Printing: Krips bv, Meppel
Binding: Litges & Dopf, Heppenheim